# Global Information Infrastructure

## The Birth, Vision, and Architecture

Andrew S. Targowski
Western Michigan University

IGP

*IDEA GROUP PUBLISHING*

**Harrisburg, USA • London, UK**

Senior Editor: Mehdi Khosrowpour
Managing Editor: Jan Travers
Copy Ediotr: Beth Green
Printed at: BookCrafters



Published in the United States of America by
Idea Group Publishing
Olde Liberty Square
4811 Jonestown Road, Suite 230
Harrisburg, PA 17109
Tel: 717-541-9150
Fax: 717-541-9159

and in the United Kingdom by
Idea Group Publishing
3 Henrietta Street
Covent Garden
London WC2E 8LU
Tel: 071-240 0856
Fax: 071-379 0609



Library of Congress Cataloging-in-Publication Data

Targowski, Andrzej.
Global information infrastructure: the birth, vision and architecture/Andrew S. Targowski.
416 pp.
Includes bibliographical references and index.
ISBN 1-878289-32-2
1. Information superhighway—United States. I. Title.
HE7572.U6T34 1996
384—dc20 96-13580
CIP

British Cataloguing in Publication Data
A Cataloguing in Publication record for this book is available from the British Library

# IDEA GROUP PUBLISHING

**Harrisburg, USA • London, UK**

## Series in Global Information Technology Management

*Information technology has become a key factor in propelling and accelerating the globalization of businesses. The recent advances in hardware, software, telecommunications, and databases have brought us into a complex, yet exciting age of true globalization. The Series in Global Information Technology Management enlightens you on various aspects of this trend by presenting new information to explore, investigate and use to prepare you for the challenges of the 21st century.*

***Books in this series:***

**Global Information Infrastructure: The Birth, Vision and Architecture**
By Andrew Targowski

**Global Information Systems and Technology: Focus on the Organization and Its Functional Areas**
Edited by P. Candace Deans and Kirk Karwan

**Global Information Technology Education: Issues and Trends**
Edited by Mehdi Khosrowpour and Karen D. Loch

**The Global Issues of Information Technology Management**
Edited by Shailendra Palvia, Prashant Palvia and Ronald Zigli

For more information, or to submit a proposal for a book in this series, please contact:

**Idea Group Publishing**
**4811 Jonestown Road, Suite 230**
**Harrisburg, PA 17109**
**Tel: 1/800-345-4332 or 717/541-9150**
**Fax: 717/541-9159**

*To Irmina, my wife and partner on four continents*

# Global Information Infrastructure

## Table of Contents

## Acknowledgments

This book was envisioned during my sabbatical in 1991 at the City Hall of Kalamazoo, Michigan. It was a great experience for me to plan some systems for the City from my perspective of the Electronic Global Village. I first chaired the Kalamazoo Electronic Global Village Committee in 1992, and one year later, I chaired the Board of Directors of the Greater Kalamazoo Telecity USA (1993-1995). It was a great field experience for me.

Formerly, during 1971-1974, I was the head of the Polish Computer Development Program and the inventor and architect of the Polish National Information Infrastructure.[1] In those years I developed a term and project, INFOSTRADA, (I associated this term with a Warsaw arterial, called WISLO-STRADA) which, translated from Italian (la strada) into English meaning INFOHIGHWAY. Today this term is a metaphor for a new information civilization. In those days, however, the horizontal flow of information could not be accepted by the Polish totalitarian regime. This forced me out of work for 6 years. I defected to United States, where I obtained political asylum and citizenship.

This book was written while I was collaborating with the Kalamazoo City executives, who gave me great support. I am grateful for the support of the former city manager Jim Holgerson, his successor Mark Ott, and MIS director Sammy Taylor, as well as to all my friends from the above mentioned committees.

I am grateful to the Business Information Department's Jon Styrd, MBA and Steven Woods, MBA, then graduate assistants at Western Michigan University who read some portions of this book manuscript and provided helpful corrections.

I address the warm thanks to Dr. Mehdi Khosrowpour, president of Idea Group Publishing, who gave me strong encouragement to start working on this book. Finally, my many special thanks go to Jan Travers, my managing editor, for her excellent job in producing this book.

***Andrew Targowski***

WESTERN MICHIGAN UNIVERSITY
Department of Business Information Systems

Greater Kalamazoo TeleCity USA
Board of Directors member and Project Director

Vice President of Information Resource Management Association

Kalamazoo, Michigan, United States, 1996.

---

[1] In Polish this NII was called KRAJOWY SYSTEM INFORMATYCZNY (KSI). A description of this project was published as "Computing in Totalitarian States: Poland's Way to an Informed Society," INFORMATION EXECUTIVE, Summer 1991, Data Processing Management Association. It was one of two such projects undertaken in the world in the 1970s. Another slightly similar CYBERSYN project was initiated by a British cyberneticien Stafford Beer for the support of the Chilean Economy in 1971-1973. Both projects have been killed before some experience could be reached. Each project has been ended for a different reason.

# Global Information Initiative[1]

We gather here today to chart a path to the future—at a time when prediction is as difficult as ever, but also at a time when our circumstances are clearly conducive to the rapid spread of a new capacity to process and communicate information that will benefit all humankind. It is a path that will take us from our shared vision to a new reality. Just as human beings once dreamed of steam ships, railroads, and superhighways we now dream of the global information infrastructure that can lead to a global information society.

But our dream today is not fundamentally about technology. Technology is a means to an end. Our dream is about communication—the most basic human strategy we use to raise our children, to educate, to heal, to empower and to liberate.

In its most basic form, communication is the transfer of information from one human being to another. Information, in turn, is the raw material of knowledge, and knowledge sometimes, if we're lucky, ferments into wisdom. And of course, in all of our countries it is by now a cliché to note that the information revolution now in its early stages will ultimately transform our concepts of both communication and information.

Last year in Buenos Aires, I attended the first World Telecommunication Development Conference to present the United States' vision of a Global Information Infrastructure that will promote robust and sustainable economic progress, strengthen democracies, facilitate better solutions to global environmental challenges, improve health care and, ultimately create a greater sense of shared stewardship of our small planet.

The Buenos Aires Conference adopted a set of basic principles we believe are the building blocks of the GII:

Private investment.
Competition.
Open access.
Universal service.
Flexible regulations.

These principles have been central to the discussions about the GII in bilateral, multilateral and regional forums, most recently at the APEC meeting last week in Vancouver, but also at the Summit of Americas meeting in Miami last December and in memoranda of understanding between the United States and both Russia and Ukraine.

The very act of holding this conference is in keeping with advice given to dreamers long ago by Mahatma Gandhi: "You must become the change you wish to see in the world."

Moreover, moving forward aggressively on a GII is the best way to deal with concerns highlighted during the G7 jobs summit in Detroit last year. At that

[1] Excerpts of remarks by U.S. Vice President Al Gore to G-7 meeting on the Global Information Initiative, February 25, 1995, Brussels, Belgium.

conference we confronted the central dilemma facing every government: how do we make sure our economies provide enough jobs?

The initial OECD jobs study outlined the connection between jobs and what we do here. Those nations best able to adopt the new technologies for a knowledge-based economy have been the best at creating jobs.

The fact is that government policies based on faulty assumptions that try to block change or protect the status quo have themselves become job destroyers. This time we have a chance to get it right. We can open markets to create job opportunities. We can use education and training to enable more workers to adapt to the new workplace.

The liberating effects of these new technologies have been clear around the world. Satellite stations brought medical advice to those tending to the suffering in Rwanda. Radio and TV broadcasts in South Africa promoted the role of voting in a democracy. Wireless technologies are allowing emerging nations to leapfrog the expensive stages of wiring a communication network—for example, in Thailand, where the ratio of cellular telephone users to the population is twice that of the US.

The effects are also visible in education. One of the biggest handicaps for those who want to learn has been distance. In Washington, the Library of Congress is a wonderful place. But we must ensure it becomes a tool for, let's say, a schoolgirl from my hometown in Carthage, Tennessee, 600 miles away.

Already, distance education is helping some citizens overcome geographic difficulties. In Japan, over 100 institutions are linked by computer and satellite, with some 150,000 students currently enrolled.

In India, there are five open universities and more than 35 distance learning programs in conventional universities.

And in Canada, the Knowledge Network delivers courses to adult students living on islands in British Columbia.

In France, the newly-discovered cave paintings in Ardeche, almost impassable to reach in real life, are accessible on the Internet to scholars, teachers, and most important, children.

The Clinton Administration is committed to the goal of connecting every classroom, every library, every hospital and every clinic to the national and global information infrastructures by the end of this decade. We must provide our teachers and our students with the same level of communications technology that shipping clerks, construction workers and government officials use every day.

Information technology is a critical element of economic policy. But there are great obstacles. How do we begin the hard work of turning the obstacles before us into opportunities?

First, by focusing squarely on those who will drive the demand for information products and services: the users. User demands will define the marketplace.

Competition to serve the users will speed up innovation and cost-effective deployment of new technologies. Private investment in diverse technologies will mean new sources of capital and expertise for rich and poor nations alike.

Computer networks have created whole new, rapidly growing markets. These networks help small and medium sized enterprises from both poor and rich countries to become more effective competitors in world markets.

In the United States, our spectrum auctions have speeded the licensing of personal communication services and are leading to the creation of hundreds of thousands of jobs in the next several years—one indication that communication is a source of economic change growth, not just the result of it. The GII will not be created in one place at one time by any one group. It will be the product of cooperation among governments, industry and citizens on a global scale.

But how do countries with widely varying needs, cultures, and technologies cooperate? First, by acknowledging that the fruits of our cooperation should be open to access to markets for all providers and users of creative content and information products, equipment and services. For the competitors in the 21st century global economy, there is no substitute for being in the marketplace and providing the users we represent with the greatest variety of products, information and services for the least cost.

Second, building the GII is going to require robust competition. And you cannot create robust competition by excluding competitors, whether those competitors are at home or abroad.

It is vigorous competition—which means global competition—that creates jobs. And so I say on behalf of President Clinton, let the message of this conference be clear: we support competition in open markets that allow any company to provide any service to any customer. What concentric actions must be taken to realize that goal?

First, we must drop our barriers to foreign investment together. For more than 60 years, the U.S. has had limited restrictions on foreign investment in certain telecommunications markets around the world in negotiations within the General Agreement on Trade and Services. The deadline for these negotiations is April 1996. Let us resolve to meet this deadline to remove our investment barriers together.

Second, let's develop and enforce effective intellectual property rights for the GII. If our content providers are not protected, there will not be content to fill the networks and give value to services.

Third, all parties should participate in the development of private-sector, voluntary, consensus standards through the existing international organizations, such as the International Telecommunication Union, the International Standards Organization and the Internet Society. The creation of truly global networks will require a high degree of interconnection and interoperability.

Governments are not the best arbiters of technology, and government intervention risks encouraging adoption of standards that are either ultimately inferior or inappropriate to demands of the market.

Our vision of an information society is one in which the most valuable resource—information—is also the most abundant. My hope is that the open exchange of ideas of all sorts and the greatest access possible for all citizens to the varied means of communication will stimulate creativity.

Global communication is not about conformity. Some fear that in losing the distance between ourselves and others we lose our distinctions as well. But communication is about bridging the differences between nations and people, not erasing them. It is about protecting and enlarging freedom of expression for all our citizens and giving individual citizens the power to create the information they need and want from the abundant flow of data they encounter moment to moment.

Communication is the beginning of community. Whether it is through language, art, custom, or political philosophy, people and nations identify themselves through communication of experience and values. A global information network will create new communities and strengthen existing ones by enriching the ways in which we do and can communicate.

Ideas should not be checked at the border. We have much to learn from each other and we should follow practices and policies that incorporate, not exclude, the greatest diversity of opinions and expressions. We all gain from the exchange of cultural viewpoints and experiences that occurs when open minds engage each other.

At the same time, users of the GII want and will demand privacy. When you ask Americans about information technology, it is their biggest concern. We must protect the privacy of personal data and communications.

Governments and industry need to work together to develop new technologies, new standards, and new policies that will provide privacy and financial transactions and ensure intellectual property rights, the GII must be secure and reliable. The OECD should continue its leadership in the area of computer security.

Fortunately, technology and human imagination keep providing us with new opportunity to enhance our communication capabilities. Take, for example, non-geostationary satellites. They hold remarkable potential, especially for remote or thinly-populated regions, and for societies eager to reap the benefits of 21st century technology even before completing expensive land-based networks. These advanced technologies can provide everything from basic telephone calls to remote medical diagnosis. Like the Internet, they have the potential to knit together millions of people in different locations and situations—and do it economically.

Our purpose in meeting here together is to advance our common goal of a Global Information Infrastructure that will bring to all countries the benefits of a Global Information Society. Our challenge today is to create the commercial, technical, legal and social conditions that will establish the foundation for the GII.

As we work across our common boundaries and oceans to build a GII, we cannot think only of today's debates about wireless or satellites; we must perform our work in the service of a global vision that can be realized in every community and village in the world.

I have outlined today the concrete steps we must take to embark on this new voyage of discovery. Empowered by the moveable type of the next millennium we can send caravans loaded with the wealth of human knowledge and creativity along trails of light that lead to every home and village.

**Al Gore**
**U.S. Vice President**

# Introduction

Futurist Alvin Toffler (1980) has written about three "waves" of civilization. In the First Wave, agriculture enabled nomadic peoples to settle into villages and cities. In the Second Wave, the industrial revolution made possible the modern industrial state. The Third Wave has begun in which computer and communications technologies will transform the national and global economies into information-driven economies. The Third Wave triggers the Information Revolution which will have political and societal impacts every bit as profound as those of the First and Second Waves.

From the 1970s - 1990s, the world created the computer and communications revolutions. They provide tools for the development of a New Information Civilization. A set of these tools includes: information technology (computers), telecommunications networks (satellite, fiber optics, telephone), and television (video). This set of technologies can be called telematic technologies. They transform the Material Civilization into the Information Civilization which becomes the dominant force of humanity. This new civilization brings about new modes of human activities and a new vision of the humanity development.

The 19th century eliminated wilderness via railroads. The 20th century developed science and technology which causes the Planet to fall apart and fragment along tribal and sectarian lines. The 21st century perhaps should implement the mass-enlightenment (the Learning Society) which will integrate us commercially and culturally as a New Information Civilization.

The purpose of the Information Civilization is:

1. To optimize operations and development of the Material Civilization in order to minimize the use of resources (including the ecology), to increase consumer choices of innovative and quality products and services, and to improve customer satisfaction;
2. To sustain the development of human cognition in order to make aware and wise decisions about; the sense of human possibility, life, education, politics, defense, business, entertainment, and leisure time.

It is important to notice that the Information Civilization does not replace the Material Civilization. Information cannot replace food, steel or plastic. Only, it can improve their utilization and support the development of innovative (green) products and services. The Material Civilization has fabricated the computer and the computer has created the Information Civilization. The Information Civilization opens opportunities for the development of the Learning Society, which constantly improves its own awareness (based on rules and laws of knowledge) of civilization threats, issues, problems and solutions.

The general architecture of Information Civilization shown on Figure I-1, is composed of the following metaphoric elements:

1. Infofactories, which generate information and seek new information, among them there are the following:

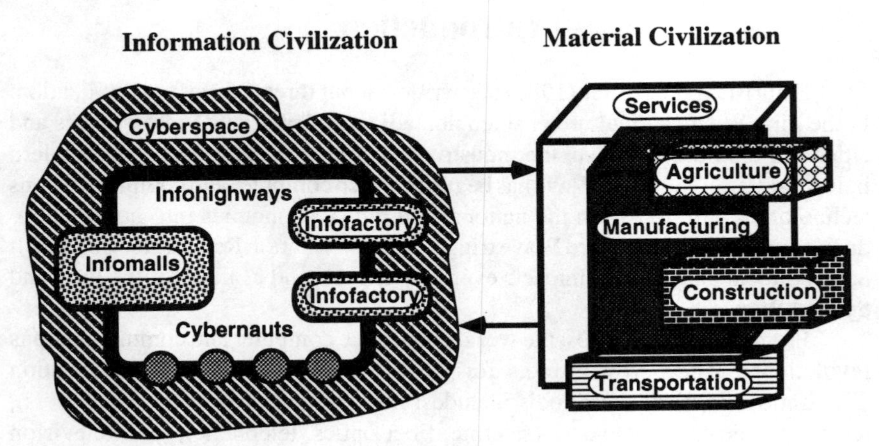


*Figure I-1: Information Civilization in Control of Material Civilization*

- virtual enterprises, virtual schools and universities, on-line governments, virtual communities, and so forth. A virtual infofactory is a geographically, dispersed organization with a minimum of building walls, and which operates by the application of telematic technology.

2. Infomall, which provides the following products and services:
   - telematic services; information services (America Online, CompuServe, PRODIGY, and so forth), home banking, home trading, value added networks (Telenet, TYMNET, infoNET), and so forth
   - electronic money is a payment system applying: credit cards, Automatic Teller Machines, Point-of-Sale machines, Electronic Fund Transfer, Automated Clearing Houses, and so forth, in which transactions processing and value distribution depends wholly on the use of telematic technologies.
   - electronic knowledge addresses an increasing impetus to automate information inquiries and digitalize libraries' depositories with subsequent electronic dissemination through: automated libraries, electronic (digital) libraries, virtual libraries, electronic newspapers, and so forth.
3. Infohighways, which transport the multimedia (voice, data, graphics, video) converged signals of computer, telecommunications networks (wire, fiber optic, satellite, wireless channels), telephone, fax, radio, and television through paths of a matrix of Local Area Networks (LAN), Metropolitan Area Networks (MAN), Rural Area Networks (RAN), Wide Area Networks (WAN), and Global Area Networks (GAN) (Figure I-2).
4. Cyberspace, which is a information-based space, that is the dispersed, infinite constellation of electronic (digital) files, databases, home pages, bulletin boards, directories, menus, where humans, with a password, interactively navigate in order to create, update, exchange, and retrieve information traces.
5. Cybernauts, who are informed telecomputer (combination of computer, television, and telephone) users with a password to access billions of information tidbits and do everything on-line from shopping and learning to working. A cybernaut can be an "electronic immigrant" who can telecommute to work over great distances.

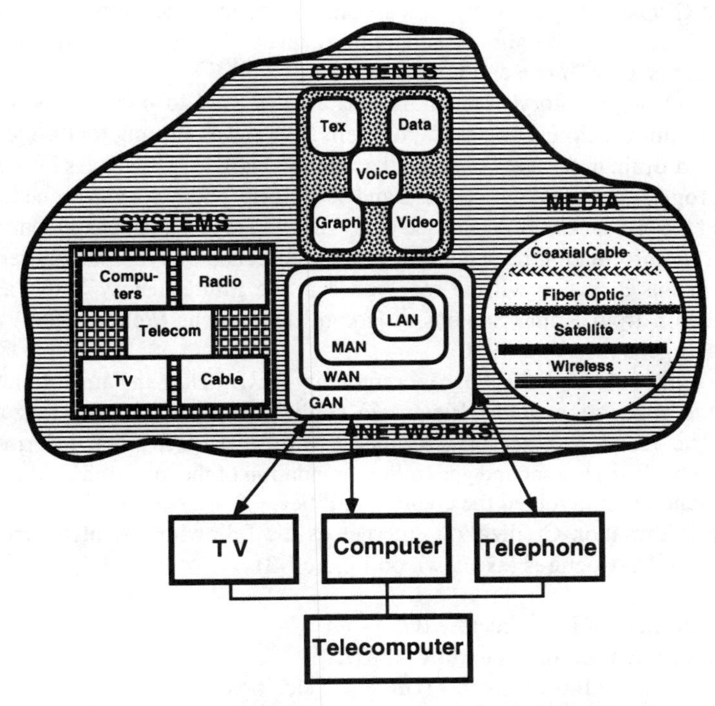


*Figure I-2: Four Views of an Infohighway and Its User Technologies*

The infohighway will connect infofactories, infomalls, and cybernauts with a speed from 100 Mega to 3 Tera bites per second.[1] One fiber-optic cable is capable of transmitting the entire Library of Congress - all 107 million volumes in less than five minutes. That is more than we want to read most evenings. But it is a metaphor of what the "national infohighway" may one day make available in nearly every home. The desktop video-conferencing and video-on demand will consume the most of the infohighway capacity.

The Information Civilization emerges as a result of massive applications of telematic technologies in the world economies (about $1 trillion spent in 1980s-1990s). Tehranian (1988) identifies four perspectives on the process of developing the Information Civilization:

1.Technophilic optimists
2.Technophobic pessimists
3.Technoneutrals (who have little theoretical pretension and considerable interest at stake not to alienate their clients)
4.Technostructuralists (technology is neutral, neither bad or good)

In this book, the author represents the pseudo-technostructuralist's perspective, which means that telematic technologies, if not guided, can be harmful. By the end of the 20th century, the relentless production and applications of telematic technology leads to the emergence of the global-green, One Human Family (OHF) living in One

Electronic Global Village (EGV), at least among the most developed 50 nations (Targowski, 1991). The aim of achieving a level of OHF in the humankind development is a legitimate and noble purpose.

Long through history, humankind has always tried to increase its mental power and culture. Almost 60 years ago, Wells (1938) was arguing for the creation of the World Brain in terms of organized academic knowledge. He was impressed with the progress achieved by scientists and educators. Today we can argue for the creation of the Global Brain, composed of all knowledge through the application of telepower by information providers for diverse audience of cybernauts. Perhaps progress in this direction will be recognized in the future history books as the transformation from Homo Sapiens to Homo Electronicus (Pelton, 1989) or to Homo Intelligens (Masuda, 1990).

The Electronic Global Village, a community for One Human Family functions through the application of the Electronic Global Machine (EGM) (Targowski 1990). The EGM is a set of information infrastructures. The information infrastructure (IFIF) is a second generation foundation of the mankind civilization. The first generation involved the creation of cities.

The Information Civilization emerges as the following set of interrelated information infrastructures (as shown on Figure I-3):

- Global Information Infrastructure (GII),
- National Information Infrastructures (NII),
- Local Information Infrastructures (LII, e.g.,Telecities),
- Enterprise Information Infrastructures (EII).

The design and operations of information infrastructures are subjects of study and professional work of information systems, services, networks developers, social and management specialists, public administrators, and politicians. This book is designed for these types of readers.

The key IFIF is the National Information Infrastructure which provides (as shown on Figure I-4):

1. Infohighways (telematic network services treated as information utility) services for every school, college, university, hospital, clinic, business and local, state, federal, government, and individual user;
2. Electronic Knowledge under the form of the National Digital Library; which is the integrated system of creation, distribution, and dissemination (common standards and formats) of books, journals, newspapers, and documents, in order to permit users to identify, locate, and access needed resources in a consistent fashion;
3. Electronic Health, which is a seamless (unified protocols) matrix of telematic networks and interorganizational computer applications of hospitals, clinics, research centers, insurers, medical vendors, and patients in order to reduce health care costs and improve delivery and quality of health care services;
4. Electronic Money, which will replace paper money so that money can be traded freely, inexpensively, and instantly anywhere in the nation and the world. This transformation will support seamless management of business transactions

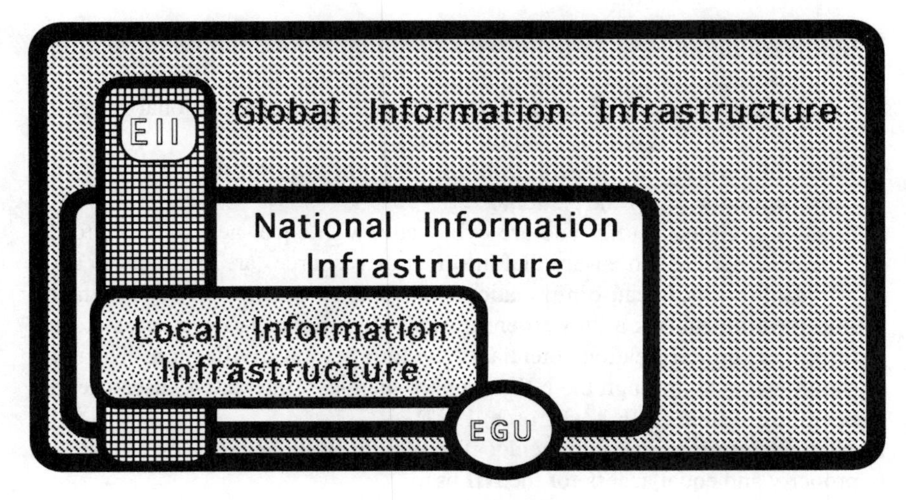


*Figure I-3: Information Infrastructures of Information Civiliation (EII-Enterprise Information Infrastructure, EGU-Electronic Global User)*

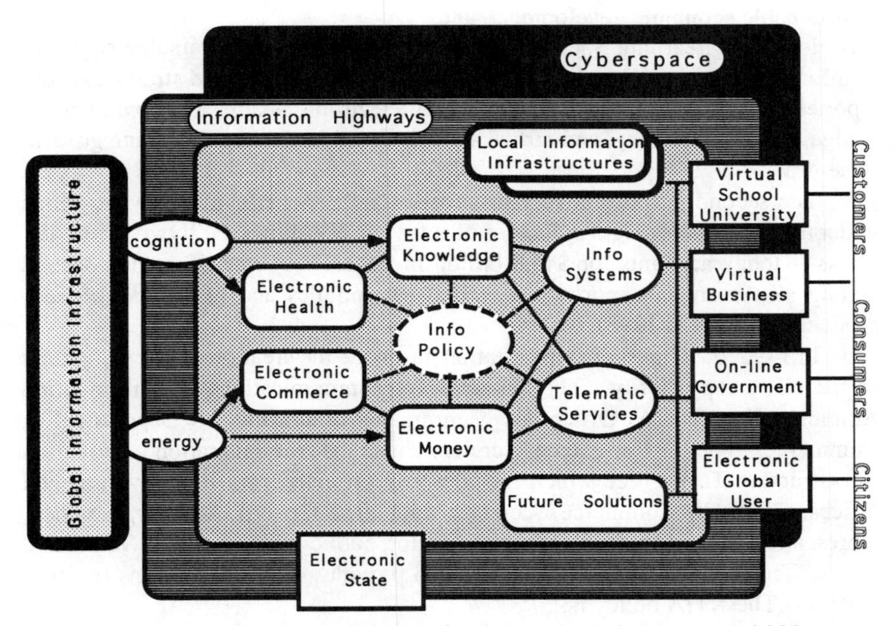


*Figure I-4: The Architecture of National Information Infrastructure-2000*

among all types of vendors and customers in order to shorten and improve the quality of money cycle in the economy;

5. Electronic Commerce, which is a seamless (unified protocols) matrix of telematic networks of businesses providing electronic marketing, sales and customer service to business customers and consumers, including electronic payments;
6. Electronic State, which either controls citizens as the "Big Brother," or is scrutinized by citizens, media, groups, and organizations as the Electronic Republic;

7. Enterprise Information Infrastructures of virtual schools, universities, virtual businesses, on-line governments that can operate within the NII;
8. Local Information Infrastructures that provide interorganizational telematic services at the level of cities and rural areas;
9. National Cyberspace which develops and operates accordingly with the constitutional rules of the Nation;
10. Electronic Global Users who are cybernauts with equal access rights to the "universal service" to ensure that information resources are available to all at affordable prices. Because information means empowerment--and--employment--the government has a duty to ensure that all Americans have access to the resources and job creation potential of the Information Civilization. The users should navigate through the NII through a seamless, interactive, user-friendly operating system (IFIF Task Force, 1993).
11. National Information Policy includes the rules for protecting privacy, intellectual property and equal access for the NII users.

The purpose of the NII is:

- To improve control of the National Material Civilization in order to support sustainable economic development, and
- To develop a Learning Society which is an innovative in sustainable solutions (information/material products, services, systems, policies, and strategies) supported by nationally integrated systems of electronic health, electronic knowledge, electronic money, and virtual schools & universities, and on-line governments.

The second purpose is very important for the American creativity, which under the form of copyright industries of software, TV shows, Hollywood movies, books— today accounts for $45.8 billion in US sales overseas, ranking second among all American exports, just behind automobiles and just ahead of farm products (Billington 1995).

In 1993, the Clinton Administration has made the creation and development of the NII a top priority. The National Telecommunications and Information Administration (NTIA) Office has been created within the US Department of Commerce.[2] The US Commerce Secretary chairs the Administration Information Infrastructure Task Force which includes key officials from intergovernmental offices. The US Commerce Secretary appointed the NII Advisory Council, representing a broad spectrum of private-sector, public-interest, and governmental views. The Congressional NII Act of 1993 provides goals and means for pilot systems. The NTIA policy is:

- To promote private sector investments, through appropriate tax and regulatory policies,
- To act as a catalyst to promote technological innovations and new applications
- To ensure the NII security and reliability,
- To coordinate with other levels of government and with other nations the standardization of interfaces and removal of obstacles and unfair policies that may handicap the US economy and society.

The benefits of the NII for the nation are immense. The advanced NII will enable US firms to compete and win in the global economy, generating good jobs for the American people and economic development for the nation. The NII can transform the lives of the American people—ameliorating the constraints of geography, disability, and economic status—giving all Americans a fair opportunity to go as far as their talents and ambitions will take them (IFIF Task Force, 1993).

According to World Futures and the United Nations report (Marien, 1995), the globalizing economy and globalizing information are two of eight key challenges facing the world by the end of the 20th century.[3]

The world economy is evidently becoming more global and more dominated by telematic technology. As international corporations emerge, they make national economies more interdependent (Cowhey, 1993). Peter Drucker (1994) notes that formal and informal alliances are becoming the dominating form of economic integration in the world economy, and that for "developed economies, the distinction between domestic and international economy ceased to be a reality."

The emerging global economy includes the globalization of services, and the agro-food industry (expansion of niche markets for fresh and processed food) (McMichael, 1994). The travel and tourism industry becomes the world's largest industry (100 million employees and $2 trillion in gross sales) with growth at 4-5% in the 1990s (Hawkings, 1991). Attali (1991) defines images of "rich nomads" and "poor nomads" roaming the planet.

In the midst of a global job crisis emerges the global labor market. Steady jobs for good pay are becoming passionate memories or just dreams for more and more people. The informatization and robotization of work means the end of jobs. The ranks of the unemployed, the unemployable, the underemployed, and the subemployed are growing so fast that the global job crisis threatens the capitalistic system itself (Barnet, 1993). The world's skilled and unskilled resources are being produced in the developing world. While most of the well-paid jobs are being generated in the cities of the industrialized world. This mismatch triggers massive relocation of people (especially young, well educated workers) with the help of jumbo jets (Johnston, 1991).

The proliferation of information technology (about 150 million computers operated in 1994) and globalization of communications are widely viewed as the key factors in the globalization of economy and in proliferation of democratizing values. Connors (1993) sees further shrinking of the world which leads to the emergence of a culturally richer international society and their households and offices interconnected by public networks. This vision is supported by a trend of the emerging one global market, driven by programmed capitalism (Eastabrooks 1988). Masuda (1990) explains that the spirit of the information society is in fact the spirit of globalism, based on global information utility with cheap information for all.

Globalization and internationalization are two different terms. The former is more complex (1.6 bigger than the US economy) than internationalization. Globalization leads to stateless consortia which do not believe in political ideology but believe in economics. Through a natural process of business evolution, a selected group of global corporations rules the world. The group of 25 top global

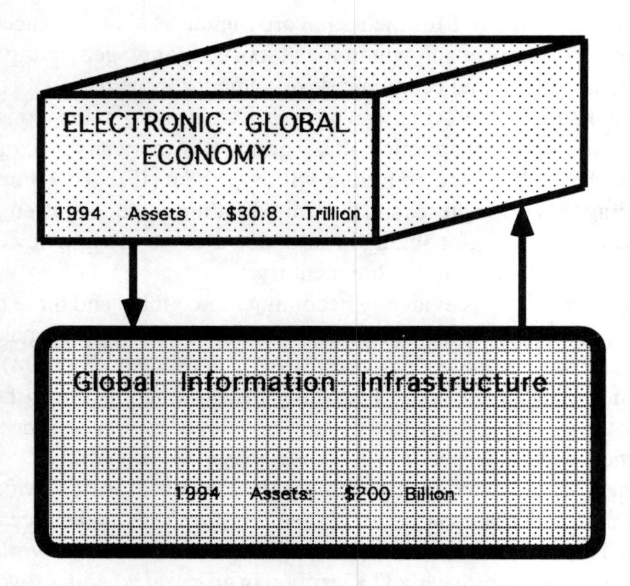


*Figure I-5: The Growth-Oriented Relationship Between The Electronic Global Economy and the Global Information Infrastructure*

corporations is called the Secret Empire[4] (Lowe, 1992).

The Fortune Global 500 largest corporations in 1994 had revenues of $10.2 trillion (28% of world GNP) and employed only 34.5 million workers. Their growth from 1993 to 1994 was at the level of 8.9% while the average growth of a developed country's GDP was at the level of 3%. In the same period of time the US Fortune 500's largest corporations had revenues of $4.3 trillion and employed 20 million workers. The comparable growth of revenues was 8.2%. If productivity (produced $ per worker) of both economies are compared, the global economy is 37% more productive than the US national economy.[5]

The growth relationship between the global economy and globalization of information (Global Information Infrastructure) is shown on Figure I-4. This relationship authorizes us to call the global economy as the Electronic Global Economy (EGE).

The goal of GII would be to transmit information with the speed of light from the largest city to the smallest village in every part of the world. The GII would be built according to an ambitious agenda that would help all governments, in their own sovereign nations and through international cooperation, take part in this revolution — a democratic effort not dictated or built by a single country.

According to Al Gore (1994), the GII will be a metaphor for democracy itself. Representative democracy does not work with an all-powered central government, arrogating all decisions to itself. That is why communism collapsed[6] and apartheid fell. Instead, representative democracy relies on the assumption that the best way for a nation to make its political decisions is for each citizen to have available the information they need, free speech and free elections. The Internet is sparking an intense debate in Saudi Arabia and other Arab countries. The authorities worry that they will lose their grip on political dialogue and public morals (Ambah 1995). The authorities of Arab countries, China, North Korea and other want to keep their citizens in informational slavery. But in the long-term there is not much they can do if the GII will be fully operational.

The architecture of the emerging GII includes:

- Emerging NII(s) of developed countries,
- Elements of EII(s), electronic money, electronic knowledge, and electronic health of developing countries
- Access points of EII(s) and electronic knowledge of less developed countries
- Private and public GAN(s)
- Internet

The relationships among the GII's components are shown on Figure I-5.

| COUNTRIES | COMPUTERS per 1000 population | MIPS* per 1000 population | NUMBER OF GLOBAL CORP. |
|---|---|---|---|
| *DEVELOPED* | | | |
| United States | 265 (1) | 516 | 151 (1) |
| Japan | 84 (11) | 139 | 149 (2) |
| Germany | 104 (10) | 141 | 44 (3) |
| France | 111 (9) | 180 | 40 (4) |
| Britain | 134 (5) | 217 | 33 (5) |
| Switzerland | 133 (6) | 220 | 14 (6) |
| Italy | 57 (12) | 98 | 11 (7) |
| South Korea | 33 (13) | 49 | 8 (8) |
| Netherlands | NA | NA | 8 (9) |
| Spain | NA | NA | 6 (10) |
| Canada | 175 (3) | 278 | 3 (11) |
| Australia | 175 (2) | 278 | 2 (12) |
| Norway | 153 (4) | 256 | 1 (13) |
| Ireland | 126 (7) | 208 | — |
| Singapore | 116 (8) | — | — |
| *DEVELOPING* | | | |
| Greece | 47 | 71 | — |
| Hungary | 24 | 34 | — |
| Mexico | 13 | 19 | 1 (14) |
| South Africa | 9 | 13 | — |
| Brazil | 6 | 10 | 2 (12) |
| Former USSR | 4 | 6 | — |
| *LESS DEVELOPED* | | | |
| Indonesia | 2 | 2 | — |
| India | 1 | 1 | 1 (15) |
| China | 1 | 1 | 3 (11) |
| *WORLD AVERAGE* | 27 | 47 | |

* Million Instructions Per Second

Source: Karen Petska Juliussen and Egil Juliussen, 6th Annual Computer Industry Almanac (Lake Tahoe, Nev. Computer Industry Almanac, Inc. 1993); population figures from Population Reference Bureau, 1993 World Population Data Sheet (Washington, DC: 1993), Number of Global Corporations from FORTUNE, August 7, 1995.

*Table I-1 Key Indicators of GII Development in 1993/1994*

The pace of the GII development will be fastest among developed countries and slowest among less developed ones as indicated by key indicators in Table I-1.

Each IFIF has the basic component as either a LAN, MAN, RAN, WAN, or GAN and open connectivity to other networks as it is defined in Table I-2. Figure I-6 defines each IFIF as a network of networks.

The impact of investment in telematic technology will have a powerful multiplier effect on the economy in coming years. Here is how: higher investment rates boost productivity. Faster productivity growth raises real incomes. Consumers spend more, companies start hiring and the economic tempo picks up. Much of the gain from telematic investment will stay in the domestic economy too, because the US producers account for a major chunk of the world telematic industry (Farrel and

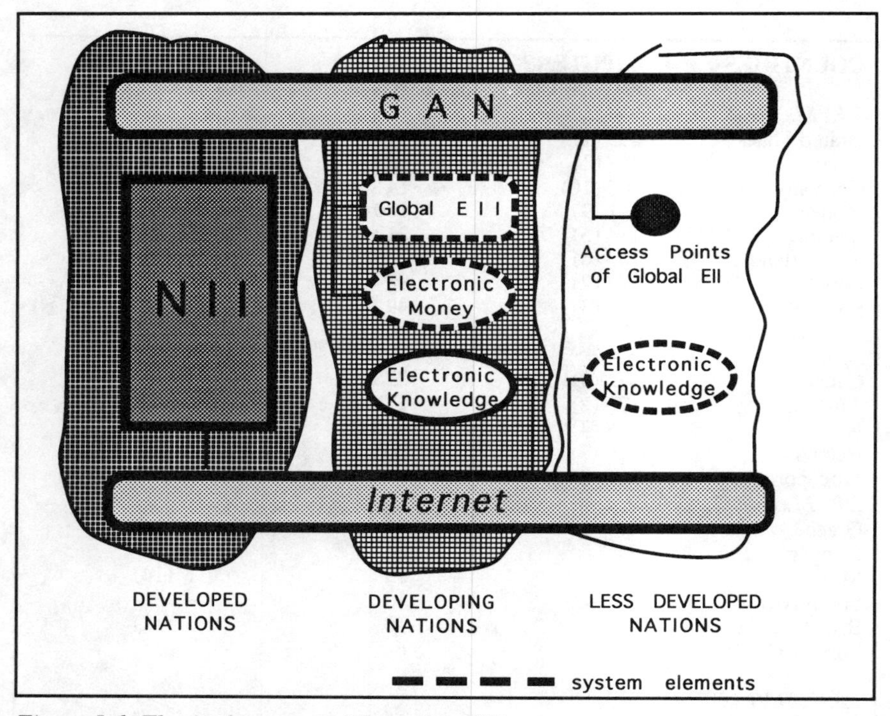


*Figure I-6: The Architecture of Emerging GII*

| IFIF | LAN | MAN | RAN | WAN | GAN |
|---|---|---|---|---|---|
| EII | B | C | C | C | C |
| LII | C | B | B | C | C |
| NII | C | C | C | B | C |
| GII | C | C | C | C | B |

B - basic network C - connectivity of the basic network to other networks

*Table I-2 Basic Networks and Connectivity of Information Infrastructures*

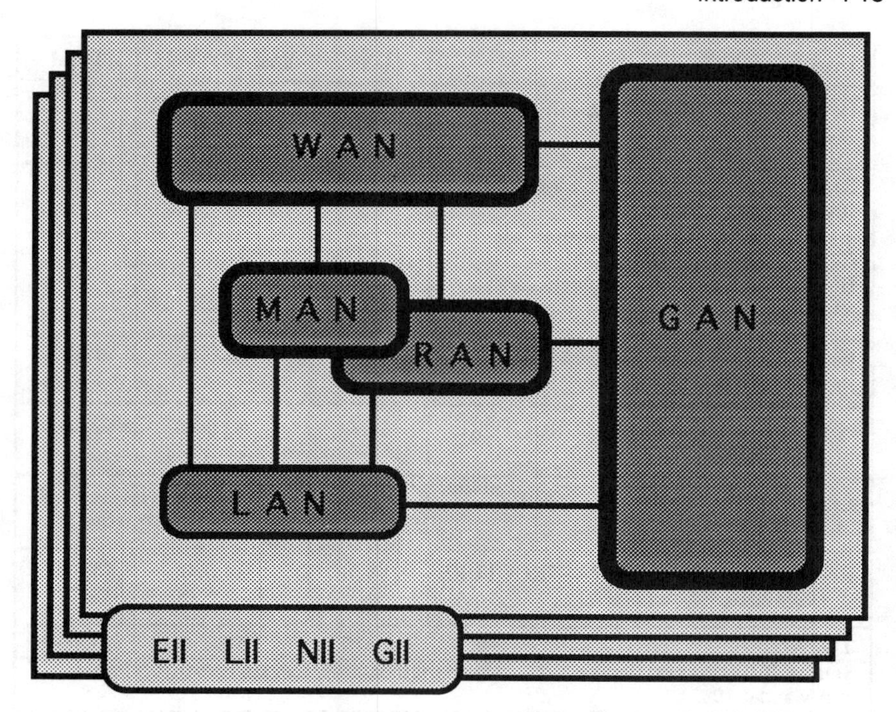


*Figure I-7: An Information Infrastructure as a Network of Networks or Matrix of Networks*

Mandel 1993).

Yet the raw investment numbers understate the dynamism that will be unleashed by building the IFIF(s). Much like the construction of the railroads in the 19th century, electric networks in the early 20th century, and the interstate highway system after World War II, the IFIF(s) will change the way we live at home and at work. It will also open up opportunities for new goods and services dreamed up by the nation's innovators and entrepreneurs—cutting-edge ideas that spur economic growth. The Table I - 3 illustrates the scope of investment in a new information infrastructure.

The information infrastructures are under construction worldwide. In 1995, the telematic industry, including software, is worth about $400 billion per year worldwide. The task of providing goods for IFIF(s) in 1995-2010 at the total value of $8.5 billion, is in the range of this industry's steady growing capacity.

If predictions are right, the gains can be tremendous, the potential limitless. Those Americans who already are active in hot cyberspace compare themselves to Lewis and Clark exploring in the American West. The potential benefits in developing and less developed countries are also immense. The GII can help in overcoming two great interdependent problems of the Third and Fourth Worlds, poverty and environmental degradation. Without the GIIs, residents of these worlds risk marginalization within the evolving EGE. The GII can provide for them the help in medical treatments, crop market predictions, weather forecasting, electronic knowledge, and so forth.

| Investments | USA | Rest of World | Total |
|---|---|---|---|
| INTERSTATE HIGHWAY (1959-63)<br>Federal gigaway grants | $38 | | |
| REAL ESTATE BOOM (1985-89)<br>Investment in office buildings | $53 | | |
| COMPUTER FRENZY (1989-93)<br>Business spending on computers, LANs, and office equipment | $22 | | $22 |
| GLOBAL INFO INFRASTRACTURE<br>in 1995-2010 | $200 | $1,000 | $1,200 |
| NATIONAL INFO INFRASTRUCTURE<br>in 1995-2010 | $400 | $2,000 | $2,400 |
| LOCAL INFO INFRASTRUCTURE<br>in 1995-2010 | $10 | $50 | $60 |
| ENTERPRISE INFO INFRASTRUCTURE<br>in 1995-2010 | $800 | $4,000 | $4,800 |
| TOTAL | $1,410 | $7,050 | $8,460 |

Source: IFIF(s)' investments estimated by the author, other investments from (Farrel and Mandel 1993).

*Table I - 3 Investment Spent for Major Infrastructures, in Billions of 1993 dollars*

For better and/or worse, the telematic technology revolution will certainly proceed to develop. Most decision-makers view technology as an autonomous force which is difficult to stop. There are a few critical voices about the euphoric opinions of the technology's social impact. Schiller (1992) argues that unfortunately, the global culture becomes the central force. Hamelink (1988) cautions that massive investments in telematic technology is a gamble. Those who are responsible for the game with technology resist to admit this. While "the problem of technology choice is, finally a moral problem." The late Ellul (1990) excoriates that the proliferation of technology creates a flood of incoherent and useless data and "enormous global disorder." He argues that each Great Design leads to the subordination of humanity to technical systems. Although he knows that the totally computerized society is inevitable--the ineluctable outcome is dictatorship and terrorism.

Kelly (1994) states in his book "Out of Control" that the new world emerges rather as neo-biological word which is the marriage of the born and the made. To some extent, cows and carrots are more indicative of the kind of inventions humans will make in the future—products that are grown (including self-made and self-repaired ones) rather than manufactured. No one can be in control—and it is better that way.

"Out of Control" is not a better way of being, argues Brzezinski (1993). He warns us that we are living in a time of fragmentation: increasing disunity in Europe, dangerous eruptions in Islamic republics, and the evils of the "permissive cornucopia" in America, which, through overstimulation and self-gratification, have led to the collapse of Western moral and spiritual values, we are living in a climate that is "out of control."

The challenge of solving all the threats to humanity requires a global mind, that, as such, is information-driven. The better, worldwide creation and distribution of information and knowledge is perhaps one of the last hopes of humanity. The Toflers (1995) motivate us that like the generation of the revolutionary dead, we have a destiny to create a Third Wave Society. In this book we understand that this is a Learning Society, with the global mindset, which is perhaps capable of solving civilization threats.

The book analyses the birth, vision and architecture of all major components of the IFIF(s). The only component—electronic health is not modeled in the book, since during 1993-95 and perhaps in years to come, the national health care policy was a subject of strong political debate in the US. Unless the national health policy will be defined in legal terms, modeling of its electronic solutions seems to be premature.

## References


Ambach, F.S. (1995) "An Intruder In The Kingdom," *Business Week*, August 21, p. 40.

Attali, J. (1991) *Millenium: Winners and Losers in the Coming World Order*, New York: Times Books.

Billington, J. (1995) "In Defence of Creativity," *Civilization*, Sep-Oct, p.91.

Brzezinski, Z. (1993) Out of Control, New York: Collier Books.

Connors, M. (1993) *The Race to the Intelligent State: Towards the Global Information Economy of 2005*, Oxford UK and Cambridge MA: Blackwell Business.

Cowhey, P. (1993) *Managing the World Economy: The Consequences of Corporate Alliances*, New York: Council on Foreign Relations Press.

Drucker, P. (1994) "Trade Lessons from the World Economy," *Foreign Affairs,* 73:1, Jan-Feb, p.99-108.

Eastabrooks, M. (1988) *Programmed Capitalism: A Computer-Mediated Global Society,* Armonk NY: M.E.Sharpe.

Ellul, J. (1990) *The Technological Bluff*, Grand Rapids MI: William B. Eerdmans Publishing Co.

Farrel, Ch. and M. J. Mandel (1993), "What's Arriving on the Information Highway? Growth," *BusinessWeek,* November 29, 1993, p. 40.

Hamelink, C.J. (1988) *The Technology Gamble. Informatics and Public Policy: A Study of Technology Choice,* Norwood, NJ: Ablex Publishing.

Hawkins, E. (1991) *World Travel and Tourism: Indicators, Trends and Forecasts,* Vol.1. Wallingford, Oxon UK: C.A.B. International.

Gore, A.(1994) "Technology and Democracy," *Discovery*, October, p.45-46.

Johnston, W.B.(1991) Global Work Force 2000: The New World Labor Market, *Harvard Business Review*, 69:2, March-April, p.115-127.

Pelton, J.N. (1989), Telepower: The Emerging Global Brain, *The Futurist*, Sept-Oct. p.9-14.

Kelly. K. (1994) *Out of Control,* Reading MASS: Addison-Wesley Publishing Co.

Masuda, Y. (1990) *Managing in the Information Society: Releasing Synergy Japanese Style,* Oxford UK and Cambridge MA: Basil Blackwell.

McMichael, P. (1994) *The Global Restructuring of Agro-Food Systems*, Ithaca NY: Cornell University Press.

Schiller, H.I. (1992) *Mass Communications and American Empire*, Boulder CO: Westview Press.

Targowski, A. (1990) "Strategies and Architecture of the Electronic Global Village,"

*The Information Society*, 7:3, p. 187-202.

The National Information Infrastructure: Agenda For Action, IFIF Task Force, the US Department of Commerce, 1993.

Tehranian , M (1988) "Information Technology and World Development," Inter-Media, 16:3, May, p.30-38.

Toffler, Alvin, The Third Wave, New York: Morrow, 1980.

Toffler A. and H. Toffler (1995) Creating A New Civilization, Atlanta GA: Turner Publishing, Inc.

Wells, G. (1938) Doubledey, Doran.

Marien, M (1995), *World Futures and the United Nations Report,* Bethesda Maryland: The World Future Society.

## Endnotes

[1]Tera = 1000 Giga; 1 Giga = 1000 Mega; 1 Mega = 1000 Kilo; 1 Kilo = 1000

[2]For the record one must notice that similar offices were created in France and Poland in 1964.

[3]Other six key challenges are: Politics in the Post-Cold War, Democratization and Peacemaking, the Global Economy, Population Growth, Global Environment Problems, A New Path for Development. Women and Children.

[4] In 1994 the Secret Empire was composed of the following global corporations: 15 Japanese (Mitsubishi-1, Mitsui-2, Itochu-3, Sumitomo-4, Marubeni-5, Nissho Iwai-9, Toyota Motor-11, Hitachi-13, Nippon Life Insurance-14, Matsushita-17, Tomen-18, Nissan Motor-23, Nichimen-24, Kanematsu-25), 8 US (GM-5, Ford-7, Exxon-8, Wal-Mart-12, AT&T-15, GE-19, IBM-21, Mobil-22, 1 Anglo-Dutch (Royal-Dutch/Shell-10), and 1 German (Daimler-Benz-20) (Fortune, August 7, 1995, p. F1-2).

[5] $295,652 (global) versus $215,000 (national)

[6]Communism collapses in Poland in 1989 when the volume of underground publications exceeded the volume of official ones and censorship lost its power.

# Part I

# National Information Infrastructure

# CHAPTER 1

# Information Utility

## The Information Utility Concept

The influence of information utility on human performance will be greater than the combined effect of television and the telephone. The information utility will applied at home, in the office, the library, classroom, and many public locations. People will act and think differently. The information utility will interact directly with human memory and mental processes. It will be an extension of a brain/mind. A combination of computers, television, telecommunications, and information services create a new telepower. This telepower will allow for instant communications with a person wherever he or she is located. It will improve the learning, storing and thinking capabilities of mankind. As a result, it will improve action and decision-making skills. The Labor and Intellectual Revolutions in the 19th and 20th Centuries have replaced human physical effort with machines and smart machines respectively. The Interactive Age in the 21st Century may improve human mental effort by telepower, "Anytime, Anyplace, Anywhere."

The first use of the term information utility appears to have been made by Martin Greenberger in his paper on "The Computers of

Tomorrow" (Greenberger 1964). He noticed, "Barring unforeseen obstacles, an on-line interactive computer service, provided commercially by an information utility, may be as commonplace by 2000 A.D. as telephone service is today." Three books were published on the subject by Parkhill (1966), Anderson (1967), and Sprague (1969). Sprague defines information utility as a class of on line-real time systems in which a large number of individual users from many different organizations share a central data processing and memory complex. Each user will be supplied with a data terminal, or input-output device, connected directly to the center at the time of use. Table 1-1 lists information utility characteristics.

Some examples of uses of information utility (IU) are: the Post Office, library, a tax service, saving accounts, stock brokerages, travel services and so forth.

Information utility, being a combination of computers, television, telecommunications, and information services, supports operations of electronic organizations, telecities and the Electronic Global Village as well as the Electronic Global Person.

## The Information Utility Technology

Information technology and telecommunications have a long history. The first telecommunications systems was the telegraph, developed by Samuel Morse in 1837. By 1851 over 50 telegraph companies operated in the United States. Western Union Telegraph was founded in 1856 and operates yet today. In 1876 Alexander Graham Bell invented the telephone and a year later in 1877 the Bell

1. Central on line-real time facility
2. Many subscribers at remote locations
3. Information storage, retrieval, processing, and computing provided
4. Services provided at subscriber's own location
5. Service simple to understand and use in subscriber's own language
6. Service—fast, immediate, and reliable—two way communication
7. Service relatively inexpensive
8. Charges for service on a base plus unit transaction
9. Terminal devices tailored to subscriber's own requirements
10. Knowledge of computers, systems, or programming not needed by subscriber
11. Question and answer mode available—subscriber-system partnership
12. Availability in some utilities for subscriber to set up, write and check programs
13. Utility responsible for equipment, reliability, file protection and control

*Table 1-1 Information Utility Characteristics (Sprague 1969)*

Company was formed. In 1885 American Telephone and Telegraph Company (AT&T) was created to provide long-distance communications services. In 1946 the first electronic computer ENIAC was in operation; in 1951 the first commercial computer was in use to automate data processing. In 1964 the third generation computers (IBM 360) were introduced which could process data from remote locations via data communications line and Remote Job Entry terminals. Four years later in 1968, the ARPANET was launched to interconnect national large-scale computers via packet distribution technology. As a result of it, the first local area network was developed in 1971 in Hawaii. Based on this experience a strong wave of LAN developments took place in the 1980s and 1990s. The enterprise-wide computing caused the development of wide area networks (WANs) and metropolitan area networks (MANs) to expand user's LANs over large distances.

It is useful to classify information utility technology into:

- Information technologies
- Access and transmission technologies
- Switching and networking technologies

It is important to note, however, that the boundaries among them are not always clear cut, and are eroding in the face of technological pace [1]. Access, transmission, switching and networking technologies form telecommunication systems.

Telecommunication includes disciplines, means, and methodologies to communicate and transmit messages over distances messages in the form of data, image, audio, and video. Data communication is limited to one medium computer-generated data. Telecommunication services contain the following services:

- Telephony
- Motion videotext (multimedia)
- Telefax
- Color fax
- Low speed file transfer
- CAD/CAM
- HDTV
- Videophone
- Video retrieval
- Videotext
- Interactive data
- High speed file transfer
- TV
- Hi-Fi distribution

A speed of message teletransmission has increased by an order of magnitude every 20 years: 1950s: $10^8$ b/s, in 1990s: $10^{10}$ b/s. We could transmit the content of 250 Bibles in 1 second in 1990. We will be able to transmit the equivalent of 2,500 Bibles in 1 second in the year 2010, and 25,000 Bibles per second in the year 2030 (Minoli 1991).

A modern telecommunication network is shown on Figure 1-1.

A telecommunication networks consist of four general components:

1. Customer premise equipment (CPE) such as:

   - telephone sets
   - PBX equipment
   - Fax machines
   - Multiplexers, concentrators, front-end computers
   - Local Area Networks
   - modems

2. Local loop between a central office and subscriber

3. Switching systems between central offices (LATA) and interchange carriers (IXC)

4. Interoffice trunks such as:
   - satellite system
   - microwave system
   - coaxial system
   - fiber optic carrier system
   - dedicated (private) line

Telecommunication networks can be public or private. Both data and telecommunication networks have the same same functionality and capabilities.

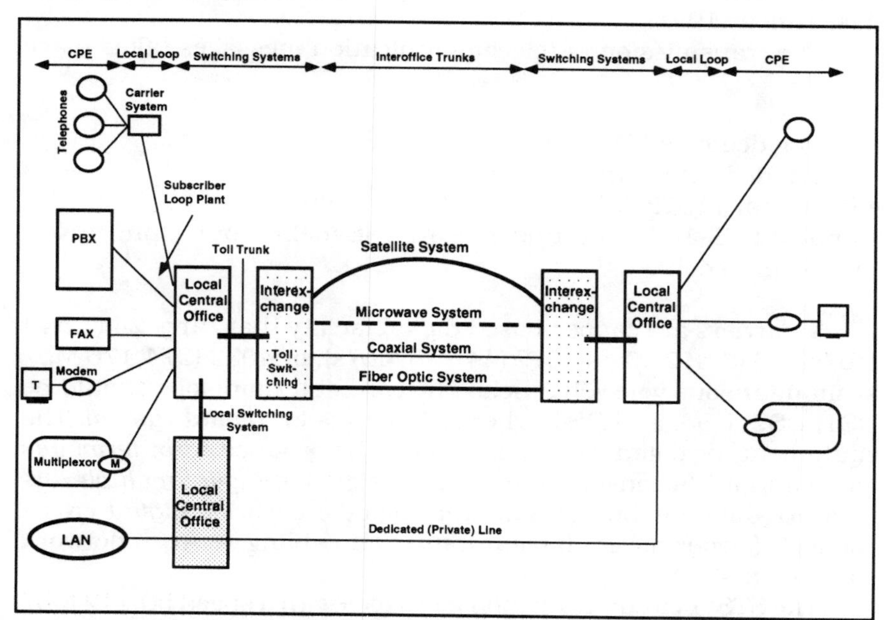


*Figure 1-1: A Modern Telecommunication Network Architecture*

Two significant developments in loop technology—the introduction of loop carrier systems and digital remote electronics and switching technology—helped implement the telecommunications network. Loop carrier systems concentrated access lines by combining many customers into one or more shared trunks. Previously each customer needed a dedicated (nonshared) loop. The introduction of digital switching reduced the amount of dedicated loop plant by allowing remote nodes to be connected to the host digital switch.

The deployment of fiber optics generally provides more capacity, reliability, flexibility, and functionality than existing metallic cable. With minimum transmission loss, fiber allows more signals to travel over longer distances using a smaller number of repeaters as does copper wire. Commercially available fiber optic technology operates in the 500 million megabits-per-second range. This speed will further increase in the future. Fiber loop systems, however, are still uneconomical for most of the residential communities.

How does fiber optics work? Phone conversations and computer multimedia information are converted by an optical transmitter to a series of light pulses. The light pulses are sent through glass fiber wire by a laser flashing on and off at very high speed. At the receiving end, a light sensitive receiver changes the light back to electric pulses. A half-inch-thick fiber optic cable with 72 pairs of fibers can transmit 3.5 million conversations. This technology is better than the traditional one since light can travel much farther before a costly amplifier is needed to boost the signal. Optical fibers are not affected by electrical or radio inference or lightning (Office of Technology Assessment 1991).

The transmission of telecommunication messages takes place via:

- analog dedicated circuits
- digital dedicated circuits
- fractional T1, T2, T3, and T4 digital circuits
- optical STS-1 circuits under SONET (Synchronous Optical Network) protocol.

T-carriers are copper-based digital facility that carry 24 (1.544 Mb/s), 96 (6.312 Mb/s), 672 (44.736 Mb/s), or 4032 (274.176 Mb/s) simultaneous voice channels. The correct terminology should be DS1, DS2, DS3, and DS4, where DS stands for *digital system*. The equipment for digital transmission can be grouped into: *terminals* (analog input is converted into digital signals), *digital multiplexers* (convert different bit rates in the digital network), *digital cross-connects* (electronic equipment frames for cabling between network components).

The STS-1 circuit transmits messages with a speed 50.112 Mb/

s via SONET. SONET network elements are: a)multiplexors (converting STS-1 signal into a grouped STS-N envelope and performing the electrical-to-optical conversion), b)convertors of networks signals into STS-1 payloads (783 bytes in length).

## Information Technology

Information technology allows individuals and organizations to store, handle, retrieve, and process data, information, knowledge and wisdom into more useful forms. Examples include computers, modems, facsimile machines and answering machines as well as system and application software. Information technologies encompass a vast array of computer system configurations such as:

- The Independent Computer System (home, personal computers, workstations,...) (Figure 1-2) is based on one central processing unit (CPU) which reflects computer power concentration in one place.
- Computer System Interconnection (fault-tolerant, non-stop processing) (Figure 1-3) is based on two or more CPUs with a network control unit which identifies a computer power reliability solution.
- Computer Networking Multisystem (Figure 1-4) is the interconnection of application CPUs to provide a single-system image of an organization's information resources. The system is based on a computer power sharing idea through applications of LAN, MAN, WAN, and GAN.

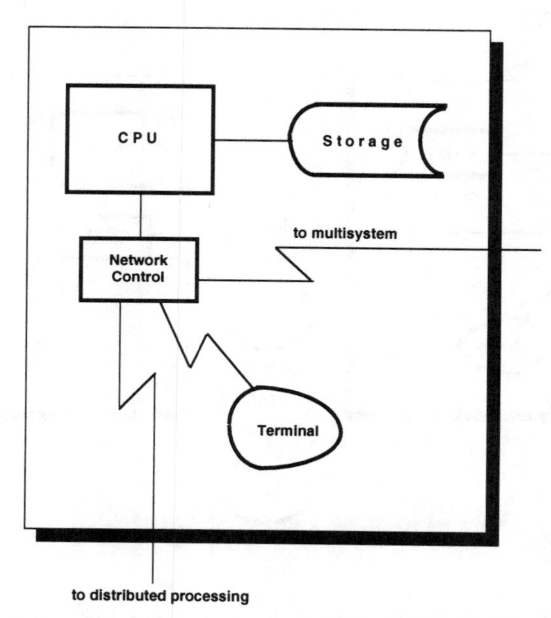


*Figure 1-2: The Independent Computer System Configuration*

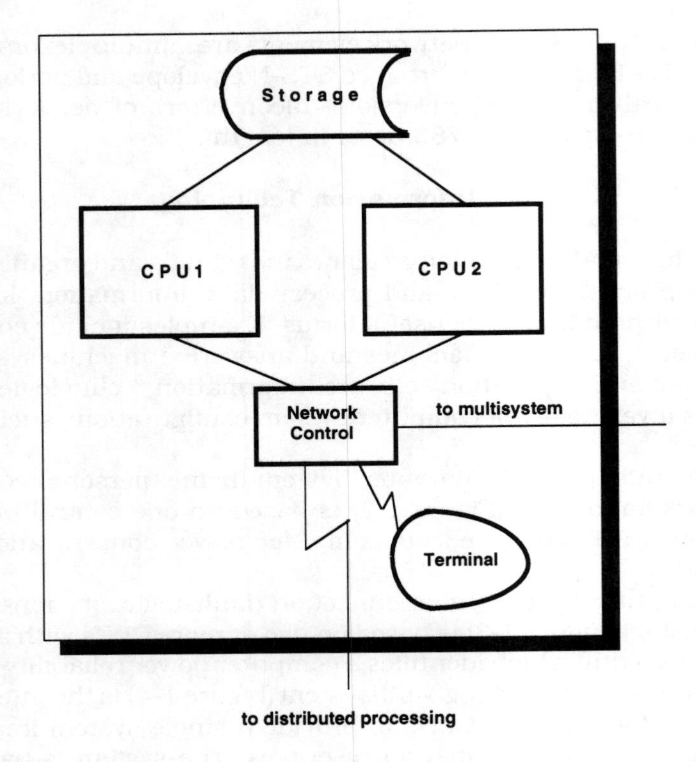


*Figure 1-3: The Computer System Interconnection Configuration*

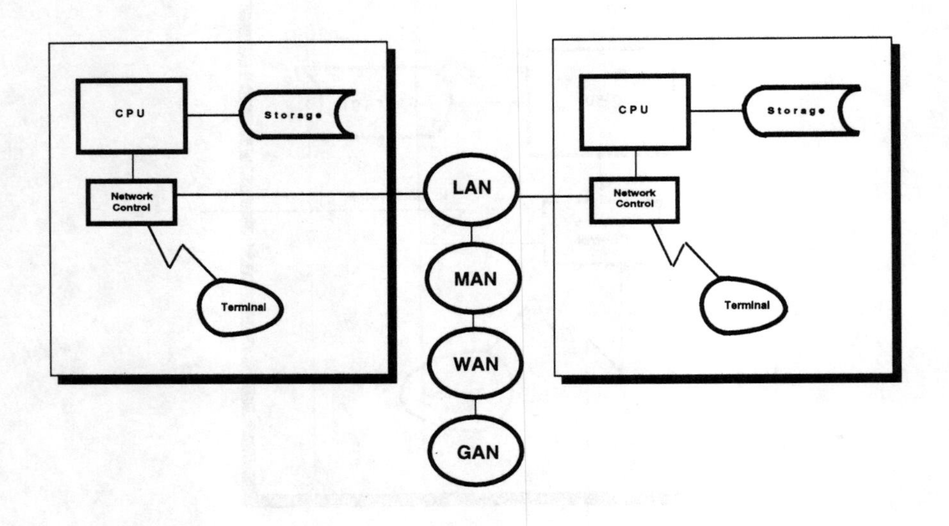


*Figure 1-4: Computer Networking Multisystem Configuration*

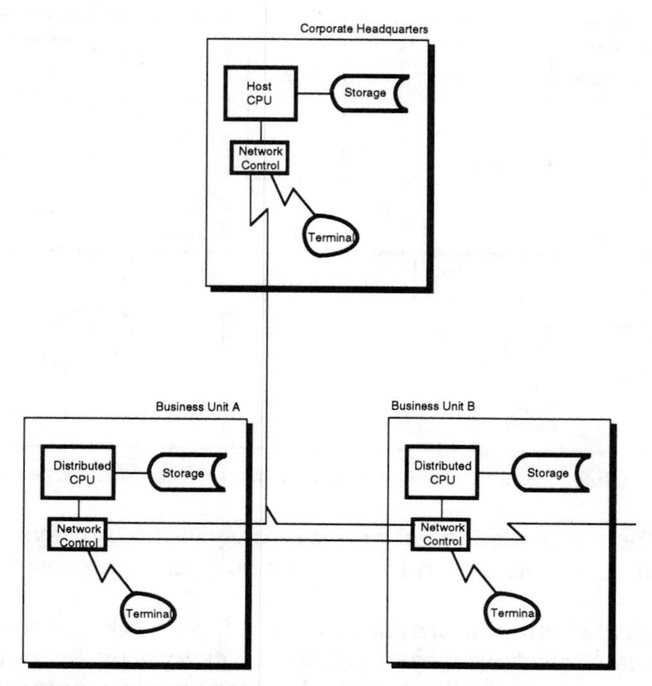


*Figure 1-5: Computer Networks Distributed Configuration*

- Computer Networks Distributed Processing System (Figure 4-5) is based on distributed CPUs to "departmental users." The idea of this system is to process internal data (about 80% of the total volume) locally and send out 20% of data volume to external users (headquarters and cooperating departments). The departmental users are in charge of data quality and software development or maintenance (or both). The application of proprietary networks or LAN, MAN, WAN and GAN is obvious in this case.
- Client-Server Computing Environment (Figure 4-6) is a style of computing in which portions of applications are distributed across a network on server systems, which clients access to obtain needed computer-based resources. This style of computing can best be defined as a series of requests and responses. Clients make requests for services, and servers respond with services.

Unlike the Computer Networking Multisystem configuration, where the goal was to get as many clients attached to the host clusters as possible, this configuration brings as many servers to the client as needed. Client/server computing focuses on the function of the computer servers in a network rather than on the size of computers. For example, display services are best provided by PCs, workstations, or windowing terminals (cooperative computing). On the other hand, computations, database, and communication ser-

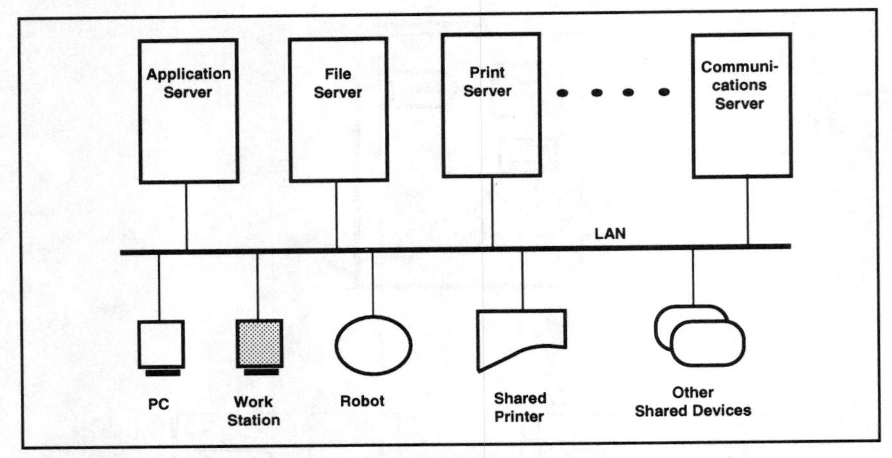


*Figure 1-6: Client/Server Configuration on a Bus or Ring of a LAN*

vices are best provided by more powerful systems. Any system may be a client or server, and in some cases both may exist on the same platform.

A typical client platform may include PCs, workstations, banking terminals, electronic cash registers, reservation terminals, robots, factory cell controllers, and laboratory instrumentation.

Servers are configured to support organizational informational needs. Servers are systems optimized to perform the following tasks: database, transactions processing, file storage, communication interfacing, printing, system management, and network management.

Client/server computing provides high-speed, low-cost (incremental) solutions to organizations. Server functionality and capacity need to be added only when necessary. Servers can off-load from more costly mainframe, or host systems. The accelerating trend of moving applications to lower cost platforms is moving many applications from host systems to servers.

Servers provide the ability to share resources among diverse client systems, improving productivity of individual clients and reducing the need to have the most powerful workstation on every desk. The pioneer of this configuration is Digital Equipment Corporation with its VAX clusters [11].

Computer applications are multiplying rapidly due to increased storage and processing capacity [2], the dispersal of intelligence throughout communication systems [3] as well as digitization and convergence of media [4]. According to Gilhooly (1989), by 1993-95, office workstations will be able to handle 32 million instructions per second, have 16 megabytes of random access memory, and cost approximately $350. Given such performance/price improvements, one can predict that by the early 2000s, the total number of computer workstations in Europe, Japan, and the United States will surpass 200 million. The majority of these will soon provide multimedia access at the desktop level.

An example of this is Computer Aided Design (CAD). According to Bruno (1990)—CAD is becoming the norm for businesses to efficiently design and test products and design and monitor machines and factories that make and assemble the products. CAD/CAM (computer aided design/manufacturing) replaces the construction and testing of actual models. Because designs are stored in electronic form rather than on paper, they can be quickly and easily updated and transferred. Intergraph Corporation recently introduced a product, the CAD Conferencing Module, that permits people at as many as eight different locations to simultaneously view and edit computerized blueprints and designs. The electronic blueprints are transmitted and loaded into each of the separate workstations so that they can be called up during the networked conference to be viewed and edited. Modifications to the still-frame image, such as zooming or highlighting features, by one person are registered at all the participating workstations. This conference arrangement expedites the design and engineering process by minimizing the encumbrance of mailing blueprints back and forth and increasing the communication between the firms various departments.

Although information technologies are becoming more user-friendly, the difficulties entailed in linking technologies and systems will likely continue to slow down the dissemination of applications.

## Access and Transmission Technologies

Access and transmission technologies transport information among and between other users and networks. They may provide:

- point-to-point interconnection, as in the case of the telegraph and the telephone
- point-to-multipoint interconnection, as in the case of radio and television
- multipoint-to-multipoint interconnection as in the case of bulletin boards, electronic mail systems, and local area networks (LANs).

These technologies can have either one-way or two-way capabilities (Office of Technology Assessment 1991).

### Telephony

Telephony so far is the most important two-way medium for transmitting information. There are today, in the United States, over 1,500 telephone companies with a total of 130 million access lines.

The top 25 companies account for 90 percent of the access lines. The Bell telephone companies serve about 85 percent of the market with about 50 percent of all central offices. The remaining companies are quite small by comparison.

A wide variety of new and more specialized service providers have emerged since the divestiture of AT&T in 1984. For example, some providers such as Telenet and Tymnet, sell packet-switched data communication services; other carriers specialize in high performance, digital point-to-point T1 service (1,5 Mbits per second). The Bell companies have on average 130 subscribers per route mile. A subscriber loop or wire between the central office and the user's premises is about 10,787 feet. Revenue per line is about $757 per year. In the United States common carriers can be classified as follows:

- Exchange-access carriers (intra-LATA: local access and transport area), such as the Bell Operating Companies (BOCs) and independent telephone companies; these are collectively known as Local Exchange Carriers (LECs);
- Interexchange carriers (IXCs), such as AT&T, US Sprint, and MCI and 200 others;
- Specialized common carriers (SCCs) such as Telenet and Tymnet which are a value- added carriers, or value added networks (VANs), or Teleport which sells fast-multimedia intra-city services and access to satellite communications.

**Cable Television**

Cable television began in the late 1940s as a technique of improving television reception and bringing in channels that could not be received locally for people in big metropolitan and rural areas. The first cable television system was installed for subscribers from a mountainous part of Pennsylvania to pick up signals from Philadelphia. By 1950, there were seventy five such systems around the nation and the industry was named as CATV: Community Antenna Television.

In 1975, RCA launched a commercial telecommunication satellite—SATCOM I. It allowed for the satellite delivery of specialized channels such as HBO, ESPN, Showtime, Cinemax, MTV, WTBS Atlanta, CNN, C-SPAN and so forth. Cable programming became a new competitive communication service in the market of "pay-per-view cable." Many additional channels with sport, entertainment, and education can be brought to a subscriber for a charge beyond the basic monthly fee.

By 1990, 12,000 cable systems had been built. The cable

industry is evolving from a technique of improving television reception to a telecommunication system providing over a hundred channels and two-way interactive television combined with computers and other information technology products. Figure 1-7 illustrates a basic cable television system.

The cable mechanism of the signal delivery consists of four steps:

- the program-supply service over the satellite or wire from a local studio
- the cable headend and a studio as a cable central office, where a set of electronic devices prepares a signal for the distribution
- the distribution system composed of trunks and hub feeder cable
- Subscribers drop cable.

Commercial cable systems are grouped into three major categories:

- community access channels (up to 12 channels)
- standard cable systems (13-35 channels)
- interactive cable systems (50 to 110 channels)

Citizen groups press hard for services they desire, and competition among applicants for franchises involves expensive publicity campaigns and sometimes attempts to discredit other applicants (*U.S. News and World Report* 1980). The Cable Communication

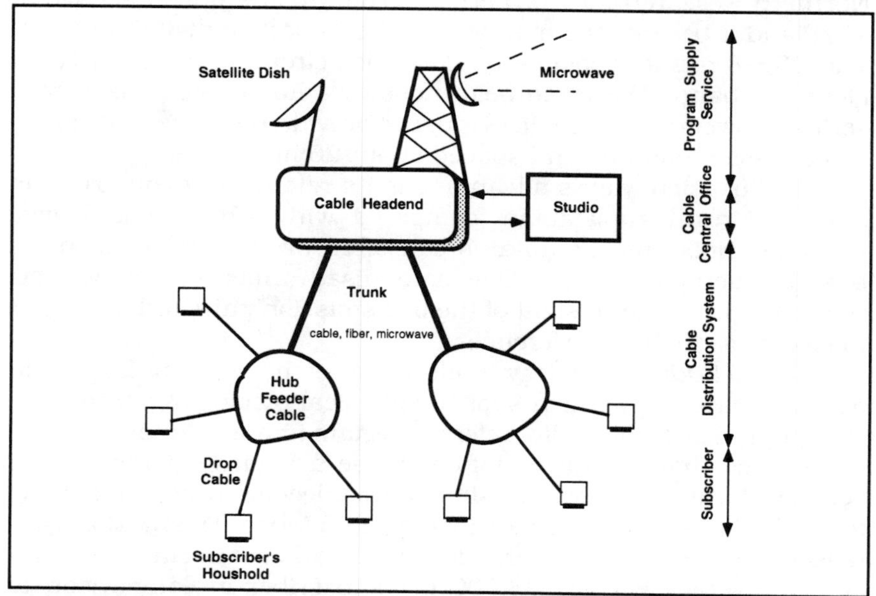


*Figure 1-7: Basic Cable Television System*

Policy Act of 1984 and the Communication Act of 1934 regulate the franchising process and the culture of communications.

The Community Antenna Television Association (CATA) estimates that cable penetration (home passed) in rural areas is higher than in urban and suburban areas (about 60 percent). Cable television can be accessed using wire or satellite. In urban areas it is available primarily through a coaxial cable provided by a cable company; in rural areas it is available primarily through coaxial cable, with 10 percent of it provided by satellite dish (cost between $750 and 1,800).

### Nonwireline Media

In remote areas, where the cost of providing wireline service is prohibitive, *microwave radio (long-haul)*, *digital radio (short-haul)* and *satellite technologies* can be used to provide less expensive access to communication services. With technological advances, these technologies provide services that are increasingly comparable to wireline service.

*Microwave radio* has long been a mainstay in telecommunications network technology. Historically, its primary use was high-capacity, long-haul toll service, and it will likely continue to be important in such markets.

Canada turned up its first coast-to-coast microwave radio system (TD-2) equipped with 180 voice circuits in 1958. The world's first long haul digital radio system, the DRS-8, was designed by Bell Northern Research for Northern Telecom. It was put in service in 1978 and it became the mainstay of the Canadian digital network. The DRS-8 has a capacity of 1,344 voice circuits that apply the 8 gigahertz band. The radio beam follows a line-of-sight path so a series of towers with amplifiers must be provided over the route of the transmission channel and spaced about 26 miles apart.

One of microwave's advantage is its relative low construction costs for rural applications compared with other technologies. Rooftops, hills, and mountains can often provide an inexpensive base for microwave towers. One major disadvantage of microwave is that it requires line-of-sight of the transmission path and is subject to electromagnetic interference.

*Digital Radio* technology is referred to as the "wireless loop," and does not have requirements for the physical placing of a transmission plant. The term "radio" refers to certain frequencies assigned to the service that are distinct from those assigned microwave toll service. Rural radio provides short-haul telecommunications using four DS3 lines—each with a capacity of 45 Mb/s. The most advantageous feature of radio technology is its low cost in rural settings. On the average it costs $3,000 per subscriber in comparison to

$10,000 to provide an access lines via copper wire.

Radio telephony is particularly advantageous for rural areas where the expense of extending wires to the customers—which may involve digging trenches, clearing rights of ways, or crossing difficult terrain—can become prohibitive. It is thus especially useful when extending service to only a few, widely dispersed customers. Another inherent advantage of wireless technology is that telephone companies have much greater flexibility in adding on additional customers and reinforcing their facilities than with conventional cable routes/ land lines.

Digital radio also has a number of advantages over analog radio:

- greater degree of security because of more complicated encoding schemes for the digital transmission,
- digital transmission is inherently better suited for handling data transmission,
- greater ability to operate in the presence of interference,
- higher capacity,
- time division multiplexing conserves spectrum and reduces cost because less base-station hardware is required to support a given subscriber population,
- ultraphone is software-based and thus more open to further technological improvements (Calhoun 1990).

*Satellite Communications*. The concept of a geostationary satellite was first published in October 1945 in an article entitled, "Extraterrestrial Relays" by a British scientist Arthur C. Clark. Twenty years later, on April 6, 1965 the first commercial satellite, named Intelstal i or Early Bird, was placed in geostationary equatorial orbit over the Atlantic Ocean—by the American Communications Satellite Corporation (Comstat). Comstat represents the United States in the International Telecommunications Satellite Consortium (Intelsat). From 22,300 miles above the Atlantic, Early Bird linked North America and Europe via 240 high-quality voice circuits and made live television available for the first time. Very soon six larger satellite were placed over the Atlantic, Pacific, and Indian Oceans. However, the first American communications satellite *Score* was put into an orbit in 1959. President Eisenhower broadcasted a Christmas message via this satellite.

If the United States were the pioneer in the launching the first international satellite, Canada was the leader in the domestic geostationary satellite system. Canada launched on September 23, 1962 the first satellite Alouette I. Telesat Canada launched its first commercial satellite Anik A1 in November 1972. Later satellites have been launched to carry telephone traffic and video services across the huge territory of Canada.

The U.S. Federal Communications Commission (FCC) allows firms to put about 30 satellites in parking orbital slots 2 degrees apart. A satellite has electronic circuitry called a transponder. It interexchanges and amplifies signals between the earth station to satellite to the earth station path. Modern satellites typically have 24 transponders carrying one color television channel and 1,200 voice/ data channels at the speed of 50 Mb/s.

Satellites have to stay in one position on the orbit. Otherwise, they would not be within the line of sight of their earth stations at all times. To remain stationary, the satellite must have a period of rotation equal to the earth's period of rotation. This takes place at an altitude of 22,300 miles (35,880 kilometers) with a speed of 6,870 miles per hour. At this altitude, three satellites are enough to cover almost the entire Earth.

Applications of satellite technology include: video teleconferencing, news gathering, point-to-point and multipoint data transmission, and direct broadcast satellite (DBS) to households. "Sky Cable" aims at delivery of digital audio and HDTV to 12-18-inch receivers ($300). In 1990, the Europeans had two functioning DBSs, Astra-1 and TDF-1; the Japanese broadcast HDTV on three DBS (BS-2a, BS-2b, BS-3). In the future video will be the most broadcasted, data transmission and will be very popular; voice will be in "the third strong place."

The development of the Very Small Aperture Terminal (VSAT) reduced costs and forced businesses to shift from wireline to VSAT technology. VSATs are particularly cost-effective when businesses need to communicate with remote sites. Thus, many major corporations—e.g., Chrysler, Nissan, Toyota, K-Mart, Thrifty Stores, and Frito-Lay are using VSATs to develop private wide area networks (WAN). On the other hand, Hughes Network Systems provides public leased WAN service to businesses from a shared hub facility (Office of Technology Assessment 1991).

Smaller companies that cannot afford to install their own systems can take advantage of shared-hub networks. For example, Terra International, Inc., based in Sioux City, IA, recently chose a shared-hub VSAT network to connect its outlets spread throughout the country with its centralized computer center, which provides an array of business support applications. Terra, which manufactures and sells agricultural products, cites a reduction in communication costs in excess of 25 percent since VSAT was installed in 1989. Another benefit is increased reliability of data transmission. The VSAT network eliminates the need to deal with many phone companies, each with different pricing structures and technical capabilities (Khan1990).

The advantages of satellite communication include: the superiority of transmission to dispersed locations, easy reconfiguration of satellites to cover different areas, ease of installing satellite antennas, and the unlikelihood of failures of the total network.

*Cellular Communications.* From microwaves to microcells. The solution is just in the name. Cells are the defining characteristic of cellular technology. Take them away and you are back to the conventional mobile telephone service of the 1970s. Throughout most of cellular's early development, from early 1950s through the mid 1980s, it was viewed from a "stand-alone" perspective. Separate companies operated separated systems. Until cellular reached the market in 1983, the only mobile telephone service available was improved mobile telephone service (IMTS), which covered its market from a centrally-located antenna and a main base station interconnected to the landline network. IMTS technology had more in common with dispatch and paging than telephone switching.

Operators now realize that cellular is the wireless extension of the public network. Customers expect to use their car phone or portable the same way they use their car phone or portable the same way they use their home or office phone. They demand features like call waiting and call forwarding. They expect to transmit facsimile and data over the airwaves with the same easy as the landline network. If they travel, subscribers want their phones to work the moment they step off the jetway onto the airport concourse. They look for itemized, timely and accurate billing (Titch 1991).

Cellular communications is something of a Cinderella story within the telecommunication industry. Although cellular companies were long regarded as a hot investment, the service they provided was viewed as an interesting, but non-essential, adjunct to the future of the public network. The current vision for the public network is of a seamless, intelligent infrastructure that provides voice, data and image communications to customers wherever they are, in a form determined by that customer. The end user who wants to be in constant communication with the rest of the world can do so. The user who wants a certain list of critical calls to follow on vacation in the Michigan woods can arrange that.

Fundamental to this process is the ability of the network to find the user and route the calls to the appropriate access vehicle, whether it is a portable hand-held phone in a Wall Streeter's pocket, an office desk phone or the lone pay phone at John's shop.

Providing that level of connectivity requires a network infrastructure that encompasses massive databases, sophisticated switches and the high-quality transmission systems that will link them to each other and to their customer base. But it also requires access to the customer in places where no landline phone can go. The wireless and LEC industries needed a market hook to justify the expense of building such an infrastructure. Personal communication service (PCS), which combines customized call routing with mobile access to the network, provides such a hook. Cellular carriers need an intelligent infrastructure to stretch the boundaries of what can be achieved today in finding the customer with any incoming call.

In the coming years landline and cellular technologies will converge. Now that Cinderella cellular has come to the ball, her choice of dance partners will be interesting to watch (Wilson 1991). In 1992 McCaw Cellular Communications, Inc. selected AT&T; Centel Corporation selected Sprint. Immediately, MCI and Bell Atlantic signed cooperation agreements with the cellular companies to create a competitive mobile infrastructure that perhaps may replace the landlines of the LEC industry. In the mid-1990s, AT&T will re-enter the local market and MCI and US Sprint will be forced to follow the case. But, it will require significant capital infusion to develop local services.

What makes cellular so popular? Simply put, people want to be on the move, but they do not want to be out of touch. They want to be available to anyone, anywhere. This is the direction cellular communications is moving. From the first active cellular system in 1983, the use of cellular has grown phenomenally. Starting with zero subscribers, there are now cellular companies building networks across America and throughout the world.

Cellular communications is the telephone in modern radiotelephone service. At the heart of every cellular system in the world is a mobile telephone switching office (MTSO). On the most basic level, it manages the cellular network like any other wireline local exchange manages the network in its territory. It routes calls, logs call completions for billing purposes, maintains a database of subscriber profiles, and connects and interfaces with other segments of the public network (with central offices, PBXs, Centrex,..).

The hundreds of cell sites checkering cities everywhere make up the true network infrastructure. As in the landline network, much of the cellular plant is geared toward transmission. But instead of digital loop carriers and local loop distribution systems, cellular deals with base stations, antennas, signal amplifiers and FM propagation methods—all of which are geared toward reproducing landline-quality service for thousands of individual mobile and portable phones in a given area.

On the broadest level, cell sites contain three components:

- Antenna, omnidirectional. As traffic in a particular area grows, operators can either create sectors within the cell, or create up to six new cells from one location by replacing an omnidirectional antenna with directional ones.
- Base station, it is a radio set which transmits and receives messages.

- Interface, connects to the rest of the network, using high-speed fiber optics and digital microwaves. In early days, cell sites were connected directly to MTSO. Now, more cells are interconnected via switching public networks. In most cases, cell interfaces (links) use four, eight or twelve DS-1s, up to DS-3. Once a level of DS is reached, it is necessary to split a cell. About 70 percent of the cellular links in the country are based on microwave. About 25 percent apply leased lines (mostly fiber), from the LEC, and the remaining 5 percent use dedicated facilities built and owned by the carrier. Microwave is used in the West and Northwest, rather than in Southeast, because drier weather conditions make for better radio transmission. Transmission equipment vendors are also looking ahead to the advent of wide area microcell deployment as an opportunity for growth. Already DMC and TeleSciences have 38-GHz microwave radios, designed for hops up to two miles. The compact 38-GHz radios, featuring antenna dishes about a foot in diameter, can be used to link microcells with larger standard cells.

The architecture of cellular communications is shown on Figure 1-8.

In Los Angeles, home to nine of the 10 busiest highways in the country, PacTel Cellular in 1991 had 235 cells covering 10,500 square miles and supported by three MTSOs. New York has 130 cells, also supported by three MTSOs.

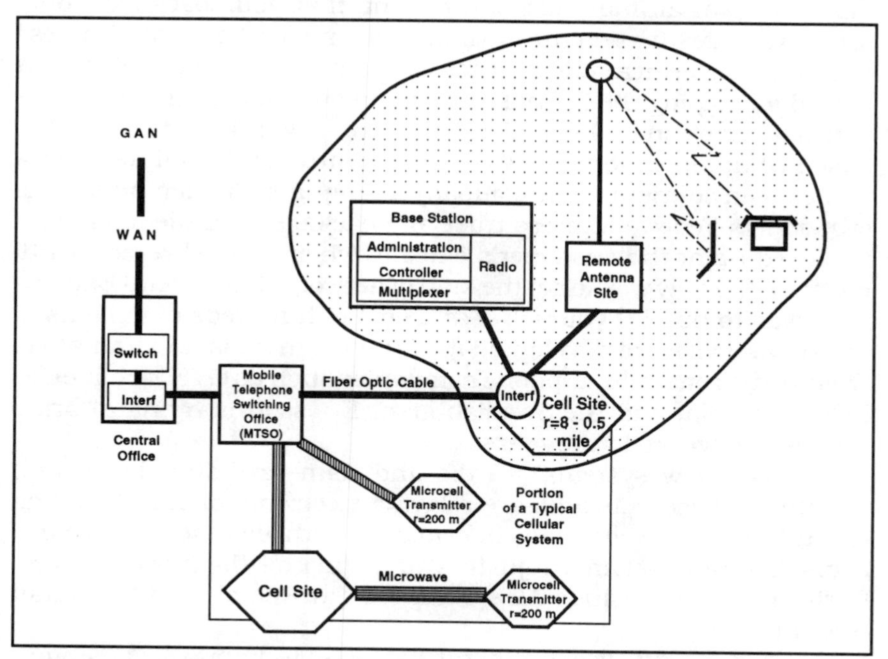


*Figure 1-8: The Architecture of a Cellular Communications System (r-radius)*

In the generic architecture, the MTSO and the Central Office's switch are tied by means of a fixed voice circuit backbone (via fiber optic cable or microwave) to multiple base stations, covering the city. The original cell sizes decrease down from a starting radius of some eight miles to about a half a mile limit. In congested cities, a microcell may have a radius of 200 meters. The number of such microcells in a Personal Communications Network can grow to over 5000 units. Today's cellular systems have capacity for about 200,000 users; however, several developments indicate that if the capacity will exceed 10 million subscribers, costs of service will go down from $80 to $25 per month.

The potential of strategic applications of cellular communications is the tip of the iceberg. As cellular systems become more a component of the global telecommunications network, it is imperative that they communicate with one another. The vision for advanced mobile communications is the concept that someone should be able to begin a car phone conversation on the George Washington Bridge in New York and not have to hang up until reaching the San Francisco Golden Gate Bridge or even moving to an airplane and traveling to Japan or Singapore, while being on the phone with the same person.

*Data Through the Air.* Frederick W. Smith became a business legend by creating the overnight package-delivery industry with Federal Express Corporation in the 1970-80s. He did it by assembling his own airline and guaranteeing that your package would arrive at its destination the next morning, no matter what. But less known is the critical role that telecommunications technology has played in keeping the 7.7 billion company the dominant force in its market. Back in 1977, Federal Express began building a radio-based communications network that eventually tied all its offices, planes, and couriers to the company's computers in Memphis. This system lets Federal Express trace any package from the moment it is picked up at the customer's office to when it is delivered. Until recently, the only way to get the kind of network that Federal Express runs was to build it yourself— at a cost of hundreds of millions of dollars and years of development. But now there is an alternative. Everybody from cellular phone companies such as GTE Mobilnet, to IBM, Motorola, and the Swedish giant Ericsson is trying to bring wireless networks to the masses.

The new systems —Ardis and Ram—are already up and running. These networks are relaying electronic mail to traveling executives and exchanging information with field service technicians, insurance-claims adjusters, and other on-the-move workers. Early customers include ADP, National Car Rental, and ICL Retail Systems.

The payoff, as at Federal Express, is improved customer satisfaction and greater operational productivity [13].

The mobile data business is a major part of an even bigger market for all kinds of wireless data communications, including office computer networks that use airwaves instead of wires. In 1992, sales of all kinds of wireless data equipment reached $450 million while revenues from wireless data services hit $160 million. Within the next 10 years, equipment sales could more than quintuple, to $2.5 billion and service can surge to $1 billion [20].

There are four ways to move data through the air, as it is explained in Table 1-2 [14].

*Global Development of Wireless Communications.* The 1992 World Administrative Radio Conference (WARC 92) of the International Telecommunications Union (ITU), which ended March 3, 1992, in Torremolinos, Spain, resulted in a worldwide allocation for mobile services in the 1.7 to 2.69 GHz band. This change in status, which brought all three regions of the world into a common standard, took place under the name of Future Public Land Mobile Telecommunications Systems (FPLMTS) for regional and global services. The conference also allocated spectrum for Low-Earth-Orbit (LEO) satellite services that may provide Personal Communication Service (PCS) to remote areas worldwide (149.9 to 150.05 and 315 to 387, and 390 MHz). The following review examines the national efforts to promote wireless communications (Wimmer and Jones 1992).

## Switching and Networking Technology

The value of information and communication technologies is greatly enhanced to the extent that they can be networked together,

| Data Networks | Fleet Dispatching | Cellular Networks | Paging |
|---|---|---|---|
| Technology: Radio-frequency for data only | Shared voice network that dispatches taxi, vans, etc. | Data is carried over cellular channels. | Advanced pagers that handle both numbers and letters. |
| Use: For mobile workers who need to communicate frequently with a corporate database | For companies that need mobile data service but do not require national coverage. | For mobile workers who need to send occasional long files. | Primary for one-way transmission of electronic mail and bulletins to people in the field. |
| Suppliers: Ardis, Ram Mobile Data | Motorola, Racotek, Fleet Call | Any cellular operator. | Motorola, Skytel, BellSouth |

*Table 1-2: Four Ways To Move Data Through The Air [13]*

allowing messages to be efficiently routed from place to place. A number of technologies support networking by performing interconnecting, switching, routing, and signaling functions. Included among these, for example, are:

- telecommunication (Telco) switches that can be defined as a means of allocating resources—space, bandwidth, and time—to people or machines that use the resources to communicate across distances (Frisch 1988),
- Telco switch networks, that organize message signal transmission, such as:
  - integrated digital networks (digital switching and transmission) (IDN)
  - intelligent networks (IN)
  - advanced intelligent networks (AIN),

- user computer networks, such as:
  - local area networks (LAN)
  - rural area networks (RAN)
  - municipal area networks (MAN)
  - wide area networks (WAN)
  - global area networks (GAN),
- repeaters, bridges, routers, and gateways these are devices to interconnect networks,
- intelligent peripherals.

Network technologies have advanced greatly over the past several years as a result of digital processing. The first computer-controlled switching systems were developed in the 1970s with the application of integrated-circuit technology. In the 1990s, approximately 98 percent of all AT&T switches are digital (Ross 1988). With respect to the regional BELL operating companies (BOC), approximately 27 (Southwestern Bell) to 66 (Bell Atlantic) percent of central offices are digital.

With the deployment of even more powerful microprocessors, faster computing speeds, and larger memories, it is possible to locate artificial intelligence not just in the central office switch, but also at nodes throughout the network. Because these "intelligent" nodes can communicate in real time with one another, as well as with other networks, communication based on this kind of architecture offers greatly enhanced flexibility—they can respond quickly to network problems and to changes in user demand, optimize network capacity, ensure greater system and service reliability (Boese and Robock 1987).

**Network Evolution**

Telco and computer networks are composed of application services, signaling and control management, switching and distribution facilities to support message movement and management. Some of the goals of networks are:

- to provide user-appropriate communications services
- to resource sharing (end of the limits of geography)
- to provide an environment to support new applications and services
- to support optimum control of communications.

The evolution of networking is shown on Figure 1-9. In the 1970s and 1980s, separated networks have evolved for:

- voice (circuit—WATS) and data (slow packet X.25)
- local (data-LANs, voice-PBX) and wide areas (data-SNA, X.25)
- public (value added networks-VAN for data) and private users.

In 1990, two trends evolved. The first trend is a further integration of networks, the second trend is associated with the discontinuity of old networks. The later is caused by:

- new bandwidth intensive services (video)
- explosive technology, such as:
  - SONET—Synchronous Optical Network
  - ATM —Asynchronous Transfer Mode

  both are used for broadband switching of integrated, large size and high speed traffic via a single network interface to communication channels for voice, video, image and data.

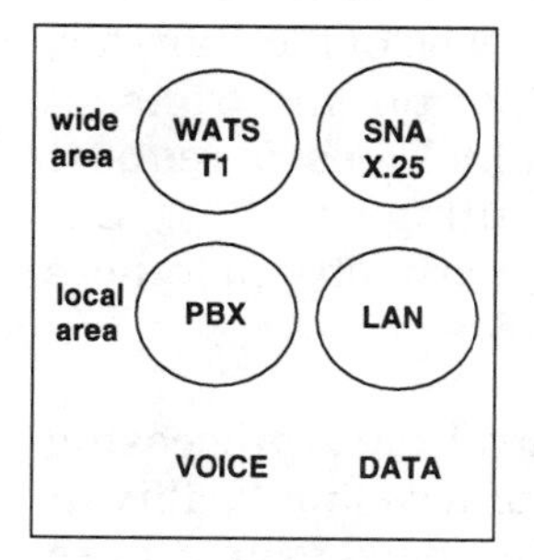


Before Integration
1970s-80s

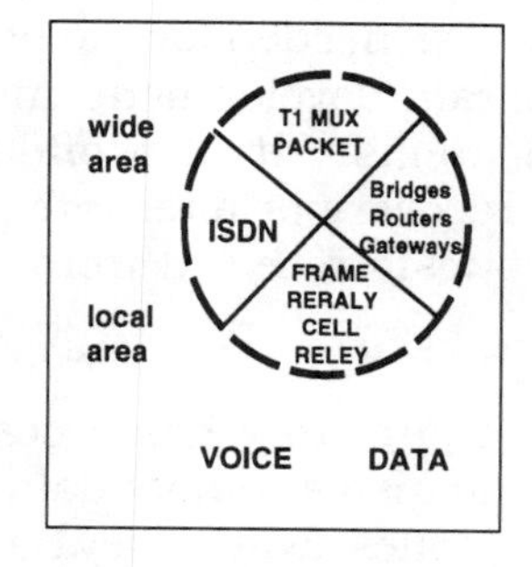


During Integration
1990s

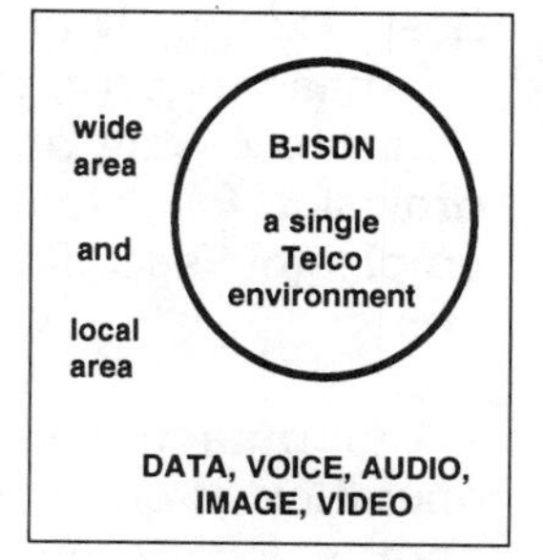


After Integration
2000s

*Figure 1-9: Networking Evolution*

In the 2000s, all network technologies will integrate into a single telecommunication environment based on the broadband Integrated Services Digital Network (B-ISDN) architecture. B-ISDN will support the following telecommunications applications:

- LAN connectivity

Examples of specific requests received from Ameritech (Midwest Bell Companies) customers are:

- A large hospital wants 100 Mb/s LAN connectivity so they can run an in-house developed, image-based record keeping system. They currently have 300 workstations on the server, and want to extend it to five additional locations.

- A large, sprawling urban university has requested a private fiber ring network to link 80 buildings which are spread throughout their campus in order to be able to transport data/video campus-wide.

- A large pharmaceutical company is looking to interconnect LANs located on their campus, and also to connect various locations with their headquarters. The pharmaceutical company sees this as a strategic move designed to improve their competitive position through the sharing of information with decentralized divisions. This pharmaceutical company also wants to deploy LAN technology to improve customer service for prescription processing (order taking) and claims filing (inventory).

- Host connectivity

Host connectivity provides communications capability between two hosts' locations or the connection of a remote location and a host. Connectivity may be needed for the purpose of file transfer, database inquiry, program development and/or on-line transaction processing applications. It also often includes a request provision for disaster recovery or alternate routing services. Examples of specific requests received from Ameritech customers are:

- A manufacturing firm currently has a central data processing operation at one site and also a remote data processing facility at another location several miles away. They have a need to link these two data processing facilities in order to balance the data processing load and resources between the two centers. In addition, the way their current operation is configured, if the main data processing center goes down for any reason, their entire data processing operation is lost. Therefore, they also want a contingency plan inherent in the linkage to mitigate the effects of a disaster at either of the locations.

- An engineering school has a need to connect user LANs to single main data processing resource center in a campus environment. This allows professors and students to perform program development and computer aided engineering (CAE) activities.

- A financial institution has requested a special technology to connect two IBM 3090 mainframe computers to a centralized disk storage. This allows for automatic backup of the day's transactions without the need to physically transport the disks.

• Video and Imaging Integration with Data and Voice and Audio

Examples of video and imaging applications are: remote computer-based video training, interactive videoconferencing, medical image-driven files, general industrial image processing, image-based publishing graphics CAD/CAM/CASE/CAE/CIM design imaging. Examples of specific requests received from Ameritech customers are:

- A large, urban hospital has expressed an interest doing teleradiology (medical imaging) over a public data network. This will improve the diagnostic process in two ways: it will speed up the process, thus providing better customer service to individual patients, while at the same time allowing the hospital to process more patients, and it will make the testing procedure safer for the patient by reducing the number of repeat tests that have to be performed due to faulty image transmission.

- A public school system wants to use distance learning applications and reduce administrative staff meeting costs. Distance learning improves a school's efficiency by providing a cost effective way for both rural and urban schools to share resources, in order to supplement their curricula with needed instruction. Students also have access to a wide range of classes that might otherwise have been unavailable due to low enrollment or lack of specialized teachers in the area.

- Three large hospitals are members of a consortium which purchased a $10 million high-speed, high-definition diagnostic device which is capable of creating a three-dimensional image of a patient's internal system. The unit is installed at one hospital, and the customer needs an extremely high speed data link to transmit the diagnostic test results to the other two member hospitals and their various branch locations. The member hospitals benefit from this arrangement by optimizing their investment resources while at the same time serving customers that they otherwise might not be able to reach.

- Route diversity/disaster recovery.

Route diversity or disaster recovery services are most often accomplished with network redundancy or an alternate routing plan. These routes support disaster recovery for the following customer business applications: Data center "hot site," LAN/MAN/WAN technologies, workstation to host on-line transaction processing. Electronic Data Interexchange (EDI) and Electronic Funds Transfer (EFT), point of sale terminals, automated teller machines, credit authorization terminals, teller window transactions, and publishing graphics in manufacturing. Some examples of requests received from Ameritech customers are:

- A financial services institution has requested disaster recovery services because their core business involving fund transfers would grind to a halt if their network went down, resulting in loss of revenue. Additionally, this customer must meet some federal recommendations for implementing a disaster recovery plan.

- A major retailer generates 35 percent of their revenues through catalog sales. They have asked for network redundancy or alternate routing solutions as a feature of a new proposal to link several new remote computer sites to the computer center. In the event of a network failure, the recovery plan would ensure that the catalog operation, and thus the customer's core revenue stream, continues without interruption [5].

This B-ISDN architecture will provide the user with standardized interfaces, digital transport, distributed processor control, flexible digital message signaling, common service structure by a Telco company, and an integrated operations and network management. In 1987, AT&T announced the Unified Network Management Architecture (UNMA) as it is shown on Figure 1-10.

Figure 1-11 illustrates a network operations center.

UNMA provides a framework for integrated end-to-end management in a multivendor telecommunication network environment. In other words, it will integrate LANs, MANs and WANs into one recoverable and controllable system within the AT&T boundaries. Something like the National Air Traffic Control System.

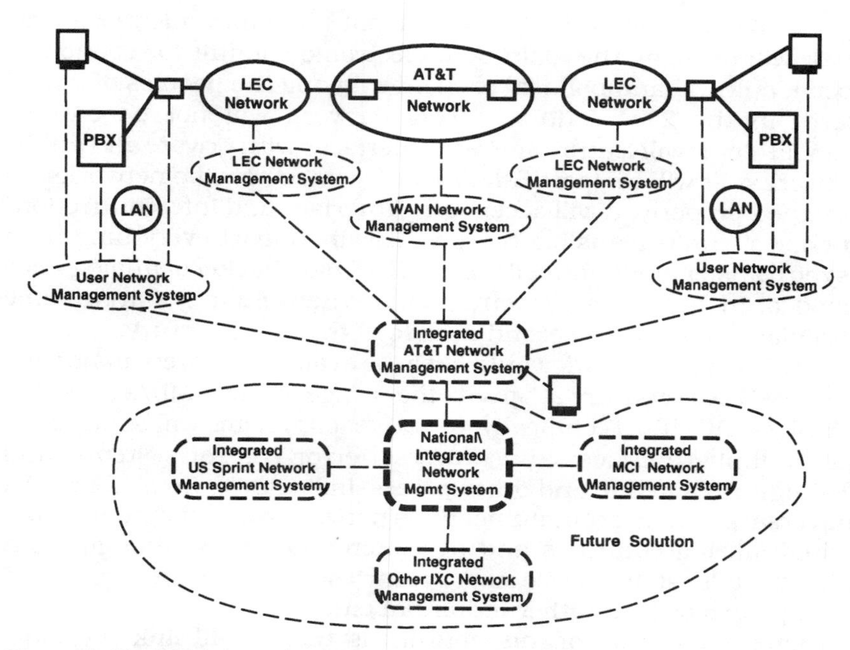


*Figure 1-10: A Framework of the AT&T Unified Network Architecture and the Future National Integrated Network Management System*

*Figure 1-11: The AT&T Network Operating Center in Bedminister, New Jersey. It controls the AT&T Worldwide Intelligent Network, the most advanced telecommunications network in the world. CNNs Terry Keenan is giving the business report from the New York Stock market. State-of-the-art computer software technology speeds the flow of voice, data, and image transmission over 2.3 billion circuit miles worldwide, handling more than 195 million calls a day. The center, known as the NOC, has a screen video wall, two stories high and 60 feet across, which continuously monitors AT&T customer requirements, 24 hours a day, 365 days a year. (Courtesy of AT&T Archives)*

The universal (integrated) global telecommunications network will serve as the main opto-electronic conduit for economic, social, cultural and political exchange among the peoples of planet Earth in the 21st century. This network will not be entirely monolithic. It will feature agnostic interfaces with private and public networks. It will support B-ISDN and service-specific networks. If designed properly, it will accept any "interface and interconnection" of choice within plausible reason. It will support everything from async transfer mode and cell relay to X.25 data packet or analog voice or video. In its early implementation, this network may start to come into place around the second decade of the 21 st century.

This global network would likely be available in every major city in every OECD country at speeds that range from 45 Mb/s (DS-3) to 2.5 Gb/s (OC-48). Technically, the biggest problems will be supporting multiple interfaces and universal error control systems, and dealing with satellite and other processing delays. Some form of a universal global telecommunications network will be found in newly industrializing countries and developing countries, though often limited at least in developing countries to the capital city and perhaps two or three other industrial cities.

One implication of this network is that it will link the most economically and technically powerful together in new and evermore effective ways [9].

## Standards & Protocols

A standard is a formally adopted and widely accepted rule that describes an agreed-upon way of doing things. Telecommunications standards are sets of rules describing how computer systems should communicate. Telecommunications or networking standards address such issues as the transmission media between communicating systems, type of interface between a computer system and transmission medium, format of transmitted message, length of transmission, and all other aspects of information exchange between two devices. Approximately 200 key standards are applied by a sophisticated contemporary network to support multimedia communication

Most large companies today use and own computer systems from many vendors. The urgent need is to link these diverse computer systems into an enterprise-wide electronic internetworking infrastructure. The enterprise-wide computer network enables companies to make maximum use of the strengths of different vendors. Companies require the flexibility to choose computer systems from several vendors so as to find the best system for the application at hand. Without communication among these systems of different vendors, however, this flexibility could lead to large-scale inefficiency in a company's day-to-day operations.

> *For example, the accounting department in Chicago may use a mainframe from vendor A; a marketing group in Seattle may use a minicomputer from vendor B; financial users in New York may use personal computers from vendors A and B. Rarely does a single vendor's equipment meet the full range of a company's needs.*

However, flexibility in choosing vendors could lead to problems. Unless the different vendor's system in a company communicate with each other, the proliferating islands of automation could have just the opposite effect.

If a company has computers from different vendors, a networking strategy should enable multivendor communication. The networking vendor that has been chosen should be committed to the national and international standards and protocols that make possible multivendor communication.

Multivendor communication is possible only different computer and communication vendors agree upon a set of conventions or standards for information exchange. Universally accepted standards in the telecomputer industry evolve slowly. Standards in the music industry enable us to buy a record anywhere in the world and play it on any turntable. Standards in the telephone industry enable us to pick up a phone set in any country and place a call without having to consult a user manual.

***Standards Architectures.*** Today, most computer networks are based or planned on the following standard architectures:

- IBM SNA (System Network Architecture)
- Digital Equipment Corporation's Digital Network Architecture (DNA)
- TCP/IP — Transmission Control Protocol/Internet Protocol developed by Department of Defense researchers who built ARPANET.
- AT&T ISDN (Integrated Services Digital Network)
- ISO OSI (Open System Interexchange)

Users who want the benefits of nonproprietary, multivendor computer networks are adopting TCP/IP protocols while they wait the arrival of commercially available OSI protocols.

Standards can be defined and established at the national and international levels. A national standard is usually developed at the national level—in the United States by the American National Standards Institute (ANSI) and in Great Britian by the British Standards Institution (BSI). An international standard is one that is developed at an international level and adapted by vendors and suppliers internationally. The International Standard Organization (ISO) based in Geneva, Switzerland, includes the national standards bodies around the world. It works closely with other important international bodies such as the European Computer Manufacturers' Association (ECMA), and the International Telephone and Telegraph Consultative Committee (CCITT). CCITT is also headquartered in Geneva and is, in fact, a Division of the International Telecommunications Union (ITU) which in turn reports to the United Nations Organization. An ISO International Standards document is prefixed by the letters ISO. A CCITT document related to data communications is prefixed by "X" — ISO xxxx.

While OSI individual layers offer the user considerable flexibility, only a few combinations are relevant for a particular application. These combinations are known as Functional Standards, Pillars, or Profiles. Several standard bodies are involved in the identification and specification of these functional standards. In the United States, the Corporation for Open System (COS) is such a body. Several United States and international manufacturers and vendors are part of the COS group. In Europe, the Standards Promotion and Application Group (SPAG) is responsible for the development of functional standards. Several major European information technology manufacturers and suppliers are members of SPAG. Several other important European organizations such as the EEC Senior Official Group on Information Technology Standardization (SOGITS), the Conference of European Postal and Telecommunications Groups (CEPT), and standards bodies such as CEN, and CENELEC are involved in the direction setting of OSI standards for real-life applications. EuroOSInet provides an ongoing demonstration of the practical applications of OSI standards.

***OSI Profile.*** The Open Systems Interconnection (OSI) model is the basis for developing a set of international telecommunications standards produced by ISO. The OSI model is the only internationally accepted framework of standards for intersystem communication. The OSI model incorporates standards developed by many national and professional organizations such as the CCITT, ECMA, IEEE, and others.

The OSI standard of networks' layers refer to standardized procedures for the exchange of information among terminal devices, computers, people, networks, processes, etc., that are "open" to one another for this purpose by virtue of their mutual use of these procedures.

"Openess" does not imply any particular systems implementation, technology or interconnection means, but rather refers to "the mutual recognition and support of the standardized information exchange procedure." In the concept of OSI, a system is a set of one or more computers, the associated software, peripherals, terminals, human operator, physical processes, information transfer means, etc., that forms an autonomous whole capable of performing information processing. OSI is concerned with the exchange of information between systems (and not the internal functioning of each system).

The OSI model comprises seven layers. Each layer contains modules that specify and define a separate aspect of the telecommunication function. Within each module are *protocols* that define message formats and the rules for message exchange between communicating systems (peer-to-peer). The OSI model was accepted by ISO in 1984 in document ISO 7498. CCITT adopted the OSI model as X.200 standard in 1988.

Figure 1-12 shows the generic OSI model. The lower layers (layers 1 to 4) are concerned with the electrical and mechanical specifics of connecting to the transmission medium, the error-free transmission of data, and transmission of data to the correct destination. Standards in these layers define the reliable and accurate transfer of data between two systems. Once the data arrives at its destination, the standards of the higher levels come into play. These standards interpret the received data and make it available to applications that benefit the end user. Without the higher level standards, transmitted data would be entirely meaningless to user applications and therefore unsuitable. In the electronic mail application, the lower layer standards would ensure that a memo written on vendor A's system is transmitted accurately to vendor B's system. The higher layer standards would ensure that received message is understood by the mail application on the vendor B's system and made available to the addressed user or users on vendor B's system.

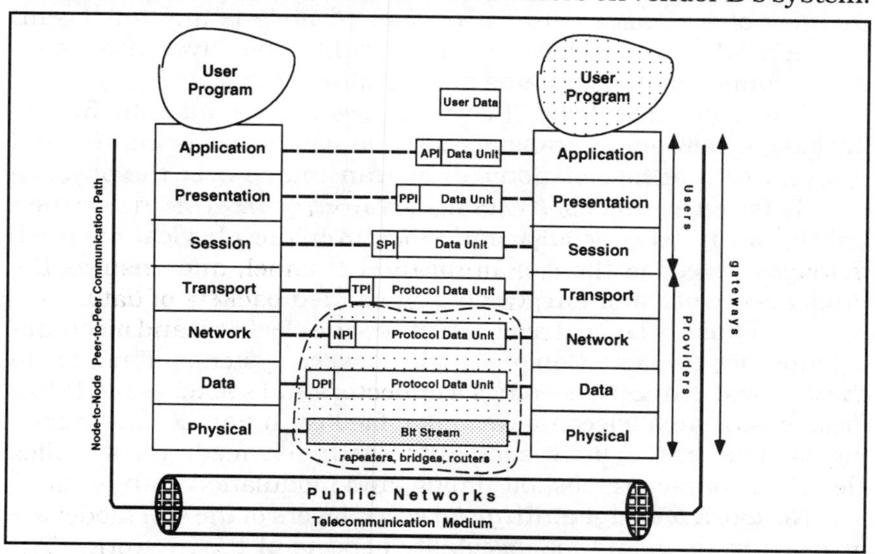


*Figure 1-12: OSI Generic Model (APIApplication Protocol Interexchange)*

The layering in OSI is as follows, in descending hierarchical order:

The *Application Layer* (Layer 7) provides services that directly support such user and application tasks as file transfer, remote file access, and database management. This layer provides facilities for a semantic exchange between applications across a network, supports user commitment, processing concurrence, recovery, remote operations, and reliable transfer.

The *Presentation Layer* (Layer 6) provides services that directly support such user and application tasks as file transfer, remote file access, and database management. This Layer controls data formats, codes, and representations, syntax selection and conversion, compression and encryption. This Layer resolves the differences of varying data formats between systems of different vendors. It works by transferring data in a system-independent manner, performing appropriate conversions at each system.

The *Session Layer* (Layer 5) is concerned with dialogue management. It establishes and controls system-dependent aspects of communications sessions between specific nodes in the network. Dialogue management contains dialogue control, synchronization, initialization, support of recovery, and termination.

The *Transport Layer* (Layer 4) provides end-to-end (source node to destination node) control of a communication session once the path has been established, allowing processes to exchange data reliably and sequentially, independent of which systems are communicating or their location in the network. This layer ensures reliability, quality, and optimization of information transport.

The *Network Layer* (Layer 3) defines routing of a message between available nodes from the source to destination and controls the flow of messages between nodes. (A node is any intelligent, uniquely addressable device on a network). This layer also establishes connections, maintains and terminate them.

The *Data Link Layer* (Layer 2) segments data into frames (blocks), synchronizes transmission and handles error control and recovery so that information can be transmitted over the physical layer between two nodes. It establishes an error-free communication link between network adjacent nodes over the physical channel, manages access to the communication channel, and ensures the proper sequence and integrity of transmitted packets of data.

The *Physical Layer* (Layer 1) handles the electrical and mechanical interface transmission of signals between systems. It relates to the physical connection—plugs, connectors and sockets—made to a local or wide area telecommunication facility. It handles the encoding of data into signals compatible with the medium, specifies electrical characteristics, bit timing, and modulation standards.

***Network Manageability.*** All seven layers of the OSI model are involved in providing manageability aspects of the network. The scope of OSI standards covers the following major management components [7]:

1. *Fault Management*: This provides support for fault detection, diagnosis, and correction. Fault Management includes the capabilities for error reporting, error gathering, setting thresholds, event-logging, and tracking communication paths.

2. *Accountability*: This allows for the resources within a network to be tracked and made accountable for the costs associated with their use.

3. *Configuration and Name Management*: Configuration management provides facilities for:

- Setting network parameters
- Initializing and changing the system configuration
- Collecting data
- Facilitating configuration management to ensure continuous operation of interconnection services.

4. *Performance Management*: This provides facilities to evaluate the behavior of OSI resources and the effectiveness of communication activities. This is done by the gathering of statistical data, maintenance, and examination of logs of the system state information. It does not include facilities for control of performance parameters in an OSI system, but merely provides for monitoring of appropriate information to accomplish its objectives.

5. *Security Management*: This provides facilities for authentication, access control, confidentiality, integrity, and parameters for providing proofs of the origin and delivery of information.

6. *Common Management*: Common Management Information Services (CMIS) provide the basis on which above functions cooperate for the purpose of network management. Information can be exchanged between management processes on different systems based on OSI. These exchanges are based on a small but powerful set of service elements. The CMIS specification provides for event notification, information transfer and control. The Common Management Information Protocol (CMIP) is the protocol used to provide the CMIS services.

***OSI Benefits.*** The benefits of OSI standards are obtained in each of the four components of the network functionality [6]:

- *Connectivity* benefits are manifested in both local area network and wide area network contexts:
  - OSI compatibility is now the expected norm, 8802.2 and X.25 standards of connection to LAN and WAN are widely accepted.

- Because OSI is accepted worldwide as the standard for public information networks, any private OSI network can make use of public services as needed.
- Different applications and types of Data Terminal Equipment (EDT) can share the same routes in and between OSI compatible networks,
- Local area networks applying 8802.2 standard can communicate with the same LANs.

• *Interoperability* through the OSI model permits all types of business equipment and information resources to interoperate, allowing the integration of multivendors networks. So hardware choices can be made purely on the basis of merit. Furthermore:
- Users of OSI networks are not aware of the inner workings of the network.
- Users do not need to become involved in the finding, retrieving or moving of information, because the network handles all routing.
- The mainframe is relieved of the task of moving information into and through the network.

Interoperability allows each user application program or user process to be accessed by others in the network. Such companies as IBM and DEC achieved interoperability among their existing networks and systems by utilizing proprietary standards and protocols. ISO and DARPA (DoD) represent user community that want to achieve interoperability between heterogeneous information technology by imposing OSI and TCP/IP standards and protocols. In 1983, the U.S. government initiated a series of workshops led by the National Institute of Standards and Technology (NIST) for organizations and vendors interested in implementation of OSI-oriented standards. A concept of interoperability is shown on Figure 1-13.

• *Distributed Applications.* The existence of an accepted international standard has paved the way for standard solutions to common needs such as File Transfer Access Management (FTAM) and electronic mail (X.400) that have been incorporated into OSI networks. This allows information to be distributed cost-effectively and to reside where it is most needed.
• *Manageability.* OSI brings a sense of order and manageability to what had been an extremely complicated network environment. It offers an orderly set of international standards on which to build now and plan for the future. New services can be added to the network without disturbing network performance or making changes. OSI users can expand facilities as needed to improve responsiveness and take business changes in stride.

OSI standards, because they are layered, can accommodate

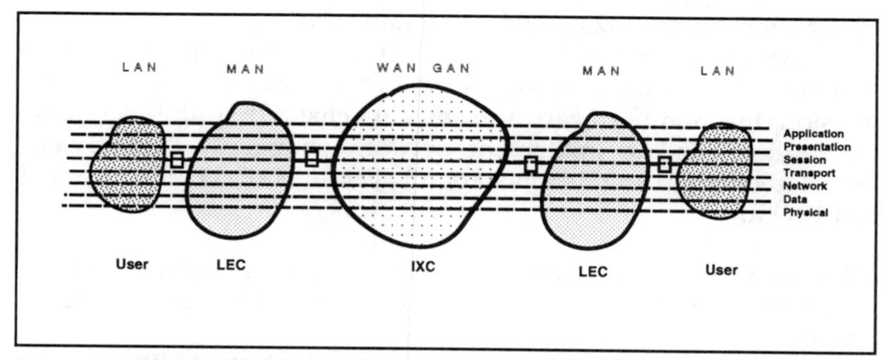


*Figure 1-13: Interoperability Through the OSI Model*

future changes and developments in technology. An OSI network is protected from obsolescence, because all OSI vendors are following the same guidelines in updating and innovating.

***NetBIOS Interface.*** This interface was designed by IBM and Sytek (manufacturer of interface cards) as a gateway between different LANs and host computers. It allows microcomputers to communicate with each other at the hardware level without going through a networking software. It is *de facto* protocol for LAN interfaces. This protocol promoted the growth of LAN applications and interconnections.

***MAP Protocol.*** The Manufacturing Automation Protocol is tailored for the assembly line and shop floor environment within manufacturing local area networks. It has roots in General Motors' problems in the 1970s, when 20,000 robots and 40,000 intelligent devices in the shop floor operations could not communicate to each other. This created "islands of automation." About 50 percent of the production automation cost was going to the communication component. The first MAP document (1984) specifies the use of a broadband cable network based on the IEEE 802.4 standard for token-passing media access control. MAP also defines the manufacturing message standard ("data packet") as well as an OSI application-layer protocol for formatting and transmitting commands between controlling software-driven programs and machine-tools.

***TOP Protocol.*** Boeing Computer Services Company has developed the Technical and Office Protocol (TOP) for engineering, accounting, marketing, desk top publishing, and other corporate functions that provide manufacturing support through local area networks. TOP networks are able to transmit and accept data with any of MAP networks. This protocol links engineering and business/management offices with the factory floor. TOP includes some OSI application-layer protocols, such as X.400 messaging protocol and the virtual terminal protocol. TOP complies with IEEE 802.3, which uses a technique to specify devices to share Ethernet and token-ring adapter boards. This technique (LLC) also defines interfacing bridges for 802.3, 802.4, and 802.5 networks.

***MAP/TOP Network Services.*** MAP and TOP are integrating into an enterprise-wide networking strategy that should save on transmission time and hardware and software cost (20%) while reducing the staff needed to implement network changes. The services associated with the OSI-Application Layer as MAP/TOP 3.0 (issued in 1987) are:

- Remote File Access to transfer in real-time contents of remote files,
- Electronic Mail from analog to postal service mode defined by X.400,
- Remote Terminal Access for workstations that would like to communicate with the centralized functions of a host computer (TOP),
- Interchange formats between users and machine-tools that guarantee the same data interpretetion of:
  - Office Documents (ISO DIS 8613/CCITT T.411 through T.419)
  - Product Definition Data for CAD/CAM applications
  - Two-Dimensional Graphics allows the interexchange of editable graphic files (Graphics Metafile Standard).

These services are designed for all major LANs and X.25 standards. The MAP/TOP 3.0 standard defines user friendly procedures of network management functions. A MAP/TOP user group with 3000 members from 1000 organizations was formed in the United States. At the international level, such countries as USA, Canada, Europe, Japan formed the World Federation of MAP/TOP users. The MAP/TOP protocols are designed for a backbone broadband network of enterprise-wide communication architecture.

*X.400 Protocol* is the international standard for global messaging as a common method for linking multiple LAN and host-based systems.

The major telecommunications interexchange carriers such as AT&T and MCI offer public X.400 gateway services for wide area messaging. The major computer vendors such as IBM, Hewlett-Packard and NCR offer gateways based on X.400 protocol. The same category of E-mail software vendors (E.G. Microsoft's cc:Mail) sell their private mail products based on this protocol. The industry group that is heavily operating via telematic networking—the Aerospace Industries Association and the Society for Worldwide Interbank Financial Telecommunications (SWIFT)—support X.400 protocol. Even the Internet is migrating from TCP/IP to X.400 in electronic mail networking among 7 million users.

The first version of this protocol was defined by CCITT in 1984. The next versions were issued in 1988 and 1992. The X.400 standard/protocol recognizes the following modules (Figure 1-14):

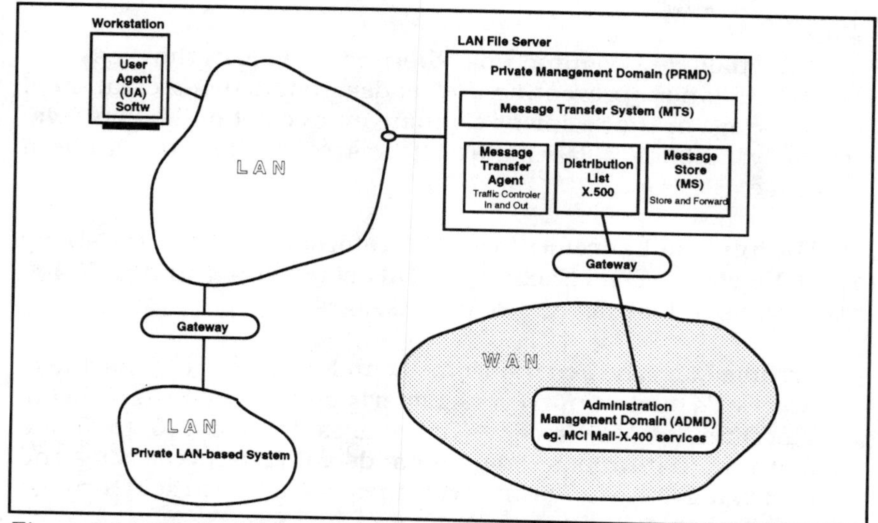


*Figure 1-14: The Architecture of X.400 Electronic Mail Protocol*

- User Agent (UA) as a piece of software that connects the end-user's workstation to the X.400 system by preparing the message in a format that the Private Management Agent can accept it,
- Message Transfer Agent (MTA) which transmits messages to another MTA in a local area network or wide area network,
- Message Storage (MS) is a staging point, similar to a post office, in which messages are temporarily held for later transmission to one or more recipients,
- Distribution List is a directory of users, complying with the X.500 protocol.

The above set of modules creates a Private Management Domain (PRMD) to which MTAs send mail. A LAN-based X.400 mail system is considered a PRMD, while, interexchange carrier's 4.00 wide area network is an Administrative Management Domain (ADMD). PRMDs usually connect to one primary ADMD, which can then route the customer's X.400 messages to multiple other ADMDs.

The X.400 message has two primary elements: the Message Transfer System (MTS) header, which includes X.400 WAN and the MTS body which includes the private mail system address as well as the message itself. The content of a user message can be text and/ or graphics in ASCII binary code.

The X.400 protocol ensures reliability and security of mailing in the following manner:

- authentication: verification of the message sender,
- integrity: ensuring that message was not altered during transmis-

sion,
- confidentiality: preventing unauthorized viewing of the message,
- non-repudiation: preventing the sender from denying creation of the message or the recipient claiming nonreceipt of the message,
- security: affixing a label that informs MTAs how to handle a message.

The future enhancements of this protocol will include the ability to mail Electronic Data Exchange (EDI) documents within an X.400 envelope, as well as fax and voice messages.

> Freddie Mac, the Federal Home Loan Mortgage Corp. is one of America's busiest purveyors of mortgages. Located in McLean, Va., Freddie Mac receives mortgages from 5,000 to 6,000 financial institutions daily and adds value by processing and packaging these funds into securities that are then sold through 200 brokerage houses to investors.
>
> Managing such high volumes of data and still providing reliable and responsive service to investors is no small feat. To achieve this, Freddie Mac implemented an X.400-based delivery system. Its first objective was to move to a single, well-defined messaging standard upon which they could deploy an intelligent messaging interface between customer systems and Freddie Mac systems. A second critical objective was to implement this upgraded service so all of Freddie Mac's 7,000 customers could potentially be switched over transparently with no changes to their existing interfaces or programs. The primary reason for selecting X.400 is that it allows Freddie Mac to deliver messages almost anywhere in the world via private and public network services.
>
> Freddie Mac replaced the proprietary delivery system with a dedicated Tandem-based Intelligent Delivery System (IDS). Messenger 400 (from OSIware Inc. with API-application program interface) is used as transport to deliver customer data to mainframe applications and vice versa. Non-stop SQL is used as the relational database and development environment. The data flow remains the same, but the interface will be via a data-driven API which triggers the IDS's X.400 communication engine. The IDS takes incoming X.400 messages, strips the X.400 headers, identifies the data and routes the message to the appropriate application for processing. Similarly, outgoing messages generated by mainframe applications are passed through the IDS. An application directory look-up is performed to identify data type and intended customer, the message is then

wrapped in an X.400 envelope, and sent on.

Freddie Mac's original goal was to provide an application-generated message delivery service to its customers while making the upgrade transparent. With X.400 an additional means of person-person E-mail will be available to customers, and Freddie Mac will be positioned to moved toward EDI. While faxes, phone calls, memos or the postal service may never be totally replaced by X.400, Freddie Mac anticipates that their business customers will eventually move to strategic messaging platforms [10].

*X.500 Protocol* specifies procedures for retrieving and the updating of information stored in the directory. This service provides support for users who require information regarding establishment of communication with another user.

The Hughes Aircraft uses eight standard E-mail packages to provide mail service to some 150,000 locations around the world. Private E-mail systems are hooked up through gateways to SoftSwitch Central Software running on an IBM mainframe. From there, the mail moves through a SoftSwitch X.400 gateway to an X.25 service. The mail is routed to MCI's public X.400 service, which can deposit it in another carrier's network, to end-users at Hughes' parent company, General Motors Corp., or to other offices around the globe.

The E-mail directories are the weakest link in messaging chain. The master directory has over 150,000 names with aliases and is easily over 2 megabytes in size. It cannot be propagated across LANs and servers, so the IS department maintains four primary directories according to domain that are updated by the master directory, which resides on a relational database. A solution to this problem will be a creep of X.500 products into proprietary systems [15].

## Message Signaling and Organization

Messages are transmitted over electronic or fiber optic communications channels in either *digital* or *analog* form. Digital signals are discrete. They consist of a choice between two possible states, namely, the presence or absence of an electronic signal (on or off), one or zero, or two different electrical signals levels. This is consistent with the basic form of communications between computers which is digital; information is conveyed as binary digits or bits (1s or 0s)

Analog signals, on the other hand, are represented as continuous wavelike signals. Each complete wavelike motion of an analog signal is called a cycle. The number of complete cycles per second at which an analog signal occurs is called the signal's frequency. The term used to express the number of cycles per second of a signal is Hertz (abbreviated) Hz. Figure 1-15 illustrates these two techniques of signal transmission in telecommunications channels.

A network is an end-to-end channel which is composed of adjacent node-to-node links made up of some form of physical media. Each media-channel has its own capacity. Capacity measures the range of analog signals that can be carried by a physical channel or the amount of message traffic that can be achieved via that channel. Bandwidth is the term to define channel capacity. In digital transmission capacity is defined by the amount of data sent over time (bits per second). Voice telephone lines with modems offer bandwidths of 2.4 or 4.8 Kb/s.

A coaxial cable has a usable bandwidth of about from 10 to 310 MHz (mega Hz). A twisted-wires medium offers 1 to 2 MGZ. A fiber optic media can transmit with frequency of several billion Hz or GHZ (giga Hz).

## Channel Allocation

With multiple nodes competing for the use of the network, the capacity or bandwidth of the network channel must be divided, or allocated in such a way as to allow the most effective use of available channel capacity. This can be achieved through multiplexing.

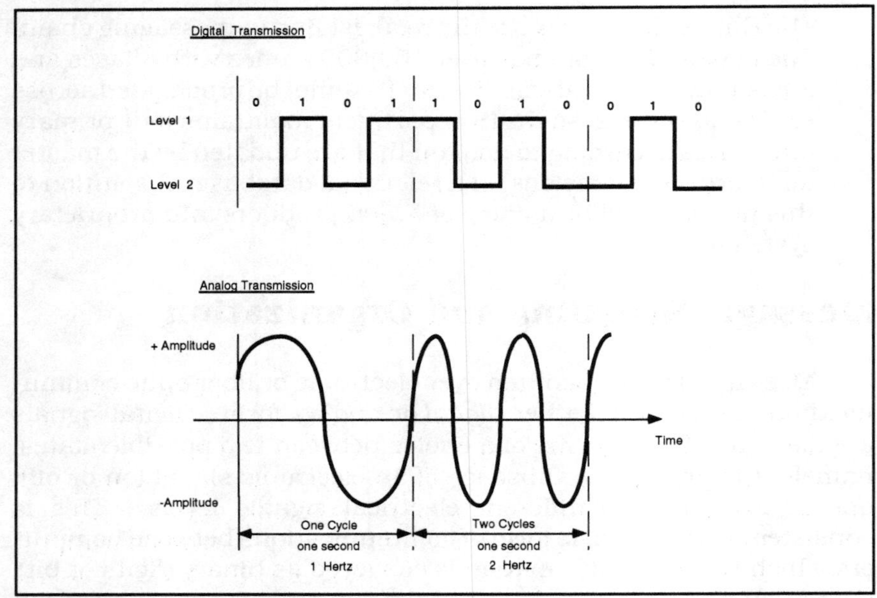


*Figure 1-15: Message Signals*

Multiplexing is a method by which a single channel is divided into multiple channels in order to transmit a number of independent signals. Different approaches to multiplexing are based on:

- space division multiplexing is created by grouping together many physical channels, like a telephone line made up of hundreds or thousands of the individual twisted-pair telephone wires that run to each home.
- frequency division multiplexing is a technique to divide the available bandwidth of a single physical medium into a number of smaller, independent frequency channels. Each of these subchannels can then be used as a separate channel for transmission. For example a channel with 48,000 MHz can be divided into 12 channels of 4,000 Hz each. Radio and television broadcasts apply this technique.
- in time division multiplexing each node is alloted a small time interval, during which it may transmit a message or a portion of a message (example: university computer time sharing systems).

## Establishing Communication Channels

For a message to travel across a network, a transmission path must be established to switch or route it to its destination. Two major techniques used today to establish a path for communication are *circuit switching* and *packet switching.*

*Circuit-switched* messages usually establish a totally new path between two points every time a call is originated. Calls are automatically directed to their destination by the most readily available route. A message does not contain an address of the destination point. This is a hard-wire solution with the customer paying for the time used.

*Packet-switching* is the process by which packets, as an envelope are placed on the channel and travel across the network to their destination. Packets contain the address of their ultimate destination. A packet can be of fixed or variable length, not less than 128 bytes. It is constructed (assembled/disassembled) by a computer according to the X.3, X.25 and X.75 protocols. It can contain bits for synchronization, control information, message number, number of the current and last packet, destination and source address, acknowledgment, error checking, and data. This is a software solution with the customer paying for the amount of information sent.

## Packet Switching Technology

In the 1970s, packet switches based on 8-bit microprocessor could handle 100 packets/s. In 1980s, when 16-bit and 32-bit microprocessors were available, network throughput increased to 500 packets/s. Multiprocessors configurations can handle even 10,000 packets/s. However, due to increased performance beyond 0.5 Kp/s, new technologies must be used.

A new technology evolving in the 1990s is based on fast packet switching through broadband networking. Fast packet switching is essentially a streamlined version of conventional packet switching (X.25) that minimizes error-checking (networks become more and more reliable) and flow-control overhead. Consequently, much less protocol processing is done on the network while traffic flows from switch to switch. The switches no longer set up connections or recover from errors because such errors are assumed to be infrequent and readily detected by the end stations and their high-level software (Savage 1992). Fast packet technology is divided into:

- frame relay (a variable-length frame formed at Level 2 of the OSI model) defines the interface to wide area network (WAN) from metropolitan area networks (MAN) and local area networks (LAN)

- cell relay (a fixed-length packet) is applied as a backbone of WAN which assures fast transmission between nodes. It is the future infrastructure of public networks transmitting voice, data, image and video.

Both technologies are complementary, hence they are designed with different objectives and applications. Figure 1-16 depicts the architecture of broadband fast packet technology, its standards and services.

As bandwidth increases, networks support increasingly complex multimedia applications. The speed of transmission depends on the communications medium:

| | |
|---|---|
| • Styled text, plain old telephone service | 10 Kb/s |
| • VSAT satellite | 15 Kb/s |
| • Basic rate of ISDN | 56 Kb/s |
| • Low quality compressed video | 100 Kb/s-1.5 Mb/s |
| • Images | 10 Kb/s-128 Kb/s |
| • Frame Relay | 384 Kb/s |
| • T-1, Primary rate of ISDN | 1.54 Mb/s |
| • Medium quality compressed video | 1.5Mb/s-6 Mb/s |
| • Ethernet LAN | 10 Mb/s |
| • Token Ring LAN | 16 Mb/s |
| • High quality compressed video | 6 Mb/s-24 Mb/s |
| • T-3 | 46 Mb/s |
| • FDDI | 100 Mb/s |
| • ATM | 150 Mb/s-1.2 Gb/s |

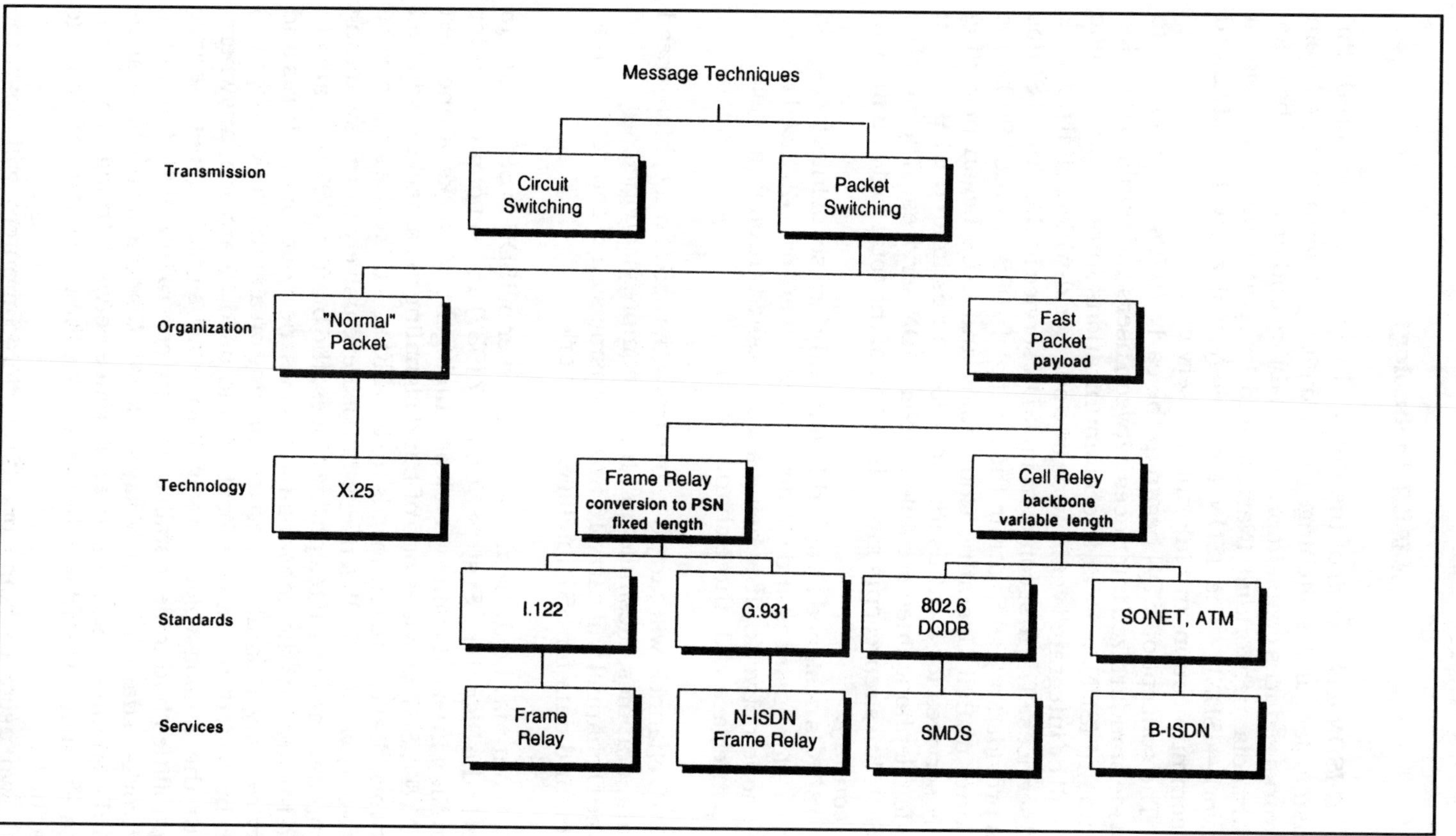


*Figure 1-16: Message Techniques Architectures*

## N-ISDN Technology

*N-ISDN technology* of the Physical Layer 1 is based on digital transmission and switching. This digital integration of telecom networks reduces costs, improves reliability and provides a base for multimedia networking (Narrowband-Integrated Services Digital Network—ISDN). The ISDN technology will allow for advanced telecommunication and information services.

The concept of ISDN began in the early 1970s. Its goal was to provide standardized interfaces between users and public networks. Islands of ISDN in local telecommunications infrastructures, one day will be integrated world-wide into one world-wide digital network and services. The first commercial ISDN service in the United States was provided by BellSouth in the Spring of 1988. In the 1990s, the majority of Bell Operating Companies and IXCs began providing ISDN services for local users. In 1994, an estimated 61.9 million single telephone lines are able to offer ISDN services. The number of ISDN lines represents more than half of the total regional lines in the country.

ISDN is designed as a world-wide standard set of interfaces and services for business and residential customers on a digitized public telephone network. It is being accomplished through a basic 144 Kb/s interface with three channels:

- Two B channels will carry 64 Kb/s transmission of voice, high-speed data, graphics, facsimile and highly compressed video.
- One D channel will carry 16 Kb/s transmission of control information and data for SS7 intelligent network.

What may be the real key to the power of ISDN is SS7. A key to this is the Signalling System Number 7 (SS7) as a set of protocols for handling internal traffic via telcom intelligent networks. Conceived by AT&T, SS7 provides out-of-band signaling via a packet switched network that can reach speeds up to 64 Kbs. Unlike the analog voice network, whose intelligence is housed exclusively in the switching computers at central telephone offices, the brain of SS7 is located in and throughout the network itself. Thus, new network features and services (say, a database service running on a mainframe computer) can be added in just one location and used by anyone, anywhere. With the analog voice networks in-band signaling, new service capabilities had to be installed in every central office switch on the network—a daunting job, to say the least. Comparing traditional in-band signaling to SS7 it is the difference between trying to communicate with a 20-word vocabulary versus the vocabulary of a college graduate.

Two users of home or office microcomputers will have the

capability to communicate via ISDN switches and network in a mode of face-to-face conversation, review a document, graphics or photos, than transmit a facsimile page in a few seconds.

> In 1991, the users at the U.S. General Service Administration office in Atlanta, Georgia had to take over a fellow employee's PC anytime they needed to use a Postscript-capable printer. Now the users can send their print jobs to John Penman's printer without forcing him to quit his own work. They use ISDN's D channel to share the printer, while dedicating the two B channels to voice. The bottom line is: they have replaced four phone lines with each basic rate ISDN line. The single 16 Kb/s D channel is shared by two PCs with terminal adapters. Staffers send jobs to John Smith's printer and he hears only an audible tone on his telephone set, then the whir of the printer. He no longer needs to give up access to his desktop PC. GSA, in downtown Atlanta, gets its ISDN through Southern Bell's ESSX centrex service. Each PC in GSA's department had a separate phone line for modem and separate line for voice. With ISDN, one line has taken the place of four. Two users, for example, use one ISDN line. Each uses one B channel for voice and they share the D channel to access printers and a modem pool. The 2400 b/s modems have been replaced by 9600 bps modems.

Basic services of ISDN can be provided via twisted pair copper wires. However, to carry a high quality video, a fiber optic cable is required. A phone set which is ISDN-ready cost between $ 600 and $1000, compared with $40 for a regular set. In 1990, four vendors: AT&T, Fujitsu, NEC, and Northern Telecom had such sets ready for delivery to customers.

Three types of ISDN services are defined by CCITT:

- Bearer services which provide the means of transmitting voice, data, image, and video on the same channel (three lower layers of the OSI model).
- Teleservices (user visible) include telephony (3.1 KHz speech communication without an echo), telex, videotex, and message handling (end-to-end)through the integration of the transportation function with the information processing function (SS7). These services correspond to layers 4 through 7 of the OSI model.
- Supplementary services associated with the bearer and teleservices, e.g., reverse charging on circuit switching or packet switching, call forwarding unconditional, city-wide centrex, call waiting, calling line identification, three party service, call transfer, credit card calling, and so forth.

*ISDN Benefits*. The main benefit of ISDN is that it allows relatively high-speed digital networking —at least ten times faster than most modems can now deliver on the analog voice network — from anywhere to anywhere.

Another key feature offered by the public telephone network, in addition to ubiquity, is bandwidth on demand. That is, rather than leasing a dedicated data link, as most large organizations currently do, with ISDN you pay only for the capacity you actually use.

The other main benefit of using the public switched network is that almost all of the responsibilities and costs of network maintenance, administration and disaster recovery are not a burden for users as they are with private networks. ISDN brings to data communications the kind of outsourcing that has always been a hallmark of voice communications. Users can dial up data and video connections when they need them and disconnect when they are finished, just as they are used to doing with voice connections. All they pay for is what they use, and they do not have to worry about network provisioning and maintenance.

> Lawrence National laboratory has the largest and most complex ISDN facility in the U.S., and possibly the world. It began using ISDN in November 1989 and it currently has over 10,000 ISDN lines going to more than 500 separate buildings, with over 160 of them connected by an Ethernet local area network (LAN). Nearly 80% of these users change locations every year. Considering that these people are data as well as voice users, moves and changes involve at least two lines, and sometimes more. It was a very messy situation,with people walking around with modems trying to get them up and running. But with ISDN, this mess has been cleaned up. Lawrence Livermore initiated a telecommuting program for 100 engineers and scientists, emphasizing remote LAN access to closely replicate their work environment at home.

AT&T Network Systems has been using ISDN since 1990. Not only is the organization is seeing reduced costs in data connections, the quality of the connections for both voice and data is significantly higher. This means much higher sound clarity and fewer dropped connections to re-dial. What may be most important, however, is that ISDN is turning out to be a big time saver in data communications, as well as a money saver.

ISDN is expected to have the largest impact on small and medium-size businesses. Firms that currently use only the telephone network for voice, fax, and some data applications will find themselves on a more nearly equal technological footing with their larger competitors. Larger companies, even those that have multi-megabit capacity private networks, will find that ISDN has a lot to offer them, too. Many companies will use it for desktop conferencing, telecommuting, remote LAN access, and LAN to LAN connections. Bandwidth on demand will lead many large users to move substantial portions of data traffic carried on their private networks to the public networks. With ISDN, these organizations can optimize their own networks around average peak levels, as opposed to *maximum* peak levels, and pay only for the additional capacity they use above average peak levels. The bulk of corporate data traffic can be handled on ISDN. Mission-critical applications, though, will be the last to be taken off private networks. The real beauty of ISDN is that you can take advantage of logical opportunities as they arise and keep your other options open.

*ISDN Applications.* Just as spreadsheet and word processing software sold personal computers, applications will also create demand for ISDN. Applications include:

- several varieties of telecommuting,
- local area network access,
- videoconferencing,
- inbound call management,
- PC screen sharing and workgroup applications,
- document image storage and retrieval,
- fax server access,
- sharing of peripheral equipment, such as printers.

A number of ISDN applications share a common theme namely, using digital networking to accomplish tasks remotely that formerly had to be done in person. The benefits are obvious: reducing travel time and expenses allows workers to devote more time to productive activities. Lets take a look at some examples of applications:

- **Videoconferencing**. While videoconferencing systems —from dedicated facilities that cost well over $100,000 in the 1980s to portable systems available in 1992 for less than $10,000—are becoming more common. The concept of personal, desktop videoconferencing has been the Information Age model ever since AT&T unveiled the fabled Picturephone over in 1966 at the New York World's Fair. With advances in the underlying image compression algorithms allowing full-motion video to be transmitted on twisted copper wire, desktop video is finally becoming available for the mass market. PC-based systems have actually dipped below $1,000 though they still resemble early black and white television image quality. But with that barrier broken, we will start to see increasingly better systems at the same price or even less. Does that mean a videophone on every desk in a year or two? Yes, it is true that desktop video systems will be affordable to many business users by the end of 1994. But it will still take some time before the budgeting process recognizes and embraces new applications of technology. This same lag occured with PCs back in the early 1980s; it was not until about 1985 that the business market really took off.

- **Telecommuting.** The notion of avoiding the daily grind of commuting without sacrificing productivity or one's career path is naturally very appealing. While "flex-time" has become accepted as a way to dodge the rush hour crush, working at home without direct supervision has a way to go before it is embraced by most managers. Yet offering people greater flexibility in how they schedule their work and personal lives has some intriguing benefits, not the least of which is the chance to attract very capable, self-motivated employees. Remote management—where direct supervisors are located in different facilities or even different cities —is becoming a more common practice in large organizations.
- **LAN Access**. A wide range of emerging applications will depend on the ability of users to access their own local area networks from remote locations (for telecommuting or distance learning, for example). ISDN works just fine as a means to access databases, peripheral equipment or other users attached to LAN. Because LAN-to-LAN connections tend to be very intermittent, basic rate service adequately handles most of these requirements as well.

People forget the evolutionary nature of building an infrastructure like ISDN. When the "golden spike" was driven to create a transcontinental railroad system, there was little more than a long stretch of track. But 20 years later, literally thousands of towns had sprung up around the transportation infrastructure. And just as the railroads carried the economic lifeblood for the Industrial Age, ISDN will carry the economic lifeblood for the Information Age. A widely available, affordable platform such as ISDN can be a critical step in jump-starting the mass-market Information Revolution (Smalheiser 1992).

On November 16, 1992, the first transcontinental ISDN-1 calls were made from Reston, Va. to sites in Chicago, Huntsville and Pasadena. This symbolic first call along with a press conference mark the beginning of the National ISDN network. Twenty two central offices equipped with National ISDN-1 software have been installed and are connected through a common signaling (CCS) and ISDN network with 64 Kbs clear channel capability. This beginning network provides National ISDN services coast-to-coast and border-to-border.

The event called the TRIP '92 (**Tr**anscontinental **I**SDN **P**roject) was dubbed the "Golden Splice" ceremony in reference to the "golden spike" driven at the completion of the transcontinental railroad 123 years ago, which heralded the beginning of the Industrial Age. TRIP '92 consisted of three distinct facets: the user open house with demo applications, the beginning of the national network, and the ceremonial event transmitted by CNN. Each of these facets involved hundreds of individuals and tens of companies. The companies who agreed to sponsor TRIP '92 were Ameritech, AT&T, Bell Atlantic, Bellcore, BellSouth, Cincinnati Bell Telephone Company, Eastman Kodak, IBM, MCI, Northern Telecom, NYNEX, Pacific Telesis, Siemens Stromberg-Carlson, Southwetern Bell Telephone, and U.S. West.

ISDN implementation has progressed to a nationally interconnected multi-vendor, multi-carrier network. The network providers participating in TRIP '92 have successfully moved ISDN implementation to this stage of maturity.

TRIP '92 certainly represents the most significant event to date in the evolution of ISDN in the U.S. Until recently, ISDN applications were limited to controlled pilot projects used by everyday workers in real world job situations. TRIP '92 marks the end of the experimental stage. It introduces the public to "real users using real ISDN in real productive activities" at over 150 sites in North America with links to cities in Europe and Asia.

ISDN is not just a sophisticated new technology, it is a core transmission medium upon which real business, education, and personal services can work. As the Electronic Frontier Foundation points out in their Open Platform Proposal; "...ISDN is a platform which could stimulate innovation in information services in a way that will benefit much of the American public that currently has no access to electronic information services" (Kapor and Weitzner 1992).

With the wide-scale availability of new services and features, ISDN will become a major force toward true public participation in the Information Age.

## B-ISDN Technology

In 1984, international effort was aimed at the development of Broadband-ISDN technology to provide greater than 2 Mb/s digital services, since LANs already provided 10 Mb/s at the time. This technology is based on SONET (Synchronous Optical Network) and ATM (Asynchronous Transfer Mode) technologies to provide multimedia communications services supporting information services.

SONET is a digital hierarchy (Physical Layer of the OSI model) based on a single optical fiber line with transmission rates at the international level starting at 155.52 Mb/s. This will retain the 50

Mb/s modularity used at some national levels (in the U.S. too). The adoption of the SONET standard will provide a complete and compatible family of lightwave, microwave, and switching products which will create an integrated global network for the delivery and management of communications and information services.

The SONET standard was developed by Bell Communications Research (Belcore) in the United States. SONET was conceived by AT&T as a backbone of public networks with standardized interfaces with MANs, LANs and other WANs (of lower speeds) to provide multimedia services. A "packet" in SONET is called a "payload." which consists of an envelop of cells with different addresses. Each payload has 756 information bytes (90 columns, each 9 rows wide = 810 bytes minus 27 bytes for transport overhead and minus 27 bytes for path overhead). Phase 1 of the SONET standard was approved in 1988.

Asynchronous Transfer Mode (ATM) is a cell-switching transfer mode. A cell is fixed at 53 octets, consisting of a 5-octet header and a 48-octet information field. These small, fixed cells can be switched more efficiently, which is important for the very high data rates (155.52 and 622.08 Mb/s) of ATM. Cells do not have to be placed in any particular sequence order in the SONET bit stream since it is the "asynchronous" mode. ATM has been chosen to organize the SONET payload ("envelope").

ATM switches with their simplified software and high traffic-handling capacity is the next giant step in the digital communications revolution. Many computer companies develop and sell office ATMs for switching data traffic from local area networks. These systems will need a public network to carry the traffic cross-town or overseas to other offices, enterprise-wide. By the end of the 1990s, the market for ATMs will hit $20 billion.

The current network of digital exchanges switch telephone calls at a steady, "synchronous" stream of 64 KBs. When a call is made the switch holds open a channel for conversation. ATMs, by contrast, take voice and data information such as video signals and computer files and place them into electronic packets. These packets are more easily accommodated by the future telecommunications systems (e.g. via SONY) that will run at many billions of bits per second and handle a cacophony of phone calls, television signals and bursts of computer data traffic all running at various "asynchronous" speeds.

By the end of the 1900s, ATM will form switching hubs for broadband communications networks. These systems will be operated by local and long-distance phone companies as well as more-nimble cable TV competitors, whose higher-capacity lines are better suited for the new traffic. These multimedia networks will carry a torrent of digital traffic, allowing computers to talk conversationally; doctors to instantly diagnose disease from electronically transmitted medical records; and consumers to see and speak with each other in high-definition television.

Switches used in the current phone network handle mostly voice traffic and cost $1 million or more. ATMs handling 10,000 data lines will cost about $10 million. However, they will handle huge amounts of new revenue-producing traffic. Large phone switches are among the most complex systems on earth, using about 10 million lines of software code to make sure calls are handled and billed properly. However, ATMs switches can be run by outside processors so they do not require the millions of lines of expensive software code that allow conventional switches to provide services such as "call-waiting" and "caller ID."

Such simplicity has raised the hopes of outside suppliers hoping to break AT&T's and Northern Telecom's hammerlock on the U.S. market. Two Japanese manufacturers, NEC Corp. and Fujitsu Ltd. are aggressively challenging AT&T and Northern Telecom. They have already supplied some ATMs switches to MCI, Williams Telecommunications in Tulsa, Okla. and to several regional Bell companies. Fujitsu thinks that ATM switching will be a centerpiece of the next generation of the communications network. This company employs 1,600 engineers to pursue the ATM technology. In 1992, AT&T shipped an ATM prototype switch called Compass to U.S.West to use in Denver's Baby Bell's broadcasting services market. In addition, Northern Telecom has a ATM switch working in its laboratory.

Proposed switch improvements are planned in the following phases:

- FIRST PHASE (beginning in 1993): Initial installation of ATM switches will augment the existing public phone network. This will allow large businesses to move whole libraries of data at high speeds and to make and receive full motion video phone calls between offices on different coasts.
- SECOND PHASE (mid-to-late 1990s): As installation of ATM switches spreads, most businesses and some homes will have access to high-speed data services, 100-channel TV programming and interactive information services. Video conferencing will be easy as dialing a phone number on the public network.
- THIRD PHASE (beginning in 2000): ATMs will begin to replace major metropolitan switching centers, taking over local phone services offering everyone sophisticated multimedia computer and entertainment lines, including high-definition TV programming. Customers will also gain access to advanced medical networks for remote diagnosis and treatment. Increased phone company efficiency could boost earnings and slow phone rate hikes (Keller 1992).

Some networking experts argue that ATM is only a transitional technology between FDDI and the next generation of multigigabit switching via electro-optical switches. Long thought to be decades away, photonic switching is coming closer to reality: MCI issued a request in 1992 for a proposal for a test photonic network.

Photonic switches may provide the terabit-per-second networks (1 terabit = $10^{12}$ bits) and 10 Gb/s nodes that scientists believe are necessary if we are to have true "telepresence" at virtual meetings, where no one is physically present, but 3-D representations make it hard to tell the difference. Such magic , perhaps, will be available in the decade of '2000.

B-ISDN is a service requiring transmission channels capable of supporting rates greater than the primary rate. It is necessary especially for image and video services. To differentiate this new network from the original concept of ISDN, original technology is now defined as narrowband ISDN.

CCITT classifies the services that can be provided by a B-ISDN into:

• interactive services (two-way exchange of information):
- conversational services (video-telephony, videoconferencing..)
- messaging services (video-mail, document-mail..)
- retrieval services (distance education, teleshopping, video-retrieval.....)
- LAN and host computer interaction
- multi-site interactive CAD/CAM

• distribution services (from service provider to B-ISDN subscriber):

# without user individual presentation control (broadcasting):
- High Definition TV (HDTV)
- pay-per-view TV
- document distribution (electronic newspaper, electronic publishing..)
- video information distribution
- high-speed medical imaging

# with user individual presentation control:
- distance education
- advertising
- news retrieval
- telesoftware
- retrieval of encyclopedia entries
- results of quality tests on consumer goods
- electronic mail-order catalog

B-ISDN technology uses compressed video techniques to accommodate high-requirements for bit transmission (up to 1 Gb/s) into a compatible bit rate for the SONET lines (150-600 Mb/s).

The deployment of B-ISDN technology will take a decade at least, since the telephone companies have huge investments in installed equipment that has not been fully depreciated. In the 1990s, only Bell Atlantic and US West will develop the SONET lines that can support B-ISDN services. AT&T and Illinois Bell will only test the SONET equipment. Perhaps after the year 2000, the availability of B-ISDN services will be on wide scale.

## Telco Switching Networks

There are over 170 million telephone subscribers in the United States and Canada. With such a big number of telephone sets it is impossible to connect them point-to-point. Telephone centralized exchange services are needed to control the telephone traffic between millions of customers. Telephone service evolved from manual switching technology to automatic telephone switching systems. As telephone traffic grew, telephone switching developed into a hierarchy of telecommunications (Telco) switching networks:

- Class 5 — at the lowest level there are about 20,000 switching centers called end offices (toll free) which serve directly to the customer via the local loop. One central office can serve a maximum of 10,000 telephone numbers.
- Class 4—at the next higher level there are about 1,300 toll centers that apply higher rates.
- Class 3— About 265 primary centers
- Class 2—About 75 sectional centers
- Class 1 —12 regional centers (10 in the United States and 2 in Canada) with approximately 7.9 million numbers that can be potentially assigned to customers.

Telecommunications traffic is routed first at the lowest level of the Telco Network and if that level is busy, higher levels are selected.

The evolution of the North-American Telco Network took place in five stages:

Stage 1 (1980s to 1950s)—analog technology
Stage 2 (1960s to 1970s) — emergence of digital transmission
Stage 3 (1970s to 1980s)— emergence of digital switching
Stage 4 (1990s ———) — N-ISDN end-to-end connectivity
Stage 5 (2000s ———)— B-ISDN end-to-end connectivity.

About 2000 telecommunication carrier companies provide services in North-America. Their progressive development goes into the integration of digital switching and transmission functions (*integrated digital networks*), the penetration of telecommunications networks by computer-based intelligence (*intelligent networks*) and the application of Computer-Aided System Engineering (CASE) to develop and implement new services via on-line intervention into telecom networks (*advanced intelligent networks*).

## Integrated Digital Networks

The term integrated digital networks (IDN) has been used to refer to:

- The integration of transmission and switching equipment;
- The integration of voice and data communications;
- The integration of circuit-switching and packet-switching facilities.

The IDN is the foundation for ISDN (Stallings 1989). The essence of IDN is digital transmission (a stream of binary digits) and switching. The digitization of telecom networks allows for the software-driven customization of networks. It leads to the emergence of private (virtual) networks defined by software in the framework of a public network.

## Intelligent Networks

The intelligent network consists of integrated hardware and software distributed throughout the telecommunications service provider's network. Thanks to software-driven technologies, telecom service providers are able to develop their own services. They can try them out, implementing those which meet their needs and discard those that do not.. From concept to implementation, a service can be introduced in a few months instead of a few years. Key elements of the intelligent network include (Figure 1-17):

- A Service Control Point (SCP), which consists of a centralized database that uses algorithms and customer instructions to route and handle messages.
- A Service Switching Point (SSP), which consists of a circuit switch (like AT&T's 5ESS)—that distributes calls throughout the network. It is a local switch designed to carry out low-level, high-volume functions such as a dial tone, announcements, and routing. The SSP performs functions as directed by the SCP. It acts as the entry

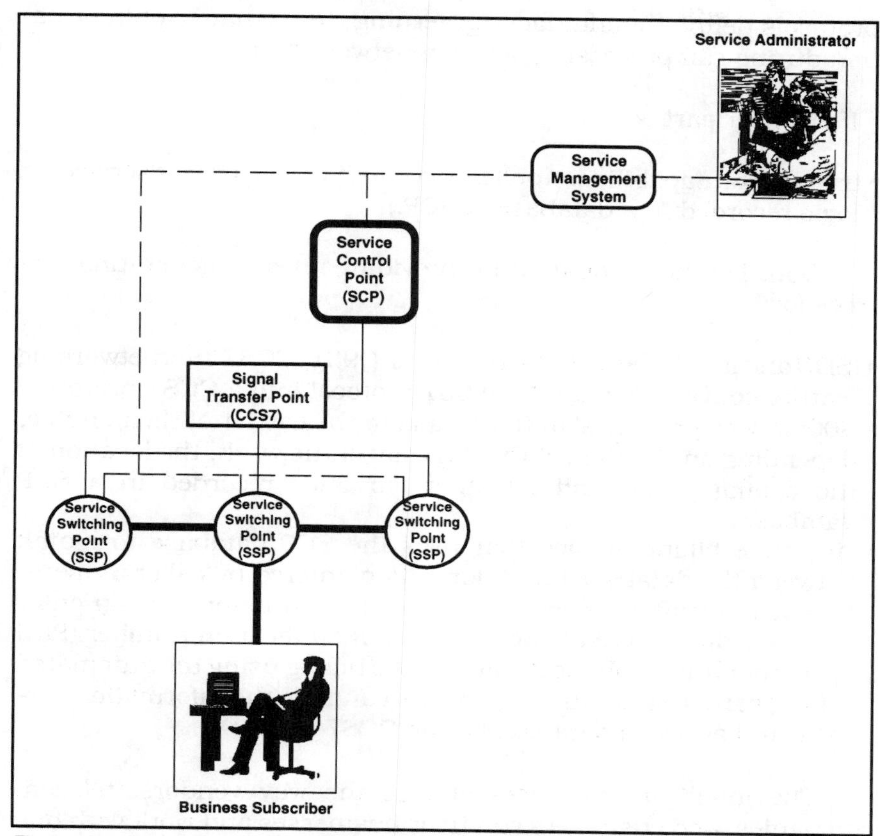


*Figure 1-17: The Architecture of an Intelligent Network*

and exit points.

- A Signal Transfer Point (STP) on a Common Channel Signaling System (CCS7) that provides out-of-band packet-switched communications among network elements acts as an intermediary signaling point, for example, routing signaling messages to SSP at the calling destination or relaying instructions from SSP to SCP.
- A Service Management System (SMS) that provides for network planning, engineering, provisioning, monitoring, maintenance, and repair.

The CCS7 separates telephone signaling from the voice path, greatly increasing the amount of signaling information that can be exchanged by switches and providing access to a variety of databases. It transmits signaling information over dedicated 56 Kb/s links that are 24 times faster than in-band systems. As a result, CCS7 reduces trunk holding time while handling more traffic through the same facility.

Essentially, CCS7 ties central office switches to STP that provides signaling distribution for the telecom network. Two CCS7 protocols define the interfaces governing the exchange of information during call processing between network points:

- ISDN user part

- transaction capabilities application part (SSP can access service via logic recorded in a database of SCP.)

Both lay the foundation for providing revenue generating services for:

- ISDN and internetworking services, the Q931/CCS7 internetworking feature converts the ISDN's q.931 protocol into a CCS7 protocol,
- 800 service providers' to route calls to the carrier of their choice, depending on the time of the day, day of the week, the location of the calling party, and on other variables recorded in a SCP database,
- alternate billing service that uses the SCP database to match between the database and information entered by callers, operators at the traffic service position system. Customers can enter their telephone credit card personal identification number (PIN) into a touch-tone phone for automatic billing using the automated calling card service. Also, operators can take the information verbally and access a databasevia the CCS7 network.

The intelligent networks change the way vendors, telecom companies, and customers run their businesses and work with one another. Vendors can develop telecommunications products that deliver certain communications service under the form of software-oriented "building blocks." They deliver these blocks to the telecom companies that assemble them into new services.

How would the IN benefit a business with many locations in a city, such as a pizza chain that makes deliveries?

One pizza chain found it could cut costs and improve service by advertising a single phone number for all its stores in town. When a customer phones in an order, the network automatically routes the call to nearest pizzeria. Here is how it works. When a customer dials the pizza company's advertised number, the local exchange recognizes the trigger — the request for AIN services, but does not know how to complete the call.

It routes the call to the nearest intelligent network switch, the Service Switching Point. The switch then queries the Service Control Point for routing instructions. This network element includes a database on how the pizza chain has set up its service. It sees that the pizza company has instructed the network to route calls to the caller's local pizzeria, based on the first three digits of the caller's phone number.

Service logic in the Service Control Point directs the Advanced Service Platform —call processing building block in the switch —telling it how to process the call. The switch makes the connection and the call is answered by the local pizza store clerk who is ready to take the order.

This service, known as Area Number Calling, could also include a time and day routing feature to forward calls if one store closes earlier than others. Should the pizza chain want an emergency feature to reroute calls if store gets too busy or if an oven fails, that is possible too (Samuelson 1991).

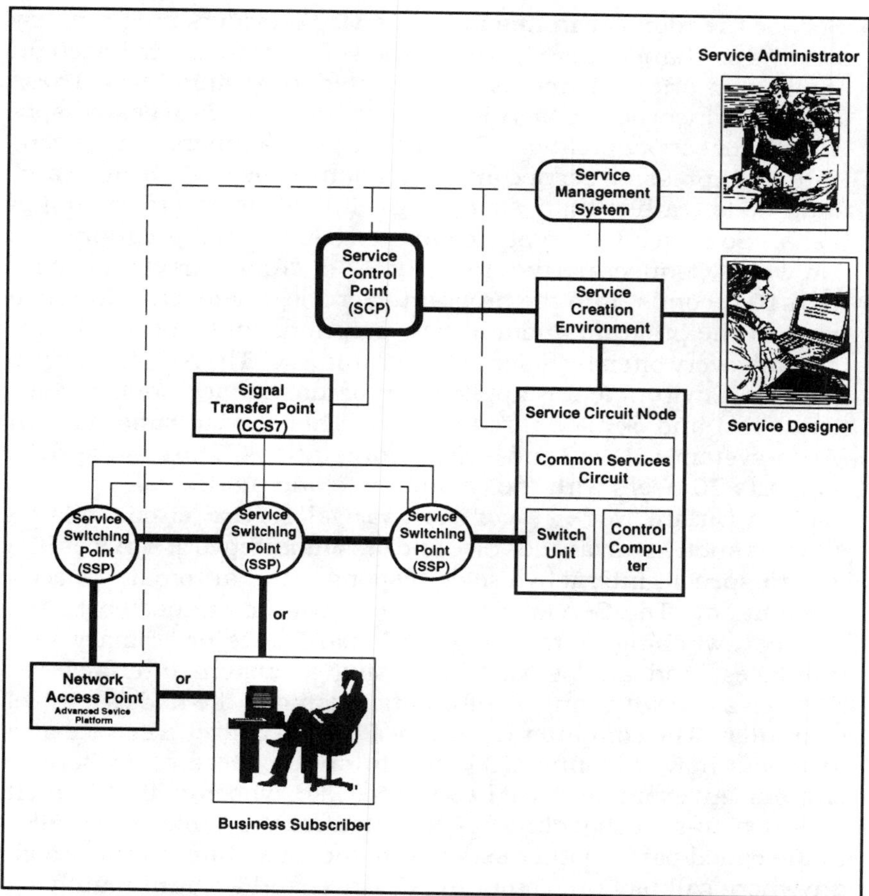


*Figure 1-18: The Architecture of Advanced Intelligent Network*

## Advanced Intelligent Networks

The Advanced Intelligent Network (AIN) is a software-driven service deployment proposed by Bellcore for the regional Bell Companies as a Computer-Aided System Engineering. This network dramatically reduces the time between conception of a service and its implementation in the public network. In such a way, new services can be tested more frequently among friendly users and focused customer groups. If a new service is accepted, the service can be quickly introduced into the complete network.

To help reach this goal, AT&T launched in the 1990s a group of products called A-I-Net™. Together the products enable the entire enterprise to be managed through a single system.

The architecture of the Advanced Intelligent Network is based on the architecture of the Intelligent Network. However, it has a few more components (Figure 1-18):

- Service Creation Environment lets service providers program new services or change existing ones. The software tools are based on application-oriented and object-oriented programming. These languages incorporate high-level commands and data geared specifically for service creation. This means programmers can concentrate on expressing service and operations logic in their programming code, rather than struggling with low-level programming tasks. So instead of cryptic computer commands, programmers can write "Connect party A to party B," or "Ring party A for more than 10 seconds" into the program. Service developers can reuse parts of the program and add new programs and features incrementally, very often by different programmers. The A-I-Net Service Creation Environment is applied to program Service Control Point (database) and Service Circuit Node. This service runs on Sun Micro-system's SUN/3 and SUN/4 workstations. The basic system supports 10 users with the capability to add more.
- Service Circuit Node — controls special service circuit-related services such as advanced voice and facsimile network services like text-to-speech, interactive voice response, and automatic speech recognition. The Service Circuit Node can be connected to the Service Switching Point using ISDN basic rate or primary rate interfaces, and analog switches using a custom interface. It receives a call with control information going to the node's control computer. The computer determines how to deliver a service and instructs how to connect the call to one or more of its service circuits. An example might be a service like "who is calling," which uses text-to-speech technology to announce the name of the caller to the called party. Other services include fax store-and-forward, anywhere call pickup, computer security, and account match.

The Advance Intelligent Network brings a rich array of sophisticated services. Here are a few of them:

Account Match—Inbound calls are routed automatically to a specific customer service representative and the caller's file is brought up on the representative's computer screen while the call is being connected.

Anywhere Call Pickup—Allows a call to be held at the Circuit Node until the called person is notified by a beeper. When a customer is away from the phone, incoming callers can leave a message or have the person paged while they wait. If there is no response, the caller can leave a message through voice mail.

Automatic Call Distribution —Allows telephone company customers to interconnect a number of call centers. Customers such as airlines or credit card companies can have traffic routed in real time to a center best able to handle the workload.

Computer Security—Those using this service will be data network and computer/information service vendors and users who want to be protected from nuisance calls and destructive data transmissions. Security screening can be performed in the network before an end-user gains access to the customer's network, systems, or applications.

Enhanced 800 Service—Routes incoming calls to specific locations based on information about each call. For example, calls could be sent to the nearest geographic location or to sites based on time of day and/or day of week.

FAX Store and Foreward —Gives customers additional control of FAX transmittal through automatic storage, retry, and confirmation of FAX transmission.

Intelligent Home Services —A network-based capability to manage on a local or remote basis the interworking of home appliances, heat, lighting, security and alarm sensors to control the efficient and secure functioning of a home.

976 Selective Blocking—Allows users to block certain outgoing calls, such as those to talk lines. Customers can determine blocked numbers; or each call can be screened through a personal identification number.

Who is calling?—A service that lets users screen their calls. Basic service includes a voice announcement to the call person before they answer, giving them the caller's name.

Warm Line — Offers security for the elderly or ill. If a receiver is taken off the hook and no numbers are dialed in a set time period, a signal is sent to another phone, such as that of a neighbor or doctor (Samuelson 1991).

## Personal Communication System and Network

Back in the 1800s, the equivalent of a network manager for the Apache, Sioux and Iroquois Indians did not have to contend with data delivery options such as radio-based cellular nets or satellites. When American Indians wanted to communicate over great distances, all they needed was a little smoke and a few animal hides (Gregg 1992). Today, we can communicate via a portable Personal Communication System (PCS).

One of the most endearing and most frequently quoted lines of any movie ever made has got to be "E.T phone home." Those of us who saw the Steven Spielberg classic in the early 1980s will always remember the plight of E.T —the lovable Extra Terrestrial who accidently got marooned on the Planet Earth and, with the help of some enthusiastic California kids, spent much of the movie trying to "phone home" for help. In the 2000s, most of us likely will be able to point out the anachronism of that film. E.T.— a being from a planet technologically light years ahead of Earth—would never have tried to phone home. E.T would have tried to call his mother, his father, his brother, or his sister. In another words, he would have tried to call a person, not a place (Brody and Roth 1992).

We Earthlings only make calls to places because directory numbers historically have been assigned to telephone jacks and other physical locations in the network. Users, in turn, have been associated with these locations. As a result, we have separated directory numbers for the terminals in our homes, cottages, offices, and cars. Faced with a choice of so many numbers, someone trying to reach us can easily become impatient, frustrated, and discouraged. This reality is changing rapidly. When PCS is fully realized, many of us will never call "home" again.

Why? Because in this new era of personal communications, we could each own a single personal directory number that could be used to reach us no matter where in the world we are located, and regardless of the type of device — wireline or wireless —we are using. A caller will only ever need to dial our personal number —and the network will automatically do the rest — find us and route the call to us. For the first time in history, the burden of locating the called party will be removed from the caller and placed squarely on the shoulders of the network. Since Alexander Graham Bell's invention first went into commercial service in a Boston bank, telephone service has been associated with a fixed place. PCS are poised to turn this traditional view of network engineering on its head.

Benefits of PCS can be illustrated by the following scenarios:

Scene 1: A marketing manager, returning to the office from a two-day off-site planning session, accesses his voice mailbox. One message is from a key client, requesting his attendance at an emergency meeting with top executives from the client's firm. The meeting was scheduled to take place two hours before.

Scene 2: A 10-year-old boy, spending a weekend at a friend's cottage, slips and hits his head on a rock, damaging his eye. The friend's mother rushes the boy to the nearest hospital. The operation consent form must be signed. The mother calls the boy's parents; the phone rings unanswered.

Scene 3: An employment agency calls a student, who has been job-hunting for weeks, to offer him a temporary job. The position must be filled immediately. No one answers at the student's home. The agency calls the next person on its list (Cohn at al 1992).

The location-independent services will create a paradigm shift in the way the world views telephony and communications. PCS will reshape the telecommunications and computer industries, our communities, and our lives. For example:

- Our view of the telephone will change. No longer will the telephone be a shared fixture, but rather a personal item that people can carry with them most, if not all of the time. At home for instance, each family member could own his or her own phone and all members could make and receive calls simultaneously. And, as telephones becomes more personal similar devices will emerge. They might be worn much like jewelry or as a wristwatch or they could be built into a portable radio or CD player.
- The concept of just-in-time communications will emerge, shattering today's statistic that 80 percent of the calls we make do not reach the intended recipient. Being able to reach people wherever they are located will end the rounds of telephone tag that sap productivity today and will accelerate the pace of business to an unprecedented level. At present, people communicate in ways that might be compared to the batch-mode communications of the early days of computing—creating stacks of voice-mail messages and "while-you-were-out" slips. Those voice messages and slips are processed in batch mode, and four times out of five the return telephone calls lead to further messages and slips, continuing the frustrating and unproductive game of telephone tag.
- The importance of call management services. It will enable users to control who reaches them, at what times, and under what circumstances (Brody and Roth 1992).

PCS will bring the full power of communications mobility to the mass market —a mobility that today is enjoyed by only a few people who can afford cellular telephones and services. By the 2000s, PCS is likely to be a $50 billion industry. The system will serve as as many as 150 million people worldwide and 60 million people in the United

States, including millions who will benefit from in-building PCS and PCS data transmission. PCS will enable poorer countries to overcome wired infrastructure handicaps affordably and efficiently (Wimmer and Jones 1992).

The key to addressing this challenge lies in the mix of technology components. The PCS network will be based largely on a hybrid use of wireline technology and three classes of wireless technology:

- high-density, on-premises, low-power wireless systems for those people in buildings (office environment);
- high-speed, wide-coverage vehicular cellular systems (cellular radio systems for those people traveling in cars between buildings, and between cities);
- high-density, wide-coverage cellular microsystems (for those people who are mobile within the office and on the road) that can inter-network to wide-coverage vehicular cellular service.

The combination of these technologies enables individuals to use the same handset in the office and anywhere in the cellular network.

On May, 1992, the first-ever fully end-to-end digital cellular telephone call between Canada and the United States was completed between Toronto (Ontario) and Fort Worth (Texas). Participating in this historic call were executives of Motorola and Northern Telecom. The call demonstrated the digital radio technology that will be supplied by Motorola-Nortel Communications Co. to Bell Mobility Cellular, a network of 400 cell sites.

> In addition to working at home, customers increasingly need to conduct business while on the move. As the largest mobile service provider in the Great Lakes region, Ameritech Mobile serves more than 326,000 customers in 22 metropolitan areas, covering more than 30,000 square miles.
>
> Ameritech Fast Track Service™ makes it easier for customers to use their mobile telephones outside their home service areas by enabling them to receive calls automatically in more than 325 cities in the United States and Canada.
>
> Ameritech Mobile Message Service™ acts as an answering machine for mobile telephone customers, storing messages until they can be retrieved from any touch-tone telephone.
>
> Ameritech mobile also is the Midwest's largest provider of paging services. For customers who need to be reachable even when away from their mobile telephones, Ameritech Priority Message Service™ combines mobile communications and paging tech-

nologies to notify users of waiting messages via their pagers.

Increasingly, mobile customers require access to information. With The Smart Call™, customers in Chicago and northwestern Indiana have access to traffic and weather reports, stock market reports and sport scores.

With the Ameritech Mobile Access Data Service™, customers can access data stored in any domestic computer that is reachable through a local or 800 telephone number.

To give auto buyers "built-in" convenience, Ameritech Mobile service and Chrysler Visorphone™ are offered on Chrysler-Plymouth and Dodge cars sold in the Ameritech region. The visorphone, shown on Figure 4-18 at Roseville Chrysler-Plymouth in Rosevilee, Michigan, has a five-year warranty, 100-number memory, one-touch dialing, and automatic radio mute. Flip it down to dial, flip it to talk. Other manufacturers are coming out with similar built-in products.

Today the average American spends more than two hours each day in an automobile. In a world that revolves around such a mobile environment, there is more of a need for mobility security. In 1990, there were 1,635,900 vehicles stolen in the United States. Ameritech Mobile Communications and METS (Mobile Electronic Tracking Systems) developed MobileVision™ in 1992. It is a mobile telecommunications network that links you, your car, police, emergency response providers, and the Central Monitoring Station through wireless technology. The system includes:

- Road assistance
- Stolen vehicle location and recovery
- 24-hour monitoring
- Comprehensive geographical coverage.

An emergency occurs while you are in your car. Whether your car is overheated or has some other problem, MobilVision has the resources to respond. You communicate the nature of the problem, e.g., stalled car, to the Central Monitoring Station, and the dispatcher at the Central monitoring Station acknowledges that the signal has been received and that a response vehicle is on its way. The dispatcher will also let you know when the response vehicle will arrive at your location. You never have to leave your car.

When your car is moved or started in an unauthorized way, the Central Monitoring Station receives a signal, and your car is tracked immediately. The Central monitoring Station contacts you to verify the unauthorized use of your car (to minimize false alarms). The Central Monitoring Station then links and assists you with the police (if you are not reachable, and you have authorized us to do so, the

Station will act in your behalf). With the help of this service, the local police track your car until it is recovered.

Whenever you leave town, carry a compact (4.11 oz) SkyWord pager of the SkyTel™ messaging system. You will be able to receive brief, complete written messages instantly, even if no one knows where you are. With its 8—character electronic display, SkyWord is perfect for getting direct answers and quick updates, like—PRADLEY MEETING MOVED UP TO 3 TOMORROW. BETTER GET AN EARLIER FLIGHT. BILL.—before it is too late.

All anyone needs to reach you is a PC with a modem (including modem-equipped laptops), any standard communications software and your personal code. They send the message to a single toll-free number, and in seconds it is relayed from a central computer to a satellite, beamed to downlinks, and sent to thousands of cities and towns via dedicated transmitters in each area.

Instantly, your SkyWord pager alerts you, either silently or audibly. You control who gets your personal code: your secretary, colleagues, clients, or family. It is literally like being in a hundred places at once (300 cities in the United States and 150 in Canada, Mexico and Singapore). For knowing up-to-the-minute details, you just cannot beat it.

As long as you are improving your communications systems, why not turn SkyWord or SkyPager (limited to numbers) into a complete communications system by adding SkyTalk. With the SkyTalk voice mail option, you get the full information you need, right away, in your caller's own voice. You are notified immediately when recorded information is waiting for you, and of course, calls to leave and retrieve messages are toll-free. SkyTel offers regional subscriptions for the Eastern and Western United States. So you pay only for the coverage you need. If things change, one phone call lets you switch to national service, which covers you in thousands of cities and towns in the United States. If you travel to Canada or Mexico, these areas can be added to your coverage. In addition, SkyTel is leading the way toward a true global messaging network. Developments in 1990s will include Singapore, Europe and the Pacific Rim. Most importantly, it doesn't matter where in the coverage area you are: a New York subway, a San Diego skyscraper, wherever. Driven by a 900-MHz frequency, it is the system best suited to penetrate concrete and steel. National rental package costs $89 per month and the regional rental package costs $69 per month.

A less expensive system is provided by BellSouth Company in 450 North-American cities. A MobileComm™ card that was introduced in 1992 is shown on Figure 1-19.

The In-Flight office is a new competition in sky-phone service at 40,000 feet. Your plane has just lifted off from the Los Angeles airport where, barely 30 minutes ago, you concluded a major meeting with your company's most important client. You are anxious to report the positive outcome in writing, but your laptop is in the shop for repairs. Even pulling together your scribbled notes for transcription by your secretary will take two days, and tomorrow is Saturday. Not to worry. From a steward, you can request a keyboard that will be installed in the back of the seat in front of you. It constitutes a word processing/ fax generating computer. In the five hours it takes to fly across the United States, you are able to compose and edit your business report, and then forward it, via phone, to your company's mainframe. You even can call up information from that mainframe to incorporate into your report. All this moving at 600 miles per hour, 40,000 feet up in the sky.

This service was conceived by John D. Goeken, the technical genius who created the long-distance telephone service, MCI, and FTD, the national flower delivery system. Goeken also conceived Airphone (now owned by GTE), the first air-to-ground telephone service to be placed aboard airliners.

Airborne telecommunications for passengers work in a mode similar to cellular phone service, except that the former's "cells" are many times larger. Two bands of frequencies are used: 849 to 851 MHz and 894 to 896 MHz. These bands are divided into 310 6 KHz channels with four to eight channels available for each airplane, depending on seating capacity. Communications from air to ground use channels in the higher frequency band, while responses from the called parties on the ground return over channels in the lower frequency band.

When a passenger initiates a call, the aircraft system automatically selects one of the 31 channels assigned to a ground station several hundreds miles ahead of it. This is to sustain the connection

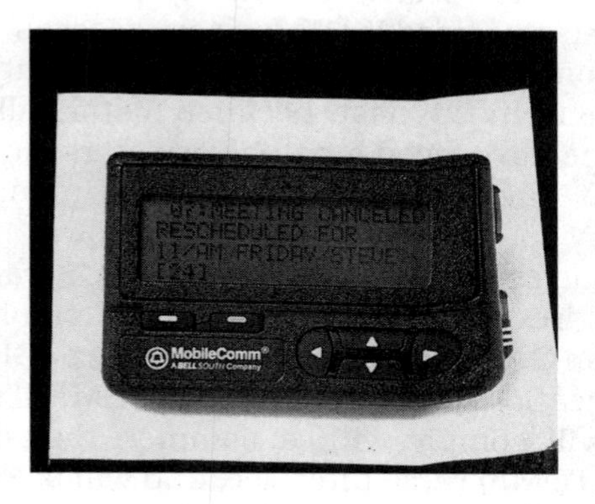


*Figure 1-19: A text pager of the MobileComm Nationwide Messaging Network supported by Bell South (Courtesy of G. Williams Associates).*

as long as possible; very rarely do conversations last long enough for the high-flying aircraft to move beyond a line-of-sight connection with the ground station. There is no system yet to pass connections from ground station to ground station. However, both the current services, GTE Airphone and In-Flight, claim they have hand-off capability based on cellular technology. When it will be established, a passenger should be able to talk, without interruption, from one coast to another. A hand-off from domestic to transoceanic service is also in the offing, but passengers engaged in a call will be prompted about upcoming hand-offs to transoceanic service because it costs about five times as much as as over-land service.

In-Flight has been testing its services on US Air, Northwest, and American Airlines. In-Flight announced its intention to offer ground-to-air passenger paging, conversations and digital transmissions upward. A passenger who wants to receive calls on the phone at his seat would run a credit card through a reader on the handset or key in a personal code so the phone system could find them. The system must maintain a constantly updated airborn phone directory through which it could direct calls to passengers (Mandell 1992).

As PCS continue to evolve, they will ultimately migrate to an intelligent network infrastructure, which can track subscribers and route calls to them anywhere in the world. For example, if a user was carrying a personal terminal or a cellular phone, the device would automatically send a radio signal to the nearest base station, informing it of the user's location. The base station and the associated Service Switching Point (SSP) would, in turn, relay that information via a CCS7 (Common Channel Signaling System 7) data channel to the user's home database which would update its records with the user's new location.

By the end of the 2000s, millions of cellular subscribers in Europe will be able to receive uninterrupted telecommunications services (including advanced ISDN-type services) as they move from one European country to another. This capability represents a radical departure from the past, because historically cellular subscribers in Europe have faced a patchwork of incompatible cellular technologies that have prevented them from using their cellular handsets in many countries other than their own.

The inadequacies and inconveniences of PCS fragmented structure are being addressed through the deployment of digital cellular networks based on GSM (Groupe Special Mobile or Global System for Mobile Communications) standard. Under GSM, all mobile stations (the handsets) will comply with a common digital time division multiple access (TDMA) radio interface, and will be compatible with all of Europe's national GSM networks. These networks will be able to roam freely within national GSM networks and from one European GSM network to another. GSM standards are also being used by the American and Japanese standards bodies as a model for their own digital cellular network architectures and standards. The GSM standards are more than 5,000 pages long and detail complete specification for a common digital cellular network (Beaudry and Parker 1992).

## User Computer Networks

A user computer *network* comprises two or more intelligent devices (computer systems, intelligent terminals and intelligent peripherals) linked in order to exchange information and share resources. A device on a network is called a *node.* Some nodes are called *hosts.* A *host* is any network node that individuals can access for resources such as processing power, information files, and user applications. Hosts are general purpose nodes that have been specifically designed to fulfill network-specific functions. They are distinguished from special purpose nodes, such as repeaters, bridges, routers, gateways which interconnect user computer networks.

From the perspective of individual users, the most apparent advantages of a network is that critical information, no matter where it is located, is available to them when they need it.

For instance, a company's financial manager preparing a profit-and-loss statement for a fiscal quarter can sit at a terminal and immediately get up-to-date sales and operating cost figures, even if data resides on two remotely located nodes.

From an organizational perspective, the immediate advantages of a network are:

- efficient use of information in enterprise-wide decision-making
- cost-effective use of expensive resources, such as storage devices (disks and tapes), computer input-output devices (scanners, printers, readers).

Consider four research engineers located in facilities in different cities. Each engineer is designing a separate part of a product. Through a network, engineers continually share their findings and even via virtual reality they can assembly those parts.

The primary advantage of the computers proliferation is that computing power is available to individuals at their work site, be it an office, a lab, or a factory floor. A major disadvantage of such proliferation is duplication of effort. Every departmental computer requires its own database of information, as does every personal computer. Multiple copies of the same data exist, and this presents updating problems. Networking helps control this inefficiency in data storage and maintenance. There is no need to create multiple copies of the same data because users on a network can share information. With a network, maintaining a database becomes simple because there is only one copy of the database to maintain .

Ultimately, the benefits of a network lie in the increased effectiveness of a company's workforce and resources.

## Local Area Networks

The history of LANs is relatively short. The first LAN—ALOHANET was installed in 1971 through radio waves in Hawaii. The Ethernet bus model was the next LAN developed by Xerox in Pal Alto in 1981 by R. Metcalf and D. Boggs. In the same year Xerox, Digital, and Intel upgraded Ethernet. It gave a birth to two LAN lines; Ether Series 3 COM was developed by R. Metcalf in 1983, and IBM PC Network was offered by IBM in 1984/85, based on CSMA/DA. However, IBM switched a year later to Token Ring Network and established a standard—IEEE 802.3. In the 1990s, the most popular LAN is Novell for its client-server architecture.

A local area network (LAN) is a telecommunications network which usually is owned and operated by the computer systems user. A LAN interconnects terminals, workstations, microcomputers, minicomputers (servers), and mainframe computers as well as other devices such as printers, scanners, high-resolution monitors, telephone stations, and so forth. LAN technology supports:

- equipment interconnectivity
- sharing common resources
- peer-level interactive communication
- enterprise-wide information management

LANs support transmission rates of 1 to 100 Mb/s and higher over distances up to 10 km. A typical LAN's area is an office building, a university campus, an industrial park, or a hospital complex.

Some organizations, such as large industrial, insurance and banking companies have a large user base and enough information traffic to develop intra-city networking. In this case several LANs can be interconnected to form a metropolitan area network (MAN). In enterprise-wide information management, a company's LANs from different cities and locations can be interconnected via wide area networks (WANs) and MANs.

Design elements for local area networks fall into several categories:

- Topology
- Channel control, allocation and access methods
- Transmission media
- Network management.

***Topology.*** The LAN concepts evolved since the mid-1970s around a group of standards defined by the Institute of Electrical and Electronics Engineers (IEEE). The following standards, topology-driven are applied:

- Ethernet (Digital)—IEEE 802.3 on a bus topology
- Token-ring (IBM)—IEEE 802.5 on a ring topology
- Dual-Ring (Telecom Australia) —IEEE 802.5 (DQDB) on rings topology
- Dual-Ring Backbone (Digital) —IEEE 802.5 (FDDI) on rings topology.

The distinguishing feature of *bus topology* is that network nodes share a single physical channel via cable taps or connectors. The bus topology has been used for distributing control in local area networks. Messages placed on the bus are broadcast out to all nodes. Nodes must be able to recognize their own address in order to receive transmission. As a result, there is none of the delay and overhead associated with retransmitting messages at each intervening node, and nodes are relieved of network control responsibility at this level. Because of the passive role nodes play in transmissions on the bus, network operation will continue in the event of node failures. This makes distributed bus networks inherently resistant to single-point failures.

The main feature of *ring topology* is that the nodes, which are connected by point-to-point links, are arranged to form an unbroken circular configuration. Transmitted messages travel from node-to-node around the ring. Each node must be able to recognize its own address in order to accept a message. In addition, each node serves as an active repeater, retransmitting messages addressed to another nodes. The need to retransmit each message can make ring nodes more complex than the passive nodes on a bus network.

The distributed queue *dual bus (DQDB) topology* uses a dual-bus to "counter-rotate" a message in a case of one bus's failure. This topology is fast and the most convenient for MANs.

The fiber distributed data interface *(FDDI) dual ring topology* provides a high-speed backbone network at a date rate of 100 Mb/s while retaining the token-passing advantage of the IEEE 802.5 standard. FDDI dual ring can accommodate up to 1000 nodes over 200 km (120 miles), including self-contained LANs.

> A new backbone from Southwestern Bell Corp. makes JCPenney one of the world's largest implementors of fiber to the desktop in the 1990s. The retail giant is installing an $11.2 million Fiber Distributed Data Interface network at its new eight-building Legacy headquarters complex in the north Dallas suburb of

Plano, Texas. The three story structure houses about 4,000 employees in purchasing, marketing, administration, finance, advertising and public relations. Southwestern Bell of St.Louis installed the network, which includes fiber, twisted-pair wire and coaxial cable. The data backbone is based on the FDDI specification, which supports LAN transmissions at speeds up to 100 Mb/s.

JCPenney's installation of fiber to the desktop is one of the largest in the world. The two sites are connected by 45 Mb/s T3 lines, multiplexers and routers to handle voice and data. JCPenney is also tapping the fiber infrastructure for videoconferencing with a system from PictureTel Corp., Danvers, Mass. The company is implementing a video library for automated training, and many users in the company can get cable television at their desktops. The video system works with JCPenney's company-wide direct broadcast system, a satellite network which lets remote sites look at merchandise or video messages from headquarters.

Typical topologies of LANs are illustrated on Figure 1-20.

***Channel Control, Allocation and Access Methods.*** Control of the channel is either centralized in a single node or distributed to all the nodes. Allocation schemes are implemented so that the capacity of the channel, which is finite, is used in the most efficient manner possible. Access techniques are the means by which nodes actually gain the use of the common channel to transmit messages. In other words, control strategies describe "Where" control of access and allocation resides in the network, access methods decide "who"

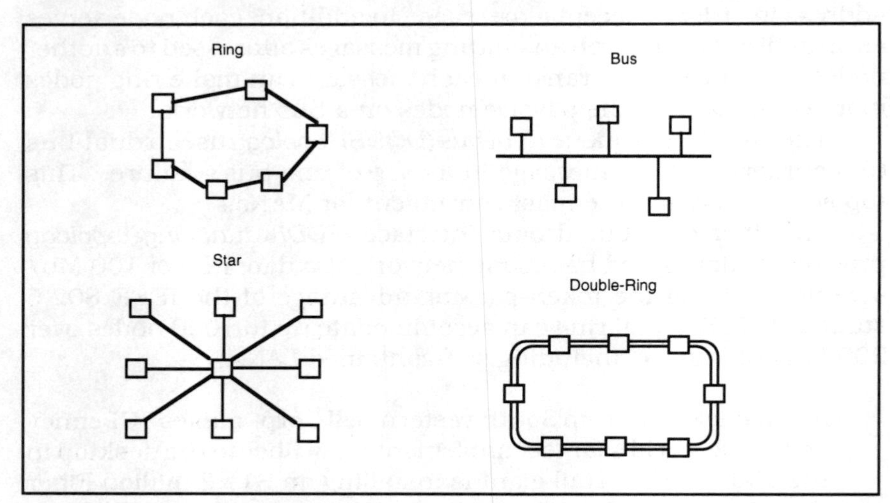


*Figure 1-20: LANs Topologies*

gets the channel, and allocation schemes determine "How much" channel capacity a node can have. These schemes and techniques are independent and can generally be mixed and matched to provide desired network capabilities. The most popular channel access techniques are:

- *Token passing*: usually associated with the ring topology but recently with the bus topology too. Tokens are special bit patterns or packets, usually several bits in length, that circulate around the ring from node to node when there is no message traffic. Possession of the token gives a node exclusive access to the network for transmitting its message, thus avoiding conflict with other nodes that wish to transmit. If a node wishes to transmit a message, it will "hold" the token and send its message, specifying the destination address. Nodes on the ring check the message as it passes by. They are responsible for identifying and accepting messages addressed to them, as well as for repeating and passing on messages addressed to other nodes. The IBM Token-Ring LAN applies this access method which allows for 1 to 4 Mb/s transmission speeds (roughly 400 to 1600 pages per second).
- *Carrier sensing* is the ability of each node to detect any traffic on the channel (called *listen-before-talking*). Nodes defer transmitting whenever they sense that there is traffic on the channel. However, because of the time is takes for a signal to travel across the network (called the *propagation delay*), two nodes could detect that the channel is free either exactly at or close to the same time, since each will not yet have detected the signal of the other. In such a situation, a collision between the two messages will occur. Upon detecting a collision, each node involved backs off and abandon its transmission. The selection of random intervals between next transmissions is more effective in avoiding further collisions. This access method is called Carrier Sense Multiple Access with Collision Detect (CSMA/CD). Owning to the ability to listen before and during transmission, the number of collisions can be quite low, and successive collisions between nodes are rare. Thus, CSMA/CD is a highly efficient form of distributed access. The Ethernet LAN applies the CSMA/CD access method which allows for up to 10 Mbs transmission (4000 pages per second).

***Transmission Media.*** Transmission media provide the physical channel used to interconnect nodes in a network. Media are classified as bounded—for instance: wires, cables, and optical fibers; or unbounded—the "air waves," over which radio, microwave, infrared and other signals are broadcast.

***Network Management.*** Network management is associated with LAN operating systems that give the network its multiuser, multitasking capabilities. The operating system's modules reside in a microcomputer (server), the printers, and other resources. LAN operating systems are built around such computer operating systems as DOS, UNIX, OS/2, MS-NET, NETBIOS.

***Client-Server Model.*** The most important aspect of LANs is the shared file system. In 1987-88 mainframe software vendors such as Oracle Corporation and Gupta migrated the client-server architecture onto microcomputer networks. This type of architecture is composed of two elements:

- client (end user) applications (such as a database, spreadsheet, or word processing data files),
- server with the database, spreadsheet, and word processing original software packages (engines).

Many vendors offer proprietary servers (minis), arguing higher performance than can be obtained from a regular workstation or a microcomputer.

A typical architecture of a LAN is shown on Figure 1-21.

## Rural Area Networks

As rural communities seek to use technology to compete more effectively with urban areas for businesses and jobs, having more choice and control over their communications infrastructures is even more important . One way of reducing the urban advantage is to deploy technologies, such as digital radio and satellite where costs are relatively insensitive to distance. Whereas many business

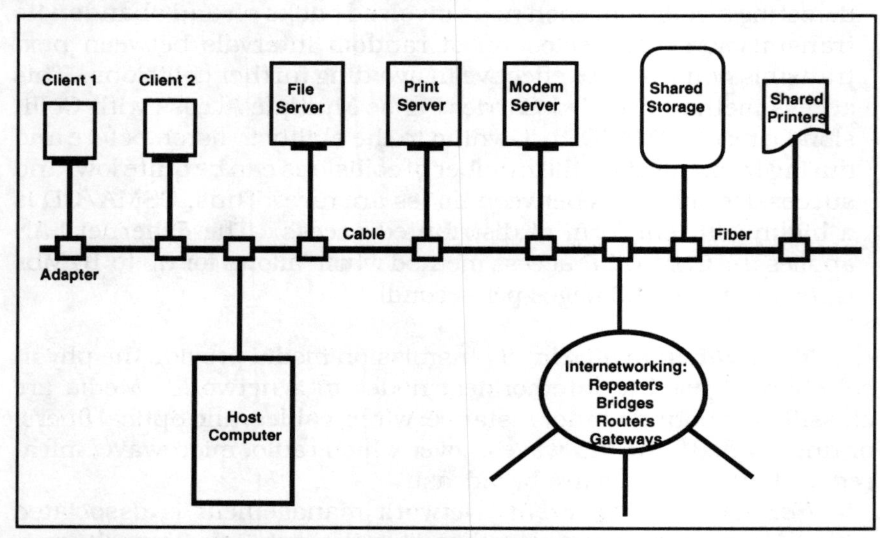


*Figure 1-21: The Architecture of LAN (either bus or a ring)*

networks are established along functional lines, Rural Area Networks (RANs) could be configured, instead, around the geographic boundaries and needs of an entire community. Designed on a topology of a ring, or a campus type, a RAN would link up as many users within a community as possible — including among them businesses, educational institutions, health providers, and local government offices. RANs could be linked state-wide by a State educational network (Office of Technology Assessment 1991).

### Metropolitan Area Networks

A metropolitan area network (MAN) is a standardized high-speed network providing LAN-to-LAN and LAN-to-WAN connections for private or public communication systems in noncontiguous real estate within metropolitan-range distances (Valovic 1989). Within the ISDN standards, MAN can provide voice and video services too. MAN extends the scope of the LAN concept into a larger environment containing a city and its suburbs. In effect, thousands of information and communication technology devices will be feeding telecommunication traffic to the MAN. If a LAN transmits with a speed of 1 to 10 Mb/s, the MAN will operate at the speed of hundreds of Mb/s. Fiber optic technology makes MANs feasible and operational.

A LAN is usually a private network while MAN is a public one. MAN is being developed by telephone companies, mostly by Bell Operating Companies (BOC). The IEEE 802.6 project/protocol is supported by AT&T, BOC, Bell Communications Research (Bellcor) and Telecom Australia. In 1982 ANSI Committee X3T9.5 was chartered to develop a high-speed data networking standard. As a result of it, a packet-switched MAN backbone network evolved. The IEEE has delineated the following criteria for a MAN:

- It must implement a fast, robust signaling scheme
- It must provide security that permits the establishment of each user's ID and password
- It must ensure high network reliability, availability, and maintainability
- It must promote efficient performance of the MAN, regardless of size.

The application of MAN can be predicted in the following areas:

- Interconnection of LANs
- Interconnection of PBXs
- Gateway to Wide Area Networks
- Connection of host computers
- Transmission of CAD/CAM data
- Transmission high-resolution images (e.g., X-ray pictures)

Since a MAN is supposed to transmit voice and video, LAN protocols cannot be applied. These protocols (IEEE 802.2) cannot guarantee 2 milliseconds of delay which is still acceptable to minimize a voice echo. MAN's larger distance than LAN's and MAN's voice requirements generate a need for a new network architecture and protocol for MAN. The MAN technology evolves in two directions:

- in campus like environment (private)—fiber distributed data interface (FDDI)
- in public large area —distributed queue dual bus (DQDB).

As a supportive switching function of an internetworking MAN —BOC provides Switched Multi-Megabit Data Service (SMDS).

*FDDI is a token-passing dual counter-rotating ring* employing two pairs of fibers transmitting at 100 Mb/s. The ring architecture means that a designated node in the network repeats the incoming data at the other nodes. FDDI can transmit data in synchronous and asynchronous mode. The FDDI network can tolerate 2 km between stations, 100 km around the ring with 500 nodes. A packet can have 4500 octets. FDDI is a good solution for a backbone network interconnecting departmental LANs. Two station types are applied. Station A acts as a wiring concentrator to interconnect several Class B stations. The class A station is connected to both the primary and secondary rings of the network. A class B station can be a terminal, workstation or front-end computer.

*The dual bus architecture.* The IEEE 802.6 group included computer, CATV and satellite industries. The CATV companies wanted to deliver pictures from earth downlink stations to households. In 1987 a consensus was reached that a dual-bus architecture proposed by Telecom Australia will be the public MAN standard. This architecture is formally known as Distributed Queue Dual Bus. A packet was established as 48 octets that can be a payload in the ATM standards. The first pilot MAN was launched in Perth, Australia in 1990.

The dual bus architecture means that one designated node in the network does not repeat the incoming data to other nodes; this particular node serves as the logical begining and end of the two buses. The advantage of the dual bus topology is its fault tolerance. Should any node or line segment fail, the opening in the ring is moved to the location of the failure. The node on either side of the break take up the bus-end function of slot generation. Operation continues at full speed with no increase in delay. In ring systems, the second ring is a backup solution which cuts the capacity of the network in half. A second advantage of the bus topology is good scheduling of a message flow. This topology takes advantage of the bidirectional nature of the bus to maintain a distributed queue of nodes for waiting access. A distributed queueing algorithm is used for controlling access to buses. When traffic is slow, the system works with the speed of an Ethernet bus. When traffic is heavy, the efficiency reaches 100 percent as in a token access method.

The architecture of a dual bus is shown on Figure 1-22.

*Switched Multi-Megabit Data Service (SMDS)* is offered for high-speed LAN interconnection on public lines (DS1-DS2). It is a connectionless packet switched service that makes the MAN working with performance of a LAN. In its initial offerings in 1992-93, SMDS will apply the DQDB architecture. The connectionless transmission means that tables with preprogrammed addresses restrict users to destinations that are allowed to exchange packets. SMDS is the first broadband service operating at 45 Mb/s. In the late 1990s it will be supported by B-ISDN. It can also evolve in to the higher performance SONET/ATM technology. Figure 1-23 illustrates the architecture of a MAN with a SMDS at the telephone company's central office.

## Wide Area Networks

Wide Area Network (WAN) is a privately or publicly owned telecommunications system that is able to transmit information over

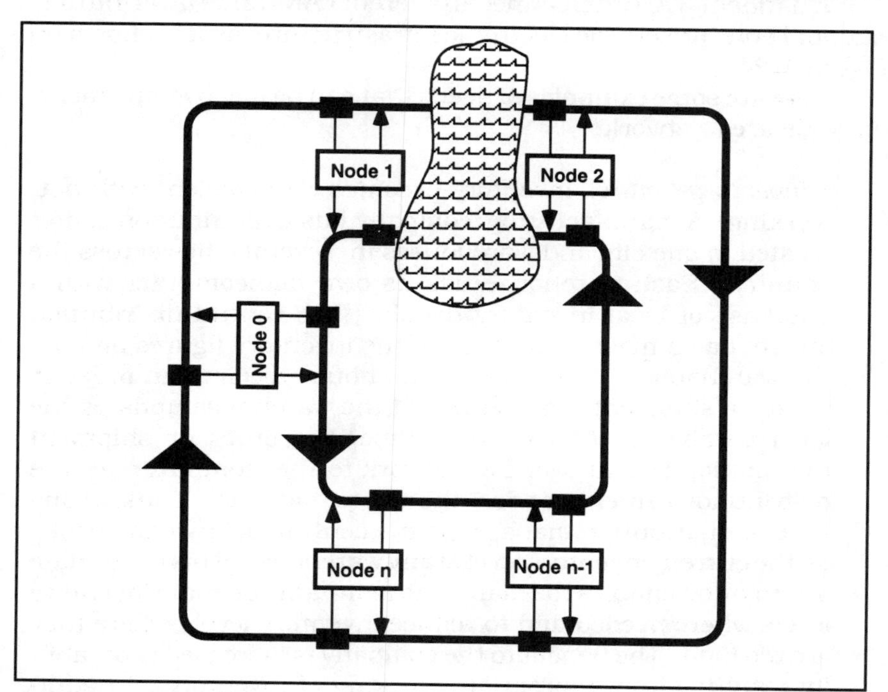


*Figure 1-22: The Dual Bus Architecture of MAN (Treatment of failures)*

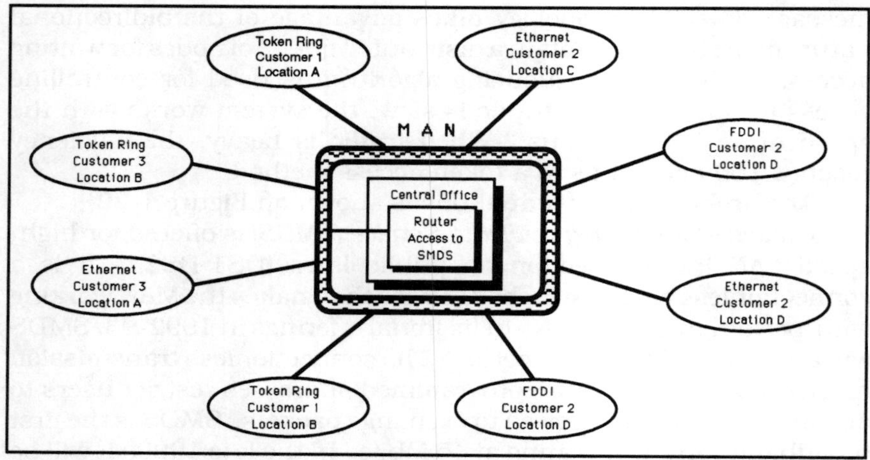


*Figure 1-23: MAN on SONET with SMDS*

a large geographic area. Mainframes, minicomputers, microcomputers, workstations, and terminals can be linked together using interexchange carriers circuits (AT&T, MCI, US Sprint and so forth), satellites, or microwave relay links. WANs transport information without any significant enhancement or change in character of information.

A national WAN interconnects different LANs and MANs into one national enterprise-wide electronic infrastructure, as it is shown on Figure 1-24.

Here are some examples of needs that can be met by implementing wide area networks:

- *Inventory Control.* Inventory becomes very efficient with networking. A manufacturing company has a distribution center located in one city and warehouses in several cities across the country. Each warehouse has its own minicomputer with a database of local inventory details. The central distribution facility has a master database of the inventory figures on all the warehouses. Whenever a warehouse receives an order or makes a shipment, an operator at the warehouse updates the local database. At the same time, the order or shipment information is sent over the network to the computer at the distribution center to update the master database. Thus, at any time, a distribution manager at the central facility can check on the current inventory level of any warehouse. This up-to-date information enables the manger to maintain adequate inventory levels where needed and to reduce inventory levels where they are too high. The benefit to the company is increased profitability resulting from appropriate stocking of inventory. In addi-

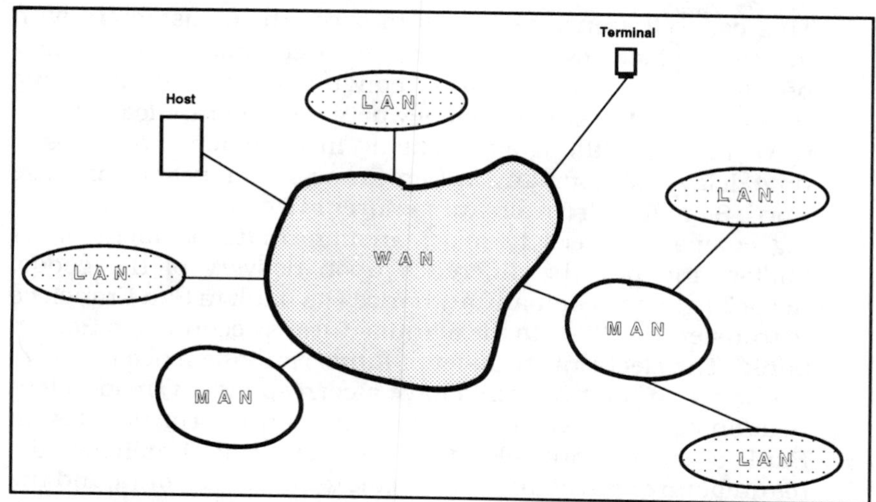


*Figure 1-24: The WAN Architecture*

tion, day-to-day inventory information enables the company to adjust the manufacturing of its products to reflect demand, thereby controlling excess production of goods with slow sales. Current inventory information also enables themanager to respond in an active manner to day-to-day problems. For example, a very large shipment would deplete a warehouse's inventory. To ensure that customers do not receive partial shipments, the distribution manager consults the master database to find a warehouse with available inventory of the product for shipment.

•*Electronic Mail.* Electronic mail helps individuals save valuable time. Employees linked on a network by electronic mail do not have to waste time trying repeatedly to reach each one another on the phone. Instead, they can sit at their terminals, type a message (of any length), and send it over the network. Within seconds, the message arrives at its destination, be it an office in the next aisle or one on the other side of the globe. The ability to communicate across time zones is invaluable, since individuals located in different countries and with different working hours could spend days trying to make phone contact. The advantage of electronic mail is that both sender and receiver can work at their convenience and still communicate with one another.

•*Electronic Fund Transfer.* Banks are linked by networks enabling them to electronically transfer funds among themselves.

This capability enables banks to serve their customers with increased efficiency. Corporate clients are among the primary beneficiaries. A corporation can conduct transactions involving transfers of large sums of money between geographically separated banks, without experiencing inconvenience and costly time delays. A company in San Francisco makes a purchase worth $4 million from a manufacturer in Boston. The purchasing company agrees to make an immediate payment of $1 million and pay the difference upon delivery of the goods. Accordingly, the company authorizes its bank in San Francisco to transfer $1 million to the manufacturer's account in a Boston bank. The electronic transfer of funds can take place in one of two ways. If the two banks have a correspondent/respondent relationship, the San Francisco bank can inform the Boston Bank, over a network linking them, of a credit of $1 million to the manufacturer's account. The credit to seller's account and the debit from the buyer's account take place simultaneously. Either bank can verify the status of the two clients' accounts and confirm the transfer. Through the network, the entire transaction can be completed in a few hours.

If the banks involved do not have a correspondent/respondent relationships, the Federal Reserve Bank in San Francisco informs the local Federal Reserve Bank, over a WAN linking them, of the need to transfer $1 million to the Bank in Boston. Though a private network linking the various Federal Reserve Banks in the country, the Federal Reserve Bank in San Francisco informs its counterpart in Boston of the required transfer. This information is passed on by the Boston Federal Reserve Bank to the seller's bank over a WAN linking them. Within a few hours, the transfer is effected through the two Federal Reserve Banks [7].

The architecture of Wide Area Networks has been evolving in two directions:

1. Circuit switching networks:

- **Proprietary** architecture developed by a number of computer manufacturers for *private* networks (leased lines), such as:

  - IBM's System Network Architecture (SNA) (1974) to connect large computer centers operating in non LAN environments. Some of these networks may have 100 host computers and tens of thousands of terminals. Several protocols for the 7-layer, pre-OSI model have been developed: CICS, IMS, VTAM,

TCAM, ACF/NCP/VS, and so forth. SNA is a WAN that uses voice-grade circuits and the synchronous data link control (SDLC).

- Digital Equipment Corporation's Digital Network Architecture (DNA) evolved since 1975. Only the lower layers of DNA are compatible with the OSI model. The node-to-node transmission relies on Digital Data Communications Message Protocol (DDCMP), which is not compatible with the OSI's HDLC.

- The Hewlett-Packard Distributed Systems Network (HP-DSN) follows the ISO recommendations.

- The Unisys Distributed Communication Network Architecture (DCA) supports the X.25 protocol.

- The NCR Distributed Network Architecture (NCR-DNA) supports the X.25 and HDLC protocols.

• **Common carriers,** such as BOC, AT&T (ACCUNET, SKYNET), MCI (Mail), US Sprint (Mail), ITT provide virtual circuits via telephone networks. Satellite transmission services are provided by RCA Americom and Western Union (also Telex and TWX services). Some services such as error detection and correction, speed and code translation between two communicating parties are provided.

* *Private* networks such as:
- SABRE, developed by American Airlines for 20,000 travel agents and 70,000 terminals via leased lines from AT&T,
- APOLLO, developed by United Airlines for the same purpose as SABRE.

2. Packet switching networks (PSNs). The CCITT X.25 protocol (1974 and with later amendments) organizes a long message into "packets," each is octet-oriented of 1024 and 2048 bytes in length. Each packet contains user data plus control information (head and tail). Packet networks can be accessed from a local telephone number. The user pays per volume of information rather than per time as in a case of circuit switching. PSNs evolves into two types of networks:

• **Public** Packet Switched Networks.

X.25 communications usually take place over public-data networks(PDNs) also called Value Added Networks(VAN) such as:

- Telenet (SprintNet) is the world's largest PSN with 800 nodes. The Telenet was developed by BBN as a copy of early DoD's ARPANET, the first PSN in the U.S. (1968). The network was sold to GTE and later to US Sprint which supports X.400 world-wide electronic mail (*Telemail*). Many commercial information services use Telenet nodes. They are *The Source, DowJones News/Retrieval,* and others.

- Tymnet is the world's most omnipresent network with 750 nodes in the U.S. and connections to 70 countries. Some commercial information services are available via this network, such as *Prodigy, Delphi, TRW Information Services, DowJones News/Retrieval, UUNET,* and other. The network was founded in 1969 and became a common carrier in 1977. In 1983, The network was bought by McDonnell Douglas.

- Uninet is the third largest PSN in the U.S. It is operated by United Telecommunications in 275 cities. The nodes transmit at the speed of 56 Kb/s. The popular information service, CompuServe, is available via this network also.

- ACCUNET packet network operated by AT&T

- EURONET was the first PSN in Europe. Later national PSNs intercepted packet-oriented traffic. TRANSPAC in France, PSS in the United Kingdom, Datex-P in Germany, PAXNET and DATAPAK in Denmark, EIRPAC in Ireland.

- AUTPAC in Australia
- PACNET in New Zealand
- DDX-P, VENUS-P, NIS/TYMNET in Japan
- MAYPAC and MAYCIS in Malaysia
- VIKRAM in India
- TELEPAC and Morelos in Mexico
- RACSAPAC in Costa Rica
- Several PSNs in the Middle East
- Several PSNs in Africa:
  SYNTRANPAK in the Ivory Cost
  GABONPAC in Gabon
  SAPONET in South Africa
  CGNET, CARINET, and Afrimail (Quarterman 1990).

* ***Private*** Packet Switched Networks (leased lines)

- ARPANET (1968) developed as a private network by DoD to

interconnect major computers at the most critical government and research centers.

- NREN National Research and Education Network, a supernetwork transmitting at the speed of 1-3 Gb/s, as a new "extension" of ARPANET.

- IBM Info-Net providing several regional information processing centers and EDI services.

The architecture of Packed Switched Networks is shown on Figure 1-25.

Each long message is divided into small units called "packets." Each packet has a "tail" and "head" between a data set. The head contains control information and an address, the tail informs a node that the packet has concluded transmission. Three components are characteristic for the X.25 network: PADs, telecom switches, and the network manager. A PAD is a Packet Assembler Disassembler which funnels information in the form of packets into the telecom switches. This feature of the communication periphery equipment assemblers a message into packets (adds a head and tail) at the beginning of routing. At the end of transmission, it disassemblers again into

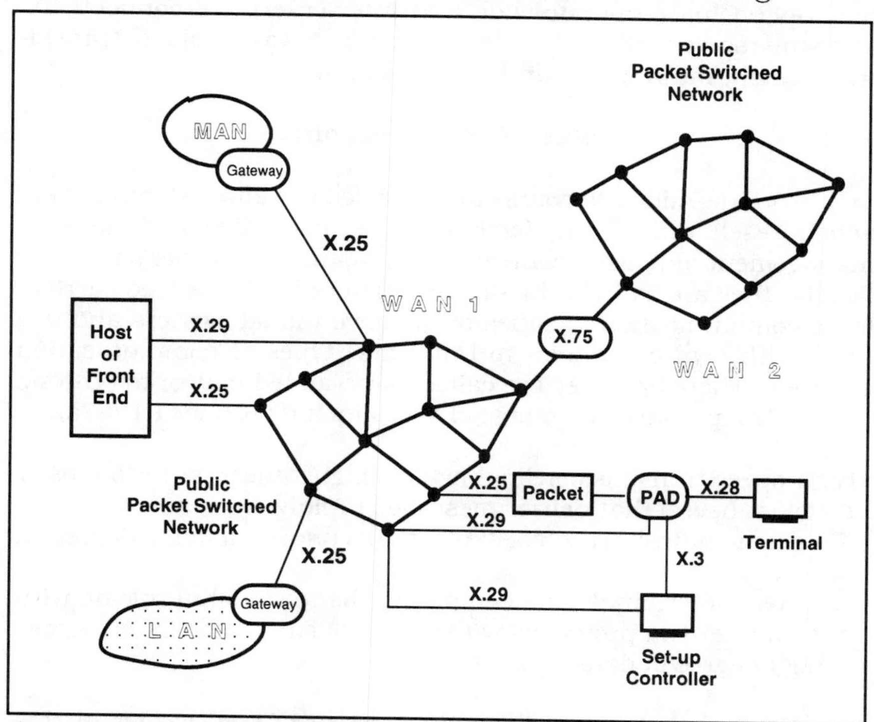


*Figure 1-25: WANs—Public Packet Switched Network Protocols*

messages. The message is passed on to the receiving host or terminal.

A set of associated "X-protocols" (CCITT) were developed to support asynchronous traffic. These include:

- X.75 for interconnection of international PSNs (X.121)
- X.3, X.28, and X.29 for interconnecting asynchronous terminals/ computers with PSNs

For example, X.3 defines 22 parameters of a terminal connected to a PSN. X.28 defines the interface between a terminal and a PAD. X.29 provides a transport layer for a X.25 PAD.

The new emerging client/server architecture of computer systems, driven by networking, is characterized by the burstiness and spontaneity of messages traveling between users, applications, and servers. Traditional, private WANs cannot handle such traffic any more. Therefore, public WANs will be configured to act like a private network—developing a "virtual-private" data network, at fees lower than those of lease (private) networks.

The ATM technology will probably integrate LAN with WAN to create true enterprise-wide network. The telecommunications environment will spin around a "superserver" that will be a decisive point for supervising corporate-wide and inter-enterprise connectivity. The superserver will play a role of a hub, gateway (protocol translator), switch, multiplexer, and session manager.

### Value - Added Networks

Value-Added Networks are provided by value-added carriers which resell raw circuit facilities. Because they add network management and error-control values to another carrier's raw line facility, they are called value-added carriers. Value-added carriers lease communication facilities from conventional carriers and use them with own computers to build new types of communication services. Therefore, they are called value-added networks (VANs).

The primary advantage of VANs include (Conard 1988a):

- High-speed transmission lines (which might otherwise be too costly for most users) that deliver messages rapidly.
- Program-control error checking that ensures a high degree of accuracy.
- A variety of terminals and computers that can communicate with one another (via protocols) without significant software changes.
- A high degree of data security.

The packet-switched data networks discussed earlier in this chapter are all examples of value-added networks. The following services are provided by VANs (Martin 1988):

- **Conversion Services.** They fall in three categories of protocol, code, and speed conversions. For example: a simple terminal that uses the ASCII code and an asynchronous data link protocol might be connected to the network via a 1200-bps data link. A computer with which it communicates might use the EBCDIC code and a bit-oriented data link protocol to communicate over a 56,000 bps data link.
- **Routing Services.** The network performs the task of deciding by what physical route the data should traverse the various links that make up the network. When one physical link fails, another can usually be substituted to make the failure transparent to users.
- **Network Control and Management.** The network controls the physical resources that implement the network and frees the user from such concerns.
- **Error Detection and Correction.** The network provides the user with what appears to be an error-free virtual channel through the network. Error control procedures implemented by typical VANs usually provide error rates as low as one undetected error bit in $10^9$ bits transmitted.

VANs apply the CCITT X.25 standard for packet switching networks. The basic solution in packet switching networking was pioneered by the Advanced Research Project Agency at the Department of Defense in 1968. However, the ARPANET was available only to universities with super-computers, government agencies, and government contractors. Therefore, private telecommunications corporations, led by former developers of ARPANET, initiated the same services for private subscribers. In the mid-1970s such second-tier carriers as Telenet, TYMNET (owned by British Telcom) and GRAPHNET began providing value-added communications services.

The *First Computer Inquiry* initiated by the FCC was completed in 1971 with a result of distinguishing data processing services from telecommunications services. Furthermore, hybrid data processing service, combining message (or packet) switching and remote access data processing to form a single integrated service was deregulated. This fact has increased the diversity and competitiveness of the telecommunications industry in the United States.

In the 1980s many new VANs started offering their services: SprintNet, InfoNet, AlaskaNET, ConnNet, PC Pursuit, DataPac. Based on VANs services, information services providers have been developing, such as CompuServ, GEnie, DIALOG, America Online, Prodigy, Sierra On-Line, DELPHI, Dow Jones News and so forth.

## Global Area Networks

To compete successfully in the 1990s and 2000s, companies must meet the following four challenges:

- be *responsive* to customer needs, *no matter where in the world a customer is located* and regardless of any constraints imposed by their own business structures;
- pursue *innovation* on a global scale, leveraging the best talent and the most creative ideas *from many markets*;
- achieve global *efficiency* in supplying quality products at reasonable prices. Economies of scale achieved only *at national levels are simply not competitive* when applied internationally;
- achieve in service, as well in production, unsurpassed *quality*. The only acceptable standard in the *global marketplace* is zero defects.

As a result, companies of all types are being forced to rethink their strategies as they reach for global scale and presence. Leading multinational corporations and consortia are finding that flexible global area networks (GAN) are an essential infrastructure tool for meeting these challenges on a world scale. Although they evolved originally as private networks, GANs today are, for most part, hybrid public and private systems that take advantage of the explosive growth in undersea fiber, intelligent gateway switches, and highly featured private virtual networks. In most cases, GANs are extensions of domestic applications developed by large corporate users in advanced industrialized countries.

The most common applications of GANs to date have focused on two of the four challenges —*efficiency* and product *quality*. In the area of organizational efficiency the following applications have been undertaken:

- global purchasing
- global inventory management
- global distribution
- global trading
- global fund transfer

In the area of product quality the following applications are the most frequently developing:

- CAD/CAM (computer aided design and manufacturing)
- global electronic mail

to link independent functional units in different countries and time zones. The site-to-site global networking of these applications expedites basic operations and business processes.

Leading enterprises, however, are discovering the other two challenges —*responsiveness* ("competing in time") and *innovations* — lie in GANs that extend beyond operations and delivery systems to encompass personal networking and *personal communications*. In other words, there is need for GANs that facilitate information and knowledge sharing among an organization's *individuals*, and not only sites.

> At Northern Telecom and BNR alone, employees use the corporate communications network to make more than 10 million telephone calls a year to other employees within the company, no matter where in the world they are located.

In short, companies need to integrate the desktops of employees on a worldwide basis— through uniform:

- voice networks
- LAN-MAN-WAN data and image communications
- multimedia messaging system (ISDN)
- EDI—Electronic Data Exchange
- dial-up videoconferencing
- mobile communications systems, that extend access service to individuals regardless of their location at a given time.

In the residential market, the following global services are expected:

- pay-per-view video programs
- high definition television (HDTV)

The enablers of GANs include:

- broadband access and transport (e.g., SONET)
- ultra high-speed telecom switching technologies (worldwide standards)
- intelligent signalling systems (e.g., CCS7 and ISDN)
- mobile communications networks.

The key to unlocking the power of global networking lies in the intelligent signalling capabilities offered by CCS7 and ISDN. These signalling systems are needed to achieve feature transparency across multiple, national and private networks and to enable deployment of private virtual networks on a global basis [8].

As the demand for global networking explodes, the twin drivers of the Industrial Age—power and transportation systems—are giving way quickly to their Information Age successors: computers and communications (C&C). Today, this global communications infrastructure forms a new kind of ecospace—the Electronic Global Village. This new ecospace challenges the world and on one hand it promises enriched lives, on the other hand, it promotes the division into information-rich and information-poor.

Two of the world's largest and most important GANs are:

- S.W.I.F.T (Society for Worldwide Interbank Financial Telecommunications) network handles electronic fund transfers for nearly 3,000 banks in more than 70 countries via a GAN since 1989,

- SITA (Societe International de Telecommunications Aeronautiques) routes airline reservations information, flights plans, and lost baggage reports to 460 airlines in 187 countries and territories. Requests for these crucial services may originate virtually anywhere in the world, and must be processed within minutes on a 24-hour-a-day basis.

***GAN's Strategy.*** A strategy for the evolution of GANs is based on (Eastland et al., 1991):

- *Digitization* through end-to-end digital connectivity. It enhances the kind of services that can be delivered to subscribers. National networks are evolving at different rates throughout the world. Digitization is only well established in the networks of developed countries. By 1990, for instance, about 70 percent of networks lines in France were digital; in Canada and the United States, 50 percent were digital, and in Japan, 32 percent. Network digitization in many industrialized countries is well under way in the switching and transmission segment of the network. The next step in this evolution, already taking place in many parts of the world, is to extend digital connections to the access portion of the network. By extending digital connectivity to the end-user's terminal, the network can be used to support a myriad of multimedia services. This extension, for example, enables out-of-band signaling between the network and the subscribes set. It allows for ISDN services.

- Telco *Intelligent Networks* delivers ISDN and Enhanced-800 services using CCS7 connections that improves:
  - the speed and flexibility of call-setup,
  - allows special routing or handling by telecom processors,
  - enables telecom companies to access customer information stored in network databases on a network basis,
  - provides detailed information about the call handling as it is set up through the network.

Once the powerful switching and database elements of telecom intelligent networking are deployed, service providers will be ready for the next evolutionary step in technology, known in North America as the *Advanced Intelligent Networks* (AIN). AIN will allow firms to :
- rapidly create and prototype new services
- implement new services rapidly through some tuning of intelligent network elements,
- increase service ubiquity and uniformity
- customize features for specific marketplaces,
- offer greater control to the end user.

- *Increased Bandwidth.* The increasing decentralization of enterprises and their use of supercomputers, for example, have created a flood of high-speed data, graphics, and text to be transported from LAN to LAN, between countries, and around the world. The world LAN market has experienced an explosive growth of 42 and 49 percent respectively in 1988 and 1989. By 1992, the total number of LANs installed worldwide was 2.3 million. Services built on dedicated lines (narrowband) are evolving into public networks built on broadband which is more flexible and multimedia-oriented (such as ISDN) since they are switched services.

- *Standardization.* Today there are about 200 standard related international committees which promote such standards as X.25, ISDN, CCS7 , SONET , ATM and so forth. GANs have to be built on international standards and should be flexible enough to accommodate a wide range of national variations of these standards.
- *World-scale Network Management.* One of the key network innovations is the capability to manage a GAN from strategically located control centers. Network management is more complex for global networks than for national networks because control of these 24 hour-a-day, seven-day-a-week systems must be automatically handed off from one continent to another as the working day moves from one time zone to another. In the SITA network, three management centers, located in Europe, North America, and Asia, can each perform all network control operations in three different cycles of a 24-hour day. The global SITA network can be totally controlled from each site, providing full redundancy in the unlikely event that one site should fail. Although the SITA network has three continental management centers at the global level, it also includes dozen of lower level regional management centers. A workstation deployed at each of these regional sites sees only the data for that region (Drynan and Jeanes 1991).

***Global Corporate Architecture.*** Figure 1-26 illustrates the architecture of a global corporate network which includes: GANs, WANs, LANs, hosts, terminals, and workstations on three continents. GANs are made up of private and public networks. The whole global corporate network is also a combination of three types of gateways: international gateways, national gateways and user gateways.

*International gateways.* The international gateway function differs substantially from other switching functions because it must be able to interact with national and international networks, accepting calls from switching nodes in other countries, performing digit evaluation and transportation, and routing these calls to their destinations. To carry out these functions, the international gateway switch must be able to recognize and process a wide variety of international trunk signalling and testing protocols, as well as translate dialed digits that differ from country to country (Barnes 1991).

> In September 1991, Northern Telecom installed, for British Telecom, one of the largest international gateway switches in the world. Called the Madley B, this BNR-designed DMS-300 switch located in Madley, United Kingdom—is configured with 45,000 trunks capable of handling 560,000 calls per hour, and 248 links capable of supporting up to 10 different British and

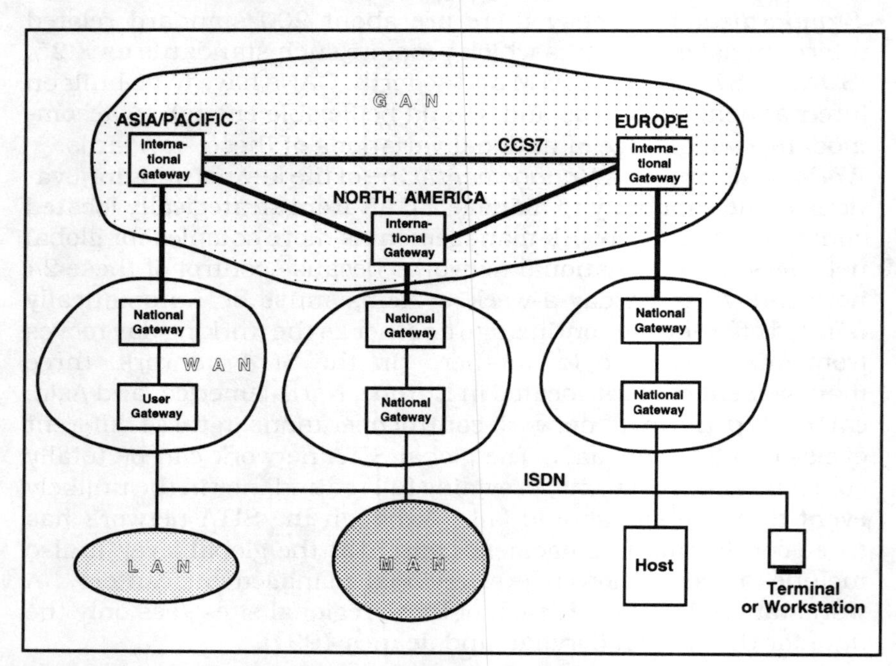


*Figure 1-26: The Architecture of a Global Corporate Network*

international common channel signaling systems. This advanced digital switch is designed to route communications traffic—including voice, data, video, and image—directly to any country in the world through terrestrial and submarine cables, radio systems, and satellite transmission facilities. DMS-300 is applied in Japan, Australia, the United States (US Sprint), Hong Kong, Canada and so forth. The switch is based on the intelligent network operations—CCS7 and ISDN. In the 1990s, this switch will support 20 different national and international trunk systems.

*National gateways* provide CCS7 and ISDN functionality for interexchange carriers in North American unregulated companies. The required backbone ISDN network is provided by national public telecommunications suppliers through Private Virtual Network (PVN) overlays on the public network. Because of the complexity of managing increasingly large amounts of customer data, these networks are evolving to centralize customer PVN network information within a network database in CCS7 system. Example of such a gateway are North Telecom's switches D100 and D250.

*User gateways* offers tremendous flexibility for internetworking with various multinational protocols and with physical user connections to different types of central offices switches. Its cards are configured for country specific ISDN implementations.

On July 8, 1991, one of the largest and most modern private Integrated Services Digital Networks (ISDNs) in Europe went into operation linking four of the BASF AG conglomerates' major locations in Germany. This fully digital network—based on Northern Telecom's Meridan 1 Communication System (user gateway) — handles more than 250,000 calls a day and provides such advanced services as call diversion, automatic ringback, voice message storage, and network-wide automatic call distribution to some 28,000 telephones. This user gateway offers business customers such powerful global networking applications as calling-line identification, and calling-name display in a given local language.

Success for global corporations is increasingly dependent on access to global communications networks that tie together an organization's dispersed operations. These networks must be easy to use, offer advanced business features, and provide call transparency to corporate employees—whether they are making a call across the hall, across the country, or around the world. Furthermore, they must include fast and efficient multimedia communication for such applications as document transfer and videoconferencing.

Where does a corporation go to find such global communications networks in today's environment? One emerging option is global private virtual networking—a service offered by alliances of national network providers. Global private virtual networking—or, more simply, global networking —represents the global extension of private virtual networks (PVNs). PVNs combine functionality of a private network —including a private corporate dial plan—with the flexibility, reliability, and the economy of the public network.

The challenge in providing ubiquitous service deployment worldwide, however, is that these PVNs—which are private network overlays on the public switched network—must be able to internetwork across national telecommunications networks having different standards, and different equipment from multiple vendors.

One of the first alliances comprises Sprint in the United States, Mercury Communications Limited in the United Kingdom, Hong Kong Telecom in Hong Kong, PTT Telcom in the Netherlands, and Teleglobe Canada and Unitel in Canada. Several other customers are currently seeking to joint this alliance. The alliance is offering a service called GVPN (global virtual private networking) (Thomson 1991).

> Faced with a lack of redundancy in its international network, Belgium Kredietbank went on-line with Infonet's virtual InfoLAN service in July 1991. InfoLAN provides global connectivity between LANs and WANs using native protocols such as TCP/IP and source route bridging. Kredietbank, world leader in European Currency Unit (ECU) clearing, is the second largest bank in Belgium with more than 700 domestic and international branches. High network availability is imperative in the global banking arena where downtime lasting more than a few hours could severely harm financial operations. Kreditbank now has one line that goes to InfoLAN and one to its private circuits. If the private line fails, the router switches over to the InfoLAN line without session loss. InfoLAN provides interoperability between InfoLAN and its own worldwide X.25 public network. Users can choose custom virtual, hybrid or private solutions with varying network topologies, geographic coverage and port speeds.

> Japan Airlines Company Ltd.(JAL) of Tokyo operates a global corporate network as a prerequisite of doing business. JAL began global communications in the 1960s using the teletype reservation system of SITA. Today, JAL still uses SITA for remote or low-traffic destinations: however, it also has its own IBM Systems Network Architecture backbone network with 3 major domestic hubs (Haneda, Tokyo, and Osaka) and 3 international ones (San Francisco, Hong Kong, and London). Some

advanced features of the JAL network are helping the company maintain a competitive edge. These include an AT&T ISDN that serves as a backup between Tokyo and San Francisco, automatic loading of flight schedules via satellite from ground operations centers to on-board computers on certain aircraft, and an in-plane fax/phone service that will soon be tested on international flights.

When it is necessary to transport information from one country to another through GANs, the services of an international record carriers (IRC) should be used. At the other line of the American source-end, user organizations will have to strike a deal with the foreign PTT (Post, Telegraph and Telephone). International carriers are licensed by national governments to offer services between countries, de facto between gateway cities. In the United States there are 25 gateway cities. The international record carriers located in the United States are: AT&T, Western Union International, RCA Global Communications Inc., ITT World Communications Inc., MCI International Inc., TRT Telecommunications Corp., FTCC McDonell Douglas International Telecommunications Co, Satellite Business Systems, CCI Inc., and Telenet Communications. These companies provide such services as telex, international telegrams, and leased lines for private lines and networks, including those of value added networks. The US-based IRC connect directly or through other countries to much of the rest of the world. Based on IRC's circuits, other telecommunications companies organize value added networks, also called public networks.

The TYMNET Global Network, with the largest market share among U.S. value added networks, supports Global Network Services (GNS) that reach 90 percent of the world's business centers in more than 100 countries. Half of Europe's multinationals use these services. GNS links corporate international divisions and departments, monitor and support those links worldwide, bill in one currency, and offer a portfolio of value added applications such as Frame Relay, Videoconferencing, ExpressLANE, EDI*Net, Messaging, electronic transaction and single vendor solutions.

***World-wide Network Services.*** In a global economy, multinational corporations see telecommunications as a key to success. Global networks transport data, voice and images around the world for these corporations. Telecommunications networks have rapidly progressed since 1921 when the first international call was placed by AT&T. Over the next seven decades, AT&T moved from this first step into the world-wide intelligent network that transmits voice, data, and images to 270 countries and territories through cables and satellites. Some 125 million calls are completed on an average day.

The worldwide intelligent network gives multinational corporations the advantage of AT&T's switched and dedicated private line services. AT&T's dedicated service offerings include the following solutions:

- International Private Line, AT&T provides a channel originated in the United States, while local service providers provide a channel terminating in their country. Either via a cable or satellite, 160 locations can be connected.
- International ACCUNET Digital Services via ultra-high capacity network of undersea lightwave cables. The 24-hour channel is available to supply virtually error-free performance and minimal signal interference. Multinational corporations applications via this network include: bulk/data file transfer, network optimization, electronic document distribution, audio and video conferencing, and computer graphics. Financial, manufacturing, transportation, and media organizations are important users of these services.
- Reservation Based Service of ACCUNET, e.g., teleconferencing during a limited time between different points on the Globe.
- SKYNET International Service via satellite operated by consortium such as Intelsat. Such applications are in use: CAD/CAM, videoconferencing, bulk data transfer, electronic mail, and graphics.

Traditional carriers such as AT&T, MCI, Sprint and TRF/FTC Communications meet challenges from a powerful global players such as British Telcom, IBM and new players offering innovative services.

British Telcom will finish building a billion-dollar-plus global network with 32 switching hubs in 2002. In 1993, the first hubs were installed in London, New York, Frankfurt, and Sydney.

IBM, together with Sears Roebuck & Co., launched a voice-data facility as the worldwide Advantis network. Advantis is the first private network which can strongly compete with the public networks. At the same time IBM teamed with Motorola to provide E-mail service through a wireless network on the Ardis packet-data radio network. United Parcel Service and Cellular Data Inc. linked their proprietary cellular data networks to the system. Such major computer vendors as Apple, Hewlett-Packard and Sony Corporation began delivering in 1993, personal digital assistance devices to run over this network.

Motorola plans to initiate in 1998 a $3.4 billion Iridium network of 66 low-orbiting satellites to connect every spot on the Earth via a wireless system.

As industry, commerce, and trade become more and more global, international global networking services are getting more developed and competitive. However, sooner or later they will become transparent and standardized for the subscribers.

## Internetworking

The first connectivity level of micro, mini and main computers as well as terminals, workstations and servers are Local Area Networks. LANs proliferate as individual electronic islands among corporate headquarters, plants, and staff-functional organizations. The need to interconnect them becomes obvious. A single enterprise-wide information utility can be built by using internetworking devices. Today a shift is observed from physical computer and node connectivity, to internetworking logical connectivity, creating one single transparent electronic environment. The following types of devices are available: repeaters, bridges, routers and gateways.

*Repeaters* are devices to connect LANs with the same protocol at the physical layer. Its function is to regenerate a signal. For example, the Ethernet LAN's bus can be 500 meters long, to extend its length, a repeater is required. In some cases, a repeater can be a short-haul microwave path which can expand a LAN's path into 6000 meters via the air. One of the newest solutions is the 1990 offering by Bellcore —Switched Multimegabit Data Service (SMDS) within the Municipal Area Network. In this case MAN with SMDS is a repeater in LAN-to-LAN connectivity. A customer' s LAN sends a packet of information via a dedicated DS3 line (44.736 Mb/s) to the local telephone company with SMDS. From that moment the packet is switched to the destination at the speed up to 155 Mb/s.

*Bridges* are protocol-independent interconnection devices. Two same types of LANs can be connected by a bridge. However, a good connectivity can be achieved only when the compatibility is at the layers, 1 (physical), 2 (data). Higher layers should be logically compatible. A "dumb" bridge must be informed (by a table) which address is local and remote to pass the information envelope. A "smart" bridge can figure out the destination address by itself. This is a learning bridge which can differentiate a mode of traffic: inter-LAN (between LANs) and intra-LAN (within a WAN). A bridge assures security of nodes' access. Bridges are recommended to interconnect:

- similar LAN networks, such as Ethernets (DEC) or Token-Rings (IBM)
- building LAN with backbone traffic on a campus electronic highway, where one standard is applied at layers 4 (Transport),5 (Session), 6 (Presentation) ,7 (Application). The bridge will isolate

intra-building traffic and when it is necessary it will interconnect with the campus traffic,

- same vendor's similar networks but with different throughput, e.g., AT&T StarLAN 1 with StarLAN 10,
- LANs with different access methods (IEEE 802.3 Eathernet and IEEE 802.5 Token Ring) but within the same framework of computer operating systems. For example, users from different LANs will use the same VAX or IBM computer via Ethernet and Token Ring LANs.

Bridges are very useful in organizing subnets. A typical bridge costs about $5,000.

*Routers* are enterprise-wide utility to interconnect same protocols-oriented networks. They are used in a LAN-to-LAN connection via WAN (X.25 packet). They provide a bridge connectivity and conversion of protocols at the 3rd-network layer. They convert LANs data protocols into WAN X.25 packet protocol. Routers are protocol-specific devices. Therefore more routers may be required to interconnect a given LAN with different WANs. However, they are multiprotocol routers that handle different protocols simultaneously. A router stores a map of paths and can decide about the optimal routing among a network's nodes. A hybrid device called "brouter" is a mix of a bridge and router. It selects the optimal routing but also assures access security, which is a bridge's function. If a backbone electronic highway is built on a fiber optic wire, then a FDDI router is required to synchronize different transmission speeds (LAN-10 Mb/s, backbone-100 Mb/s). Such a router can convert 2000 packets per second. A typical router costs about $10,000.

*Gateways* convert protocols between two different types of networks throughout all the OSI layers. For example, DECnet applications can communicate with IBM SNA applications or with AppleTalk. Gateways are more intelligent than routers and bridges and also they are slower than the predecessors. They have to translate more layers of protocols and therefore they transmit at the speed of hundreds of packets per second. Special servers under the form of microcomputers are applied as a base of a programmable gateway. Also, the gateway can serve as a network management system, generating reports on network performance. "Intelligent" gateways behave as routers, talking with each other and selecting the optimal route for a message.

The architecture of internetworking is shown on Figure 1-27.

## Transborder Data Flow

One of the dominant issues in global communications during the late 1970s was transborder data flow (TDF), which arose largely

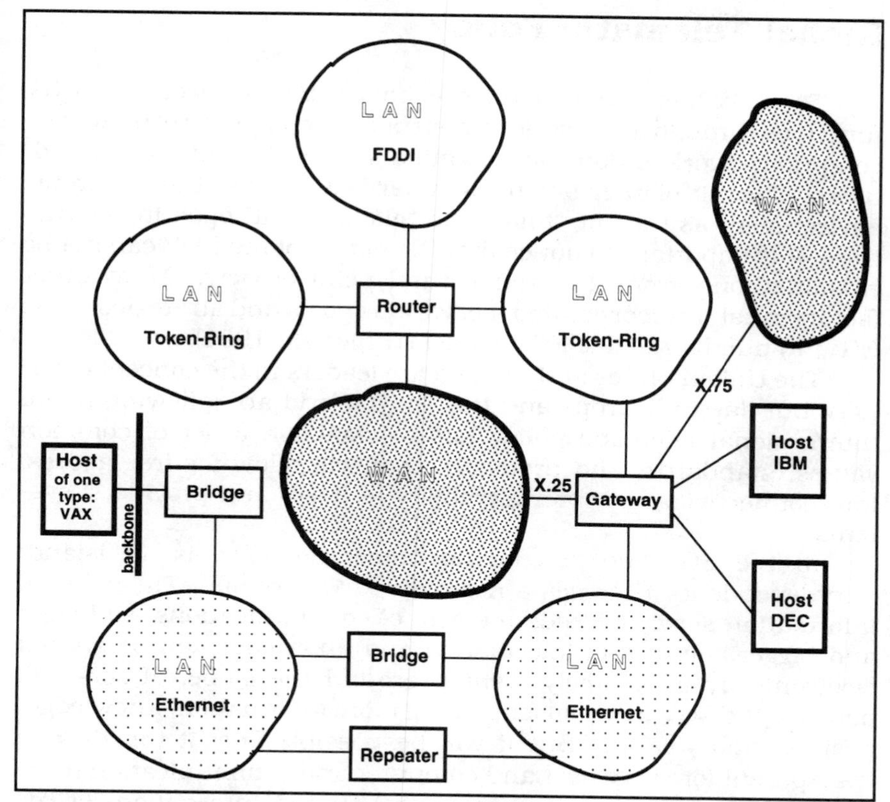


*Figure 1-27: The Architecture of Networking*

from increased data communications in Europe. Countries had to contend with the problems created by massive databases that were held in one country, used in another, and transmitted through processors in yet a third country. Questions were raised concerning which country's law would apply if an individual's or a company's rights were violated through misuse of the data. The Council of Europe and the European Economic Community (EEC), among other groups, have sought to establish international standards for legal protection regarding data privacy and to coordinate the data protection laws that several countries have already passed. Such laws are in force in Sweden, Germany, France, Austria, Denmark, Luxembourg, and Norway., and the list continues to grow. In some cases, the laws directly affect the transfer of information through data communications across national borders (Conard 1988). With the emergence of the Electronic Global Village, the TDF issue may disappear sooner or later.

## Global Telematic Policy

The EGV has to reverse the effect of the Babel Tower. The EGV functions through a telematic infrastructure, composed of networks, computers, workstations, and terminals, as well as interactive end-users. The role of international standards is to make this electronic environment as the one "total" and "global/local" open infrastructure. It is important to notice that the notion of the EGV can not be examined only from the architectural point of view. Many other factors must be incorporated (i.e. various goals and strategies of the EGV) to build a reliable EGV model (Targowski 1991).

The United States and Canada are leaders in the concept of the EGV but Japan, Europe and the Third World are following them. International telematic policy has to establish a set of common values, standards, and protocols, that will allow for free access, interconnectivity and interoperability of users and providers' systems.

In the EGV, no person, no home, no office is an island; interdependence becomes a necessity, a way of life. The value of information slowly exceeds the cost of communications, so bigger and bigger volumes of information and knowledge can be shared electronically in a timely fashion around the globe. Users will perceive the EGV as a tool for gaining more information, knowledge, position and wealth. But it will be possible only if the EGV is transparent for every user and computer and communication technology. Standards and agreements between international C&C providers slowly, but relentlessly, create elements of the international telematic policy. International telecommunication policy is evolving in the areas of:

- radiocommunication;
- switch networking (telephony, cable TV, multimedia, tariffs, etc);
- user computer networking (networks, internetworking, multimedia, etc.).

The growth of international standards has been spectacular. There are about 2000 acronyms and the same number of standards and protocols. These international rules slow down the telematic revolution and, fortunately, the telecommunication industry and users evolve their systems step-by-step and standard-by-standard. The purpose of it is to have mandatory end-to-end connectivity throughout of the world.

By the end of 1990s, the unlimited technology race in C&C will have created a sea of new opportunities. Also, many solutions of the EGV, such as telecity and/or the EGC are still unknown. Competition in this and other areas is open. New ideas, technologies and solutions will surprise the world. At the same time, the process of standardization will be delayed since the above-mentioned technological race is steered by still-active telecommunications monopolies. The international telematic policy as a vehicle of broader scope should provide a tool for social awareness of uncontrollable technological consequences.

The emerging global economy is driven by international C&C technologies. This can eventually lead towards "trade wars" and the global telematic policy must be aware of the possibility of a Global Information War, which is rather contrary to the official politics of the United States, Europe and Japan. This war would perhaps be led by multinational business organizations which would control the majority of electronic highways and processing centers. At stake will be free access and the flow of information for all eligible users of the EGV.

Global telematic intervention into the future of humankind has to take into account that telematics is a result of human practice, rather than a goal or an external agent able to act on its own and manipulate the rest of the world's social fabric. The global telematic policy has to be determined in relationship to other modes of human practice. It should give voice to those who have been silent. We still know little about how the telematic environment shapes social structures and vice versa. Social and economic sciences do not explain "the End of History". We must challenge a notion that telematics is invented to benefit only corporate profits and managerial control. Perhaps, telematic can be perceived as a new tool of democracy, freedom, and humanity (knowledge sharing) without limits of geography. From this perspective, we must know how and where to intervene effectively through the international telematic policy to build the right EGV. Today, we are at the beginning of not only a new millennium but a new telecivilization.

## Managing Global Information Technology

The emerging global economy is driven by the New World Order (post Cold War period), the World Free-Trade tendency and Global Information Technology (GIT). GIT also becomes a supportive tool as a strategic system. Its services are demanded by globally minded corporations and executives. Senn (1992) defined five areas of GIT impact on global enterprises and economy:

Strategy 1: Compress business response time
Strategy 2: Facilitate mass customization
Strategy 3: Enable local presence
Strategy 4: Aid in managing a group of states as a region
Strategy 5: Aid in driving the cost of operations.

Further, Senn also offers key issues of corporate GIT:

*Issue 1:* Firm's Ability to Interconnect with disperse key suppliers and customers to exchange appropriate information.

*Issue 2:* Firm's Reach Across Markets to acquire sales and demand information effectively.

*Issue 3:* Synchronization of Business Strategy with Technology Strategy to support business performance.

*Issue 4:* Suitability of Technology Platform that is not a set of "islands of automation."

*Issue 5:* Application Location. Where should applications be launched? At the headquarters or where distributed?

Palvia and Saraswat (1992) perceive the emergence of eight multinational categories of GIT issues:

*Issue 1:* IT Transfer from developed to under developed countries to mutually benefit from information exchange.

*Issue 2:* Cultural Differences may influence systems development, putting either more emphasis on technology or on socio-political concerns.

*Issue 3:* International Standards ensuring interconnectivity and compatibility of systems in different countries.

*Issue 4:* IT Infrastructure should be developed in some regions such as Africa, Asia, Latin America, and some countries of Eastern Europe and the former Soviet Union to support the flow of global information.

*Issue 5:* Global IT Applications to support manageability of LDCs, control and prevention of environmental pollution throughout the world, and the effective and controllable management of multinational corporations.

*Issue 6:* Global IT Policy. Supports the harmonized development and operations of systems without domination and interruptions at an efficient cost/performance level.

*Issue 7:* Global IT Market may cause some unhealthy competition from more aggressive countries.

*Issue 8:* Transborder Data Flow (TDF) causes two tendencies: (1) restricting TDF to protect a country's sovereignty, (2) more open exchange of information in the borderless world. A matter of security seems to be more serious for many multinationals than the political aspect.

It is common sense that firms with global business strategies for competing in a global marketplace must make sure their IT strategies move parallel with them (Keen 1992). The emphasis from the off-the-shelf software installation moves toward the architectural planning of a global business model, global business strategy, global systems strategy, globally integrated federational information systems, operated on compatible and globally interconnected IT platform, and managed by globally minded IT/IM specialists.

Telematic technology provides the infrastructure for the evolving Electronic Global Village. The advancement of the telematic technology and its profound and successful applications makes the Electronic Global Village more and more real. It is no longer a metaphor or science fiction —it is a *modus operandi* of the global economy and the global human family.

## Conclusions

The following chain of statements and conclusions may be formulated from the above presentation:

1. "Virtual reality", or universal networks spanning computers, phones, videos and any other information-generating devices are the next step in the paradigm of information conveying systems.

2. All developed countries have abolished or are in the process of abolishing monopolies of state-owned PTT enterprises, and are encouraging the development of competing data processing and transmitting systems. This speeds up the process of creation of the Virtual Reality.

3. Creation of the virtual reality will have a revolutionary effect on the future of civilization. Unlimited access to data will create a fascinating situation where the creativity of a human mind is limited only by its functional capabilities, rather than by lack of access to the all available data and information.

4. This bright picture is distorted by the knowledge that virtual reality is created by profit-driven organizations rather than by society in general (*vide* the deregulatory legislations). Some of these organizations still enjoy a monopoly position on their market. This may lead to a trade war with the population of the world being the primary target. To avoid this developments in the C&C field must be watched very carefully indeed.

### End Notes

[1] With the deployment of fast packet-switching and the integration of further intelligence into the telecommunication network, it will become increasingly difficult to distinguish between the functions of switching and

transmission (Kim 1987). Some examples noted by the author are statistical multiplexers, digital cross-connect systems, concentrators, and switches with built-in optical interfaces such as DS3.

[2] A critical factor in this regard has been the rapid advances in microelectronics resulting from the development of very large scale integration (VLSI). VLSI allows the placement of over $10^6$ logical operations on a single integrated circuit chip, and this number is doubling every 18 months (Estrin 1986).

[3] More and more systems are becoming defined and driven by software. This development will make future communication technologies and systems more flexible and more versatile.

[4] As Brand (1988) has described this phenomenon: "With digitization all the media become translatable into each other—computer bits migrate merrily—and they escape from their traditional means of transmission. A movie, phone call, letter, or magazine article may be sent digitally via phone line, coaxial cable, fiber optic cable, microwave satellite, the broadcast air, or a physical storage medium such as tape or disk. If that's not revolution enough, with digitization the content becomes totally plastic—any message, sound, or image may be edited from anything into anything else."

[5] Source: "Broadband Services Product and Applications Guide," Ameritech Services, Product Development, June 22, 1992.

[6] Adapted from "An OSI Guide for Management" Digital brochure EB 31339 (1988)

[7] Adapted from "Networking, The Competitive Edge," Digital brochure EB-27241 42/85A 08 65 80.0 (1985).

[8] Adapted from "Viewpoint," Telesis, issue no 93, p. 2-3.

[9] The idea expressed by Joseph Pelton, Communications News, August 1992, p. 6.

[10] Adapted from "Freddie Mac Integrates X.400 and Applications," Communications News, August 1992, p. 20-21.

[11] Source: The NAS Handbook, Digital, EC-H!!90-58/91 06 43.

[12] Source: "A Double Standard For Cellular Phones," Business Week, April 27, 1992, p. 104.

[13] Source: "Wireless Nets Aren't Just for Big Fish Anymore," Business Week, March 9, 1992, p. 84.

[14] Source: Datacomm Research Co.

## References

Anderson, B.R. et al. (1967), *Future of the Computer Utility*, New York: American Management Association.

Ball, R. (1993), "Help Wanted", *TIME*, No 1.

Barnes, T. (1991), "Gateway to the World," *Telesis*, No. 93, pp.15-25.

Beaudry, M. and J. Parker (1992), "Global System for Mobile Communications," *Telesis*, no.94, July, pp. 53-69.

Boese, J. O. and R.B. Robock (1987), "Service Control Point: The Brains Behind the Intelligent Network," *Bellcore Exchange*, November/December 1987, p.13.

Brand, S. (1988), *The Media Lab: Inventing the Future at MIT*, New York: Penguin Books, p.19.

Brock, G. (1975), *The Computer Industry*, Boston, MA: Ballinger.

Brody, G. and J. Roth (1992), "Viewpoint," *Telesis*, no. 94, July, p.2-3.

Bruno, Ch. (1990), Integraph Ushers in CAD Conferencing Era," *Network World*, Nov.12, p.37.

Calhoun, G.M. (1990), "Wireless Access and Rural Telecommunications," Washington, DC.: The Office of Technology Assessment.

Carney, J., Zarakhovich, Y., McAllister, J. (1992), "Why It Still Doesn't Work" *TIME International*, No 49.

Cohan, S., and M. Leroux, P. Luff, P. MacLaren, R. Mo, (1992), "The New Frontier: Personal, Mobile Communications," *Telesis*, no 94, July, pp. 5-17.

Conard, J.W. (1988), "Access to X.25-Based Packet Networks," in ed. J.W.Conrad: *Handbook of Communications Systems Management*, Boston: Auerbach Publishers, p.359.

Conard, J.W. (1988a) p.325.

Cordaro, G. (1991),"The Telecommunications Single Market", *Telecommunications*, No 1.

"Daily Report" (1993), Radio Free Europe/Radio Liberty, No 6, E-mail Internet, Newsgroups: misc.news.east-europe.rferl.

Drynan, D. and D. Jeanes, (1991), "Global Data Networking with DPN-100," *Telesis*, No. 93. p.37-45.

Eastland, E., and H. Lilleniit, A. Manning, J. Marson, I. Stewart, (1991), "The Globalization of Services," *Telesis* No. 93, pp.5-13.

Estrin, D. (1986), *Communication Systems for An Information Age: A Technical Perspective*, Washington: the Office of Technology Assessment.

Fitzgerland, J. (1990) *Business Data Communications*, J. Wiley & Sons, New York.

Flaherty, D.L. (1991), "The Worldwide Intelligent Network Manages the Flow of International Communications, *AT&T Technology*, vol.6, no.2, pp.24-29.Frisch, I. T. (1988), "Local Area Networks versus Private Branch Exchanges," Telecommunications, November 1988, p.24

Gilhooly, D. (1991), "Rewriting Europe: New Politics, New Networks?", *The Global Telecommunications Traffic Report*, Staple, G., ed., International Institute of Communications, London.

Gilhooly, D. (1989), "Welcome to a Future Where Less Is More," *Communications Week*, CLOSEUP, Sept.4, p.C4.

Greenberger, M. (1964), "The Computers of Tomorrow," *Atlantic Monthly*, July.

Gregg, L. (1992), "Mixed Signals from Wireless Communications," *Network World*, June 15, p.51.

Harler, C. (1992), "You Can Vacation but You Can't Hide," *Communications News*, March.

Janczewski, L., and A. Targowski (1994), "Toward a Global Telematic Policy," *Global Information Management*, vol. 2, no 2, p. 30-41.

Kapor, M., D.J. Weitzner (1992), "The Open Platform Proposal" *The Electronic Frontier Foundation*, p.2.

Keller, J. (1992), "Telephone Switching Moves Toward Increased Speed," *The Wall Street Journal*, November 25.

Keen, P.G.W. (1992), "Planning Globally: Practical Strategies for Information Technology Strategies in the Transnational Firm," ed. by Palvia, S., Palvia, P. and R. Zigli, *The Global Issues of Information Technology*

*Management*, Harrisburg, PA., Idea Group Publishing.

Khan, T. (1990), "Third Generation' Technology Fuels VSAT Growth," *Telecommunications*, September 1990, p.29.

Loberg, L. "Mobitex Brings a New Look to Dispatch Communications" (1985), *Communications International*, February.

Mandell, M. (1992), "The In-Flight Office," *Portable Office*, March, pp.33-34.

Martin, J. and J. Leben, (1988), *Data Communication Technology*, Englewood Cliffs, NJ.: Prentice Hall, p. 551.

McClelland, S. (1991), "The USSR", *Telecommunications*, No 10.

Minoli, D. (1991), *Telecommunications Technology Handbook*, Boston: Artech, p.3.

"Missing Computer Software" (1980), *Business Week*, No 46.

"Missing Networks A European Challenge: Proposals for the Renewal of Europe's Infrastructure" (1991), European Roundtable of Industrialists, Brussels.

Nagel, D. (1992), "Statement of Apple Computer, Inc." submitted to the Committee on Commerce, Science, and Transportation, United States Senate, Washington, D.C. February 28.

Nellist, J.G. (1992), *Understanding Telecommunications and Lightwave Systems*, New York: IEEE, Inc.

Office of Technology Assessment (1991), *Rural America at the Crossroads: Networking for the Future*, OTA-TCT-471, Washington, DC.: U.S. Government Printing Office.

Quarterman, J.S. (1990), *Computer Networks and Conferencing Systems Worldwide*, DIGITAL Press, p.619-627.

Palvia, S. and S.P. Saraswat (1992), "Information Technology and the Transnational Corporation: The Emerging Multinational Issues," ed. by Palvia, S., Palvia, P. and R. Zigli, *The Global Issues of Information Technology Management*, Harrisburg, PA., Idea Group Publishing.

Parkhil, D.F. (1966), *Challenge of the Computer Utility*, Reading, Mass.: Addison-Wesley Publishing Co.

Ross, I. M. (1988), Kenote Address for Publication in the Conference Proceedings of the 1988 Bicentennial Engineering Conference, Australia, February 23, 1988, p.12.

Samuelson, M.A. (1991), "Advanced Intelligent Network Products Bring New Services Faster," *ATT&T Technology*, vo.6, no. 2 pp.2-7.

Seiden, E. and J. Tarlin (1992), "Opportunities are Good for U.S Cellular Business in Latin America," *Telocator*, June, p.8.

Senn J. (1992), "Assessing the Impact of Western Europe Unification in 1992: Implications for Corporate IT Strategies," ed. by Palvia, S, Palvia, P. and R. Zigli, *The Global Issues of Information Technology Management*, Harrisburg, PA., Idea Group Publishing.

Sisodia, R. (1992), "Singapore Invests in the Nation-Corporation", *Harvard Business Review*, May-June.

Smalheiser, K. A. (1992), "National ISDN, Promise Becomes Reality," *Fortune*, November 16, A special advertising section.

Sprague, R. (1969), Information Utilities, Englewood Cliffs, NJ.: Prentice-Hall, Inc.

Stallings, W. (1990), *Business Data Communications*, Maxwell MacMillan Int., New York.

Stallings, W. (1989), ISDN, an Introduction, New York: MacMillan, p.

17.

Sutherland, E. (1992), "Telecommunications in Eastern Europe" in *The Global Issues of Information Technology Management"* Idea Group Publishing, Harrisburg.

Tanzillo, K. (1992), "LAN Revolutionizes Data Handling at Army's Tank Command," *Communications News*, November, p. 27.

Targowski, A.(1991), "Strategies and Architecture of the Electronic Global Village" in *Proceedings of the 1991 Information Resources Management Association*, Memphis.

"Telecommunication Network-Based Services - Policy Implications" (1989), Organization for Economic Co-operation and Development, Vol 18, Paris.

Thomson, K. (1991), The Importance of Global Alliances, *Telesis*, No.93. pp.61-66.

Titch, S. (1991), "The Heart of the Network," *Telephony's*, October, p.5.

Valovic, T. (1989), "Metropolitan Area Networks: A Status Report," *Telecommunications*, July 1989, p. 27

"VSATs: Far-Out Communications for Remote Sites" (1990), *Telecommunications*, September, p.37.

U.S. News and World Report (1980), "Scramble to Bring Cable TV to Your Area, 6 October, p. 47.

Wimmer, K. and J. B. Jones (1992), "Global Development of PC," *IEEE Communications Magazine*, June 1992, pp. 22-27.

Wilson C. (1991), "We're all connected," *Telephony's*, October 1991, p.14.

CHAPTER 2

# Telematic Services

## Classification of Telematic Services

The new strategic Networking Telematic Services (NTS) industry is rapidly emerging from the marriage of computers, telecommunications, and television. NTS is a combination of information services, communications, and telecommunication services. This industry permits users (consumers) to interact directly with one or more computers, databases, and problem-solving procedures from remote terminals in a multimedia mode.

The NTS industry is in the business of providing computer/telecommunications/TV-mediated information and communications services. Typically, NTS systems have the ability to handle large numbers of remote users in a conversational mode and to adapt to a wide range and form of information & communications products and services.

The classification of networking telecommunication services includes (prevailing paradigm):

- Electronic transaction processing systems (measurement)
- Electronic information services (change)
- Electronic communications services (link)
- Networking telecommunication services (area and speed)
- Cable and broadcast TV information services (infotainment).

The further explosion of the networking telematic services by categories and subcategories is shown on Figure 2-1.

# Electronic Transaction Processing Services

### Home Health Information Services

Health-care expenditures already constitute 12.2 percent of GNP, and the Federal government expects them to grow to 15 or 16 percent by the year 2000. Compared to other industrialized countries, spending approximately 7 percent of their GNP on health care, and the United States still has a higher infant mortality and life expectancy than those spending less. Over 33 million Americans have no health insurance coverage at all.

The U.S. rankings indicate that American health care expenditures could be far more efficient and effective and more oriented to preventive than to acute care medicine. Progress in medicine suggests that over the next twenty five years the changes could be profound. Taking ever more effective self-care and health promotion

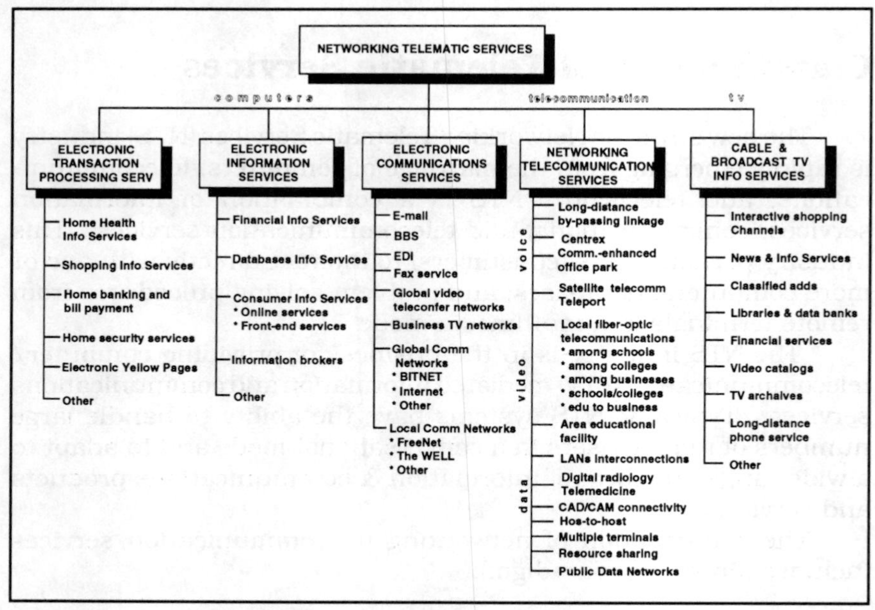


*Figure 2-1: The Classification of Networking Telematic Services*

into the home is a critical component of these improvements.

According to the Institute for Alternative Futures and the Consumer Interest Research Institute in Washington, D.C., three fundamental developments are changing the whole picture of what home based health care can be like:

• increasing attention to disease prevention and health promotion (96 percent of medical expenditures are related to treatment, and less than 4 percent go to prevention),
• the growing importance of home care (many treatments that once required hospital care can now be provided in outpatient and home care settings),
• advances in biomedical research (treatment will be customized taking into account individual genetics —"DNA fingerprint," biochemical uniqueness and health history).

As health care increasingly emphasizes prevention, care of chronic conditions, reassurance, and informational support, the delivery of health care to the home becomes an increasingly significant health care strategy.

Many of the patients and home workers' needs can be met by creating electronic links between patients and their health care professionals, between home workers and their supervisors and between patients and families. The ability to talk directly with health care professionals is important for getting medical advice, avoiding the inconvenience and expense of unnecessary office visits, and receiving assurance from authoritative sources.

Two-way video can greatly enhance these contacts, especially where emotional assurance and motivation is important. Physicians can observe their patients' body language and facial demeanor. Members of patient or family support groups can feel more socially connected. Patients can be shown visually how to use their wheel chair, how to go up and down stairs or how to take their medication. This type of interactive communication is far more effective than detailed instruction sheets.

Figure 2-2 illustrates a general model of the electronic home health care environment that will be emerging in years 2000-2010 [1].

## Computer Based Medical Records - Telemedicine

In the coming years, medical and health care organizations will electronically record and store all patient information including patient problems, tests results, orders submitted, treatment plans, X-ray and other images. Over time, they will incorporate new types of information such as personal "DNA fingerprints," providing infor-

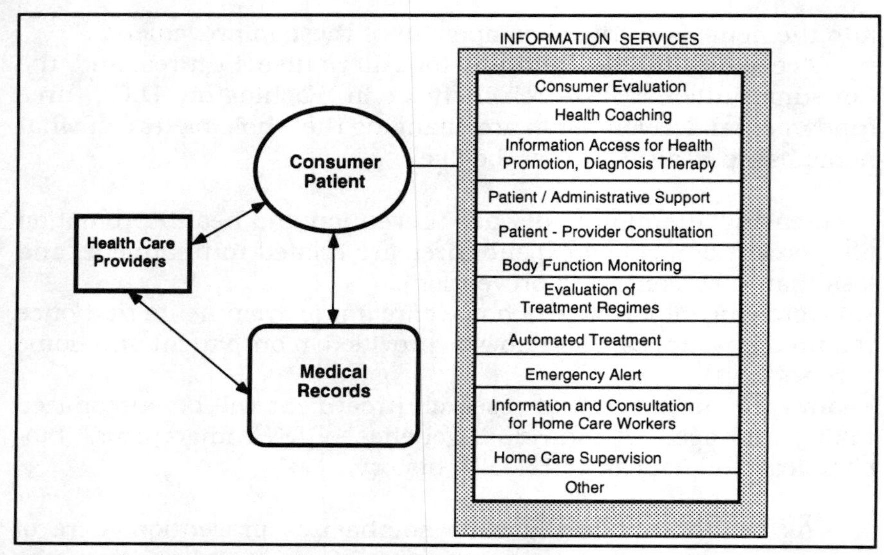


*Figure 2-2: Information Services for the Home Health Care*

mation on genetic proclivities to disease. Electronic records will be "in the home" in a real sense. They will be stored on credit-card sized "patient cards" as well as in home and health provider information systems. Individuals will have electronic access to their records and will be able to add to the records from home, for example, by entering the results of home tests or various kinds of bodily function monitoring. By the late 1990s, other home health applications will interact with electronic medical records.

As medical record information is increasingly put into electronic form, it can be readily aggregated and analyzed. The increased ability to perform comprehensive data analysis will let us identify patterns we could not possibly have seen before such as the cost-effectiveness and comparative medical outcomes of different treatments and therapies. The resulting information will give doctors a more scientific basis for assessing "what works," and helping consumers better evaluate alternative therapies and providers. This way of gathering information will require personal information privacy safeguards.

## Health Information and Communication Systems

Consumers will have greater health information available to them at home including clinical advice about specific diseases, information on their own health conditions and normal expectations, access to their own medical records, disease prevention/ health promotion information geared to their individual health status, and access to evaluations of providers and therapies. Hypertext and hypermedia will be increasingly used to allow "layered access" to information. Artificial intelligence technology will allow the applications of "knowbots" to filter information that people are interested in.

### Health Outcome Measures

In the beginning of the 21st century, outcome measures are likely to play a significant role in exposing inappropriate medical practices and directing therapeutic selection. As medical records become electronic, and large amount of aggregated patient data are analyzed, consumer rating systems will be able to include measures of specific clinical success levels of alternative treatments and providers.

### Diagnostic and Therapeutic Expert Consultation Systems

In the 1980s, numerous expert systems were developed, primarily to be used as teaching devices , or to assist physicians in clinical diagnosis. By the year 2000, one is increasingly likely to see expert systems installed on physicians' desktops workstations used for consultation and quality control purposes. A similar pattern is likely for therapeutic consultation systems. In the 1990s, expert systems will be developed as "decision support" tools for providers that will perform functions such as suggesting the appropriate work-up testing sequence and treatment patterns for specific illnesses. In the late 1990s they will be linked to the electronic medical records and knowledge bases that will advise practitioners on the logic and medical literature supporting specific decisions.

### The Health Coach

Health Coach expert systems will help to reinforce positive healthy behaviors such as following a healthy diet, maintaining proper weight, stopping bad habits, and encouraging regular exercise. They will also increase compliance with therapies, helping patients to follow physicians' instructions regarding medicine, home tests and treatments and other matters. After the year 2000, low cost voice recognition and synthesis will allow these expert systems to "talk" with family members.

### Automatic Alert and Monitoring Systems

Simple emergency alert systems are already widely available that allow home bound patients to call for help if they fall down or experience other emergency situations. By pressing a button on a light weight pendant around their neck or wrist, they are instantly connected through their telephone to their hospital, fire department, or other emergency response center. More sophisticated response systems will be able to operate automatically in health emergencies and transmit relevant information to the response center.

### Body Functioning Monitoring

These devices are increasingly used in the home to allow health care providers to monitor the progress of patients. The Holter monitor, for example, already provides 24 hour EKGs. Other devices are designed to conduct post-surgical checkups and send data on blood pressure, pulse and other vital signs. Effects of medications can be monitored in the home and dosage levels calibrated to achieve the best results for each individual. Automated treatment by drug dispensing and other means is also possible in a home setting. Already, for example, adaptive pacemakers with built-in sensors are used to stabilize erratic heart rhythms and insulin pumps automatically provide diabetics their medication.

By the year 2000, low cost biosensors will promote growth in the use of home monitoring systems linked, when appropriate, to computers in physicians' offices and managed care organizations. Wrist watches can perform a growing variety of diagnostic evaluations. They may include expert systems modules, administer pharmaceuticals, and be linked to physician offices via portable radio. The Institute for Alternative Future predicts it will be a virtual "Hospital on the Wrist." (Olson, Jones, Bezold, 1992).

## Electronic Commerce

### Shopping Information Services

Imagine a future where shopping is radically different. No more commission-hungry salesmen touting overpriced VCRs, cars, or life insurance. No more phony "Going Out of Business" ads, or "Annual Sale" signs that seem to appear every month. No more impossible choices to make between dozen of complicated and expensive products. Snider and Ziporyn (1991) in their book *Future Shop* call for a "New Consumerism" that will empower consumers, through private enterprise, instead of through government watchdogs or well-meaning nonprofit organizations.

At the heart of this revolution is "Omnimedia," a projected new telecommunications infrastructure based on existing technologies, such as fiber optics, high-definition TV (HDTV), multimedia, and wireless portable computers. Once Omnimedia is in place, privately owned "Independent Consumer Information Companies" (ICIC) should emerge to provide shoppers with customized, affordable, and unbiased advice about their purchases.

Better information about consumer choices will profoundly affect almost every aspect of Electronic Global Village life. Among Future Shop's startling predictions: traditional shopping centers will give way to "electronic malls;" retailing as we know it may disappear; the role of advertising will diminish; and there will be a multitude of new entertainment and news products.

At various times in U.S. history, we have discovered that our continued prosperity and vigor depended on building new technological and institutional foundations. Such times are brought on by technological innovation. Thus, the postal, railroad, electric utility, telephone, and highway systems were all created in response to new technological developments—inexpensively printed newspapers, the steam engine, the electric light bulb, the telephone, and the automobile respectively . In economics, these systems are called "infrastructures." For example, contemporary suburban shopping malls could not exist without the transportation infrastructure.

Today, the time is right to start building the information infrastructure to move consumer information. The authors of the *Future Shop* call for the creation of such a infrastructure composed of three components (analogous to the three components of the monetary system:

- Omnimedia circulating consumer information ("money")
- Independent Consumer Information Companies ("banks")
- National Institutes of Product Information ("Federal Reserve").

Omnimedia can do everything that all other media combined can do and more. Omnimedia would supersede the incompatible babble of today's telephones, stereo systems, newspapers, TVs, papers, magazines, cellular telephones, books, and computers.

Omnimedia might allow the following scenario:

*Newspapers, magazines, and yellow pages are available electronically. If you want to buy a bicycle, you will go to your electronic yellow pages and find out about all the bicycle dealers in your area and any specials they might be running. You might even get an interactive tour of the store, including detailed information about any particular models that strike your fancy. This will be much faster and more comprehensive that any search you could do looking in your local newspaper for relevant ads. Similarly, if you are reading about bicycles, you probably won't want to view car ads that will just have to wait until you are going to buy your next car, and when that time comes, you won't be interested in wading through bicycle articles to get to them.*

Already some VCR tapes are addressed directly to a customer. Also, the new picture-in-picture (PIP) features of some TVs allow one to skip commercials jumping to another channel until the commercial is over. The future of omnimedia is in the fiber optics lines provided to each household.

Independent Consumer Information Companies (ICIC) will provide independent consumer information only if:

- they finance their information without any advertising dollars
- they must receive no money from the owner of any product they analyze
- ICIC owners must be free of conflict ( cannot hold stock in analyzed companies).

As third-party information providers ICIC(s) share many characteristics with today's media. They simply rely on an as-yet-unrecognized medium; the computerized, intelligent database and the reports and analysis it generates. Before the full implementation of Omnimedia, many of these are being sent through the mail like periodicals. Unlike periodicals as we know them today, however, these reports and analyses are customized to the individual consumer and are thus disqualified from the postal discounts afforded to second-class (high-volume, periodic) mailings. On the other hand, any information generated without advertising support merits the status of at least second-class and possibly, even non-profit status.

A revolution in product-trucking technology makes it feasible to institute a new institutional structure: a government-sanctioned, quasi-private organization called the National Institutes of Product Information, or NIPI, from which ICIC(s) would purchase the raw factual data on which they would base their shopping recommendations to customers. NIPI could gather and disseminate four basic types of information:

- purchases (e.g. from large stores)
- customer satisfaction
- post-purchase service
- features.

According to *Future Shop*, each major consumer industry— for example: hard goods, travel, health care, legal, or finance—would have their own institute responsible for collecting and disseminating data about their particular industry. ICIC(s) will collect data from other sources too, but their more important role would be to provide perspective—to integrate and interpret NIPI and other data so to develop specific recommendations.

NIPI meets a need that sellers and buyers have already recognized for years: one-stop shopping. Vendors will be motivated to provide information about their products, resulting in the creation of a comprehensive national database, a electronic industry-wide information clearing house.

The technological foundation for electronic shopping is in:

- machine-readable universal product codes
- wireless, automatic identification of a product in a store
- computerized prize lookup (updated tables)
- electronic data interexchange
- quicker response (elimination of various mailing cycles)
- computerized UPC catalog
- telecommunication networks
- computers and terminals.

Service information would be particularly relevant for mass-produced products that run on electricity, battery power, or fossil fuels and have substantial (e.g., $20) retail values or non -eplaceable components.

Snider and Ziporyn (1991) say that our problem is not that the supply of products do not exist, it is that this does not do us any good if we cannot find them. In a modern society, information is the glue which shapes our lives; it is the means by which we choose a better life. The Age of Electronic Shopping will predict nothing less than the death of the promotional society — the death of promotional companies and personalities which now thrive in advanced industrialized countries. It is been said that today, by age seventeen, the average American will have seen 350,000 television commercials. The death of commercials will leave room for more knowledgeable information more important for the human life than the commercials they replaced. The giant media conglomerates which have grown solely because of their control over distribution channels, rather than their ability to generate quality information, will become a dying breed. The authors predict the following trends:

- Power will shift from knowledge distributors to knowledge creators.
- When promotion and reputation give way to merit, no company will be able to count on past successes to carry it along.
- Brand names will loose their market value.
- As selling costs become intrinsically inefficient, savvy sellers will focus their energy and resources on producing distributing superior products.
- In-home selling increasingly will replace in-store selling.
- Mail-order retailers will be hurt more than local retailers.
- Local retailers will have to provide in-home shopping services (previewing the store before entering it, pre-ordering merchandise, delivering merchandise from store to home, particularly from mega-

discount stores).
• Information services, not application software, will be the primary application of personal computers.
• Traditional telephone companies and cellular telephone companies will both flourish.
• Information will be seen as an increasingly important moral resource (informed citizen-consumer, not manipulated one, who is able to prevent domination by technocrats and other information elite).

## Evolutionary Steps Toward Electronic Shopping

It is yet a long way toward shopping through "data malls." The realization of data malls requires the fully automated and technologically advanced informatting of the NTS-mediated transfer of product information from producers to consumers. To achieve this level of technology, the retail industry is in a process of regrouping and reshaping its services.

One can recognize the following phases of this process:

• *Phase I — Retail Giants Rule the Marketplace.*

Giant "power retailers" using sophisticated inventory management, finely tuned selections, and above all, competitive pricing are crowding out weaker players. By the year 2000, retailers now accounting for half of retail sales today will disappear through bankruptcy, mergers, or other reorganizations. Triumphing over them are superpowers including Wal-Mart, Kmart, Toys 'R' US, Home Depot, Circuit City Stores, Dillard Department Stores, Target Stores, and Costco, among other. Consumers are flocking not only to Wal-Mart but to new retailing channels. They are now patronizing warehouse clubs (e.g., Pace) and tightly focused "category killers (e.g., Structure). Retail giants apply advanced enterprise-wide systems requiring electronic hookups from the vendors so the right products hit the shelves at the right time.

Quick's Candy Inc. a $3.5 million candy company in Buchanan (Michigan) invested $15,000 in computer systems that monitor inventories at Kmart's 12 distribution centers. When supplies run low, Quick's automatically replenishes inventories with no prior approval from Kmart. It is the only one of Kmart's 250 vendors having that capability.

• *Phase II — Megastore that Entertain and Educate.*

A three story Nike Town retail store on the famous Michigan Avenue in Chicago displays its products amid life-sized Michael Jordan and Bo Jackson statutes and glassed-in relics like Nolan

Ryan's shoes. A half-court basketball court lets visitors try out Nike shoes, which are delivered through clear, Star-Trek-like tubes. Unlike most other stores that line the Michigan Avenue promenade here, Nike Town's mission is much greater than simply displaying its wares. It is out to entertain customers and to inform them about Nike Inc.'s athletic shoes and other merchandise. This store attracts 5,000 to 6,000 visitors daily on weekends, compared with an average 3,600 people who visit the Art Institute of Chicago, also located on the same street. And its format—which combines elements found in theaters, museums and amusement parks—is cropping up elsewhere. Also on this "Magnificent Mile," Sony Corp. and FAO Schwarz have also opened extravagant retail outlets that often seem more appealing than the goods they are trying to sell. In October 1992, on New York's Fifth Avenue, Warner brothers opened its largest store next to such posh merchants as Tiffany and Bulgari. The store resembles a movie-studio back lot with giant Batman and Superman sculptures bursting from the wall.

This concept is based on the mass-media solution—pumping in the idea and motivating segmented customers. Such an approach can afford big brands with huge corporate staffs new solutions and mega-projects. Retailers, however, that do not plan for and respond to NICS developments will be the dinosaurs of the last decade of the 20th century. They may even be extinct by the year 2000.

• *Phase III — From Segmentation to Relationship Management.* Retailers look beyond the segmentation of customers into groups and apply NICS and Information Systems to track customers as individuals, responding to them on an individual basis. Tracking customers can be done using bridal, birth, and graduation registries. These events in the life of the customer are at the core of what buyer-behavior analysts would call the impetus to "need recognition." Knowing customers individually in the Information Age means keeping track of customers in a database and recording information about them that can be used to serve them better. Also, the move from customer service to customer experience (computer simulation of pre-sale solutions) is a part of the relationship management (Peterson 1992).

• *Phase IV — From Store-Based to Distance-Based Retailing.* About forty million homes now have computers. Homes are being linked either by cable or fiber-optics allowing greater information capability at home. Customers will be able to achieve full-motion videotex experience right in their own homes —probably on a HDTV screen. The Yellow Pages will be computerized so the customer can retrieve its database to find-out the best store for a given purchase. With a connection to Geographic Information System, the customer will get a map with the shortest route from his/her home to a retail store. Distance retailing can be possible either via the computer or the two-way cable TV. The consumer does not have to shop from their homes, they can do so from their workplace, car (cellular phone), or the airport. The consumer applies distance shopping from his/her location but the retailer does not have to be located in the same city or town or village. Sometimes it will be easier to shop in a retailer located in a different city than it is to shop one a couple of miles from home (Peterson 1992).

Examples of distance shopping are:

- TV Channels
- Consumer information services; CompuServe, Prodigy, America Online
- Specialized networks; Shoppers Advantage—On the Electronic Mall (3 million customers, which is 1% of the population).

- *Phase V — Electronic Shopping through "Data Malls."*

Electronic shopping via NICS will take place when all three components of the system: Omnimedia, ICIC, and NIPI work together, and when a majority of American households have easy access to either a computer or interactive TV. This phase should be in operation within two to three decades. Figure 2-3 illustrates the architecture of electronic shopping.

## Home Banking and Bill Payment

In the mid 1980s banks started experimenting, they began receiving electronic payments from home computers. It was expected that the creeping revolution of microcomputers would equip a majority of America's households with home computers in the 1980's. The revolution took place, however, the number of installed modems was much smaller then expected. Only a mass service of home banking would return investments undertaken by the banks. The Chemical Bank attracted only a few thousand users of home banking in the metropolitan area of New York City, and Citybank developed a telephone service which applied a phone handset as a data key entry terminal for the home user.

Home banking provides the following set of main remote and online services:

- Inspecting account and balance information and electronic statements
- Generating bill payments to creditors
- Transferring money between accounts
- Utilizing the bank's other information resources

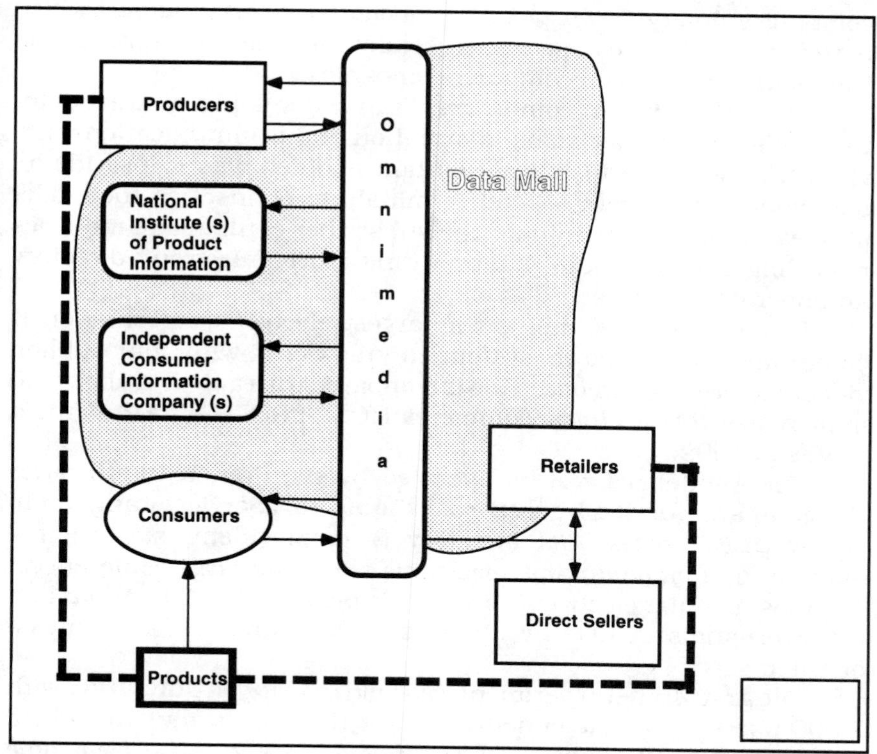


*Figure 2-3: Electronic Shopping Through Data Malls*

• Other

In the early 1990s, some commercial information services providers (PRODIGY and America Online) offered home banking through own networking telematic services.

## Electronic Yellow Pages

The electronic yellow pages is a well organized data collection of business and community services. The electronic yellow pages is a dynamic database of data, information, concepts and eventually knowledge about local services. The best known yellow pages are the French Minitel Services.

In the early 1960s, the French government, industrial circles, and academics realized that information and telecommunications technologies were going to be one of the most important infrastructure issues in the coming decades. The national "*Plan Calcul*" was defined and an office of *Le Delegue de Governement pour L'Informatique* was created to implement the Plan. The French created the national computer industry (*Compagnie International pour L'Informatique*), which however, could not compete successfully with the French-American Bull, and the American computer companies.

Instead of French computers, France developed a strategy for the application of electronic information and communications services (I&CSr). The technical solution adopted was to provide all telephone subscribers Minitel terminals to connect to their telephones. Using the terminal. the subscriber could make inquiries regarding telephone numbers, providing a sort of electronic directory and operator services.

In the 1990s, Minitel is the largest electronic information & communication services system in the world with four million videotext terminals in use. This is compared with a total of about 1.6 million users of the largest commercial ICS—PRODIGY in the United States in 1992.

The Minitel network is operated by the French PTT (France Telecom) and funded by fees from the users. Services range from yellow pages, residential directory to employment, sex-oriented databases, home banking, electronic mail, and large conferences. The most popular electronic directory is provided by PTT. About half of the terminals are used by businesses which support the business-oriented services.

The first Minitel experiment was held in Velizy in July, 1981 with 2,500 users. The Teletel network, over the European value added network, TRANSPAC was set up in May, 1992. The electronic directory service began in February, 1983. Instead of providing paper telephone directories, the French PTT distributed Minitel terminals at no cost to the user. By distributing terminals free, the French PTT hoped to increase telephone traffic generating additional revenues and income for them. The strategy worked, providing one fifth of the 1 billion dollar new industry (I&CS), because of increased telephone line usage. The French PTT (Direction General Telecommunication—DGT) organized a profit-oriented subsidiary —France Telecom (FT) which collects the Minitel fees through the user's ordinary telephone bill. FT keeps a percentage for *grand public* services and passes on the rest to the service providers. This arrangement, called service kiosk, begun in February , 1984 and was initially controversial because FT effectively acts as a banker (Delivers and Pays 1988).

## Electronic Information Services

### Financial Information Services

As early as 1850, Paul Julius Reuter first used carrier pigeons to fly stock market quotations between Brussels and Aachen,

Germany. One year later, an underwater telegraph cable opened between Dover and Calais. Reuter then began delivering news and market quotes from London to Continental Europe. Reuters is, 150 years later, still one of the dominant market information service vendors.

The market for financial information services can be broadly divided into three categories—news, data on exchange-trade instruments, and data on over-the-counter instruments. The market structure is different for each of these.

***Financial News*** . Financial news may be gathered by information vendors themselves, or they may carry reports from leading news organizations. Dow Jones & Co. Inc., is the leading provider of financial news in the United States. Dow Jones has tried to extend its dominant position in equities news to the fixed-income bond market through the Dow Jones Capital Markets Report, but in-depth news is not as essential for the bond trader as it is for the stock trader.

Reuter has an edge over Dow Jones in news that effects foreign exchange and fixed-income prices because of its vast international communications market. Reuters is also a strong competitor in delivering news about U.S. commodity markets, but Knight-Ridder is a major presence in this market through its Commodity News Service and has also made headway in supplying news concerning financial features and underlying cash markets. Other providers of online financial news include the Associated Press, McGraw-Hill Inc., Financial News Network, and Market News Service.

***Stock Quotation.*** Five companies dominate the market for securities and futures quotations in the United States—Reuters Holdings PLC, Quotron Systems Inc., Automatic Data Processing Inc. (ADP), Telerate Inc. These five companies had a total of approximately 426,000 terminals worldwide as of February 1989 (Philo and Ng 1989). For most stocks, all commodity and financial features, and all options, the market data— bids, offers, last-sale prices, and volume information—are generated by exchanges and over-the-counter markets and delivered to vendors. In foreign exchange and fixed-income markets, where there is no central exchange, price information is contributed by banks and securities firms to vendors.

Quotron Systems Inc. has long dominated the market for U.S. stock quotations, but this market is now in ferment [2]. ADP is a strong competitor. Outside the United States, the leading position is held by Reuters, which recently entered the market for stock prices. In the past, Reuters supplied quotes and news for foreign exchange, money market instruments, and commodities in this country, but not equities.

The internationalization of the securities markets has promoted foreign vendors such as Reuters and Telekurs of Switzerland to enter the U.S. market, while American companies such as Quatron and ADP have been expanding their operations overseas. The growing links between the equities, futures, fixed-income and foreign exchange markets have also led to diversification among vendors who traditionally specialized in one market. Telerate Inc., which holds a near monopoly in the market for U.S. government securities prices, has entered the equities market through acquisition of CMQ Communications Inc., the leading stock Quote provider in Canada. It remains to be seen whether Reuters and Telerate can replace Quotron and ADP, or will merely add equities quotes to their existing terminal base. There are about 200,000 terminals receiving real-time quotes from U.S. stock exchanges, and some industry observers are skeptical that the pie will become any bigger with the entrance of new players.

Nevertheless, the relative ease of acquiring and distributing prices for exchange-trade instruments has attracted several new competitors in recent years, including PC Quote Inc., and ILX Systems, a new venture backed by International Thompson Organization. Despite the competitive conditions in the securities quotation business, there is always room for new "niche" companies offering innovative products, such as proprietary analytic.

***Value-Added Products***. The relative ease with which any vendor can obtain data from the American stock markets and many of their foreign counterparts has made the market for exchange-trade data into a "commodities" market, in the sense of highly standardized products competing on price or value-added features. In order to maintain their profit margins, vendors are trying to add value through new technology or exclusive products, trying to generate as much revenue per terminal as possible. This has encouraged third-party suppliers to offer historical information, research, analytics and tailored news services through the terminals of vendors such as Quatron, Reuters and Bridge Brokerage Systems. Vendors that control the distribution network typically keep 30 to 40 percent of the revenue generated by the third-party products.

Among companies successfully exploiting demand for third-party services is MMS International, which delivers analysis and commentary on Telerate, Bridge and Reuters. MMS was recently acquired by McGraw-Hill Inc. Another third party provider is First Call, part of International Thompson's InFiNet group, along with ILX Systems. Jointly owned by Thompson and a group of securities firms, First Call is a leading provider of on-line research produced by Wall Street analysts. Both Quotron and Reuters have tried to compete against First Call's research distribution service, but Reuters recently discontinued its own service and signed an agreement to offer First Call to its subscribers.

***Foreign Exchange Data.*** The commoditization of exchange trade data has no parallel in markets where there are significant barriers to entry for vendors. Reuters created the market for real-time foreign exchange data in 1973 when it first put computer terminals on the desks of traders and convinced them to enter their rates into the system. Reuters charged subscribers a flat monthly fee but did not pay banks for contributing their quotes to the service. In 1981, Reuters launched the Monitor Dealing Service allowing traders to negotiate transactions over their terminals instead of telephones. This system has been successful in part because of its built-in audit trail. In 1989, between 30 and 40 percent of the $640 billion traded each day in the interbank foreign exchange market took place on the Monitor Dealing Service [3].

While Reuters is the best established in the foreign exchange market, Telerate is a competitive alternate service. Traders like having a backup quotation system, and like the idea of Reuters having competition. It was nevertheless difficult for Telerate to gain a place in a foreign exchange ("forex") until Reuters agreed to permit its subscribers to install "binco boxes"—bank in-house computers—that let them simultaneously update their rates on Reuters and Telerate. Until then, Telerate's forex market coverage was often slightly behind because dealers posted their rates on Reuters first. Other reasons for telerate's success in penetrating this market was the availability of AP-Dow Jones foreign exchange news and the traders' need for U.S. interest rate data, which Telerate's system provided.

Telerate did not until recently offer dealers a transactional system such as Reuter's Monitor Dealing service. It has now launched a foreign exchange conversational (on-line) dealing system through a joint venture with AT&T. Known as The Trading Service, this service allows dealers to talk to several dealers at once, unlike the Monitor Dealing Service. Now Reuters in turn is taking another step forward with an enhanced version of the Monitor Dealing Service and a centralized order database facility. While the original Dealing Service facilitates one-on-one negotiation between two traders, Dealing 2000 will emulate an auction market where bids and offers from multiple parties are exposed. This is designed to replace "blind" brokers, who act as middlemen in foreign exchange trading. The system will display the aggregate size of all bids and offers at each price, but will not disclose the identities of the dealers participating.

***U.S. Government Bond Data.*** Telerate is currently the only vendor broadly distributing prices in the government securities market. Under an exclusive agreement scheduled to expire in 2005, Telerate disseminates bids, offers and last-sale prices from Cantor Fitzgerald Securities Corp., the only major inter-dealer broker serving both primary dealers and retail customers. Other brokers provide price information only among the primary dealers, those who are authorized to deal directly with the Federal Reserve Bank of New York. In a 1987 study, the General Accounting Office encouraged brokers to distribute quotations to non-primary dealers within 2 years. In April 1989, major government bond dealers reportedly pressured a large government bond broker into abandoning a controversial effort to broaden access to bond-trading information by offering its electronic trading information screens to a wider group of customers (Herman 1989).

Reuters, Quotron, and Knight-Ridder have periodically held talks with individual brokers about disseminating their quotes, and three inter-dealer brokers have discussed distributing consolidated last-sale prices, but none of these efforts have reached fruition. When they do, "commoditization" will probably also occur in the market for U.S. government securities prices. Vendors would have to compete by providing proprietary analytics, news, or by specializing in a particular area of the Treasury market.

Reuters and Quotron are likely to try to expand into the fixed-income information business. Since its acquisition by Citicorp, Quotron has been developing information and transactional services in both foreign exchange and fixed-income markets. However, Quotron is facing the same obstacles here as do Reuters and Telerate in equities—lack of critical mass and a shortage of space for terminals on the already crowded desks of traders.

## Competition and Technological Change

Technological change is creating upheaval and uncertainty among financial information vendors. As recently as 1985, an equities trader typically had one terminal on his/her desk — probably a Quotron—which carried Dow Jones News Service and gave the trader access to prices for U.S. securities only. In the fixed-income department of the same firm, each trader would have a telerate terminal. In the foreign exchange area, each desk would have a Reuters terminal and perhaps one from Telerate. Because markets did not greatly affect one another, there was no need for the traders in one market to be watching other markets. However, fixed-income traders always needed to follow the foreign exchange markets since currency prices and interest rates are closely linked.

The technology used by the vendors was essentially the same, a dumb terminal connected to a host computer by dedicated telephone circuits. But as a number of niche services grew, traders ended up with more and more dedicated terminals on their desks. The use of single dumb terminals declined sharply when the PC permitted local storage and manipulation of price information. Now, because of digital technology, the way vendors transmit the data is becoming less important than what data they transmit.

Several other technological advances in the early and mid-1980s also irrevocably changed the delivery of financial information. The video switch, long used in the broadcast industry, reduced the clutter of terminals on traders' desks by allowing several screens to be controlled by a single keyboard. These screen then became an important part of trading rooms and were also responsible for the rapid rise of two companies that installed thousands of new system integrators for trading room worldwide. There were also rapid changes in the manner in which stock quotations were transmitted from vendors to customers. In addition to delivering prices over dedicated telephone lines, vendors began exploring other alternatives, such as broadcasting data by FM sideband and satellite. Midwestern commodity market data vendors began in 1981 to use small, low cost, receive-only satellite dishes which were particularly effective for one-way broadcast communications such as financial quotations. They now distribute financial data for vendors such as ADP, Dow Jones, Knight-Ridder, PC QUote, Reuters, and Telerate. Although dedicated interactive networks remain the primary delivery mechanism of financial information vendors, financial data accounts for 63 percent of the 114,000 data broadcasting satellite receiving sites currently in operations [4][5].

## Databases Information Services

Database and information retrieval services offer subscribers the ability to tap into more than 5000 commercial databases covering thousands of subjects provided by hundreds of companies. Dial-up access is provided via WATS lines or public networks such as TELENET, TYMNET, SprintNet, InfoNet, AlaskaNET, ConnNet, PC Pursuit, DataPac and others.

Database information services are organized into six functional levels:

1. The information originator, a publication like the Washington Post, the Wall Street Journal or the New York Times.
2. The database maintainer, an agency, company or private individual in charge of loading and updating a computer-based database system and leasing telecommunications lines to end users.
3. Service vendor, a company or agency that provides marketing and sales of a database or a federation of databases to information consumers (e.g., Lockheed with some hundreds of databases).
4. Packaged analysis creators who provide analysis of the raw data

provided by the databases and sell them to the information consumers.
5. Search services, intermediary information brokers who provide industry-wide search of all electronic storages for information consumers. In such a manner, the end user is not limited to the subscription of one or more databases.
6. The information consumer.

## Sets of Databases

The databases are organized into categories (sets), semi-categories (subsets) and subcategories (fields). One can distinguish the following examples of databases sets:

- DIALOG (owned by Knight-Ridder) offers over 300 databases to 100,000 customers in 100 countries. Such fields are covered as:

- Agriculture
- Arts
- Automotive
- Books
- Business & Corporate Education
- Chemistry
- Conputers & Electronics
- Consumer Information
- Corporate News
- Current Events
- Economics
- Education
- Energy
- Engineering
- Environment
- Food
- General Reference
- Government Publications
- History
- Specific Journals/Dbases
- Law
- Literature & Language
- Magazines
- Mathematics
- Medicine
- Patents
- Physical Science
- Psychology
- Social Sciences
- Humanities

- Knowledge Index is a subset of DIALOG with 100 databases that can be retrieved with simple queries.
- DOW JONES NEWS/RETRIEVAL (DJN/R) (owned by the Dow Jones company which publishes the *Wall Street Journal*) is a collection of several databases in business, economics, financial, investment, general-interest news and information services, and about 50 other information services.
- DATATIMES collects full the text of 600 newspapers, magazines, trade journals, news services, and other financial data sources worldwide (USA, Canada, Europe, Asia, and Middle East). The databases are concentrated on an eclectic scope of business

information.

- NEWSNET is a full-text service in the area of business newsletters (500), general news, and other specialized services. Newsnet specializes in current news and information with approximately ten years of issues on-line. Gateways to AP DataStream Business News Wire, Jiji (from Japan), PR News, Reuters, UPI, Xinhua (China) are also provided. Such fields are covered as:

- Advertising and marketing
- Aerospace
- Automotive
- Banking and finance
- Corporate communications
- Defense
- Electronics and computers
- Energy
- Entertainment and leisure
- Environment
- Farming and food
- Government
- Health and Hospitals
- Insurance
- International
- Investments
- Law
- Manufacturing
- Management
- Metals and mining
- Public relations
- Publishing & broadcasting.
- Real Estate
- Research and develop.
- Social sciences
- Taxation
- Telecommunications
- Tobacco
- Transport and shipping
- Travel and tourism

- BRS/AFTER DARK offers about 100 databases. This is a non-prime-time service provided by Bibliographic Research Services (BRS). The dependence between BRS/AFTER DARK and BRS is similar to that between Knowledge Index and DIALOG. The BRS's collection of data, however, is smaller than that offered in DIALOG.
- EASYNET is a set of 700 databases on almost any topic. It offers abstracts and even the full text of books. It is accessible through some consumer information retrieval services.
- LEXIS and NEXIS are owned by Mead Data Central. LEXIS is a huge database of federal and state laws, specific cases, and references in abstracts or a full-text. NEXIS is a full-text database of legal magazines, newspapers, and newsletters.

## Consumer Information Services

**CompuServe** is the largest consumer online service today. CompuServe began in 1969 when a insurance company, based in Columbus, decided to use its excess computer capacity to enter the remote time-sharing market, at first on a local basis in around the Columbus area, later expanding state-wide and nationally. In 1979 CompuServe provided access to its machines during the evenings

(MicroNET service) to computer hobbyists. This was the beginning of the largest commercially operated videotext service in the world. In 1990, CompuServe, a H&R Block company, had about 500,000 subscribers. In addition to its public services, CompuServe provides communications services, database maintenance, and private network services for a variety of businesses (Fortune 500) and governmental agencies. This gives the information service room to develop with the market, free from the pressure that have bankrupted so many other on-line services. You can access CompuServe via the CompuServe Network, SprintNet (Telenet), TYMNET, DataPac (Canada), ConnNet (Connecticut).

CompuServe main menu is called TOP and is composed of the following entries:

CompuServe TOP

1 Subscriber Assistance
2 Find a Topic
3 Communications/Bulletin Bds.
4 News/Weather/Sports
5 Travel
6 The Electronic MALL/Shopping
7 Money Matters/Markets
8 Entertainment/Games
9 Home/Health/Family
10 Reference/Education
11 Computers/Technology
12 Business/Other Interests

Enter choice number !

CompuServe is a gateway to over 700 major-league information databases, including those on Dow Jones News/Retrieval DIALOG, NewsNet, BRS, QL Systems, VU/TEXT, and MEDLINE. It also provides a gateway to MCI Mail and the Internet.

**PRODIGY** is a consumer front end information service with 1.7 million users in 1992. The service was developed and is owned by IBM and Sears. The users pay a flat fee for an access, and are not billed for the time users are connected online via the local telephone number. The main Menu looks as follows:

**AT YOUR SERVICE**
BOARDS & MESSAGING
BUSINESS & FINANCE
COMPUTERS
ENTERTAINMENT

FOR KIDS & TEENS
FUN & GAMES
HEADLINES
INDEX OF ADVERTISERS
NEWS & INFORMATION
SPORTS
SHOPPING & SERVICES

E-mail, weather, and travel services are the strongest features of PRODIGY. There are plenty of information providers and advertisers, who are not available online anywhere else. About 100 topics are covered ranging from advice on financial matters to pets. PRODIGY does not require users to be computer literate; it is a "plug-in-and-go" online service.

**America Online**, provided by Quantum Computer Services since 1989, uses a graphic user interface (Ensemble) with icons and pull-down menus for the following categories:

List Unread Mail
Compose Mail Form
News & Finance
People Connection
Lifestyles & Interests
Entertainment
Learning & Reference
Travel & Shopping
Computing & Software
Members' Online Support
Stock Quotes Area
Top News of the Day
Directory of Services
What is New Area
Keyword Window
Download Manager
File Search/ Software
A Real-Time clock
Display Print Options
Save the Current File

The most popular service is "Forums", which provides message bases and real-time conference areas for the special-interest groups. America Online access is through SprintNet and TYMNET.

### Other Online Services

Other online services one can find providers such as:

- BIX (Byte Information Exchange) is operated by General Videotex Corp. It is designed for communications among the BYTE magazine readers and information technology vendors. Its main services are computer conferencing, electronic mail, real-time conferencing, and software and file databases.
- DELPHI is provided by General Videotex Corp. since 1982. It evolved from a provider of an online encyclopedia toward a full-service network. Besides its consumer and business services, it

provides private network services for businesses and smaller special-interest groups (SIG). Each SIG may have their own conferencing area, separated from the general conferencing area. DELPHI also provides a gateway to DIALOG.

- GEnie, operated by General Electric Information Services since 1985, offers the same range of services as CompuServ and DELPHI. It has an easy system of Round Tables for discussions, and a clipping service that supplies new stories to the users' e-mail boxes, based on selected keywords (semi "knowbots"). The service has gateways to Dow Jones News/Retrieval and Internet.
- The Source, provided by Source Telecomputing Corporation, a subsidiary of the Reader's Digest Association, Inc., was developed with the participation of the Control Data Corporation. The Source is a full information service offering personal computer, investor, communication, news and information, and travel and leisure services, including accu-weather. The CHAT interactive communication is particularly interesting. It is available with a local call in over 500 major metropolitan areas.
- The Well is a "people's network" designed to provide local services to the San Francisco area. The Well has operated through CompuServ since 1985.

## Information Brokers

Information brokers act as intermediaries between information resources and the people who use that information. It is a fee-based service to gather the information for a customer. According to Rugge and Glossbrenner (1992), the heart of the information broker's job is:

- information retrieval based upon what the client's informational needs are (e.g., searching of online electronic databases).

- information organizations which
  - assemble and prepare bibliographies
  - catalogue book and materials collections
  - provide book indexing and library management
  - consult on library or information center design.

The information broker is a knowledge service worker who adds value to the client's performance by increasing awareness about available data, information, concepts, knowledge, and wisdom. The art of asking for and using information becomes the critical skill in surviving in the Electronic Global Village. The information brokers becomes the facilitators of this art. They offer clients their skill of goal-oriented traveling through the electronic universe, where hundreds of systems, thousands of databases, and thousand of electronic libraries and publishers provide answers to any dilemma, issue or question that any customer may have today.

## Electronic Communications Services

### Electronic Mail, The Global Postman

Electronic mail is a communications service that is converted and interconnected almost universally throughout the whole Globe. This is the most popular electronic communications service worldwide. It is a glue of the Electronic Global Village. Electronic mail is not an "end to end" service. The sending and receiving machine does not need to communicate directly with each other to make it work. It is known as a "store and forward" service. Just as in the Postal Service, if the destination and source are not on the same network, then an "application gateway" handles the interconnectivity between machines.

Electronic mail has a simple format:

- the body which contains the actual text of the message,
- the header which contains the address of the sender and recipient, a subject line, a message identifier and date provided by the local mail system,
- the envelope which is used by the intermediary mail delivery systems in routing from one point to another one. This component is not seen by the users.

Mail can be applied to carry binary files. E-mail must be addressed to a person not just a machine. E-mail is very similar to the telephone and to the postal mail. If the telephone offers immediate delivery, e-mail is delivered in a range of seconds to a day. In contrast to the synchronized telephone call, e-mail is asynchronous. The receiver reads e-mail at his/her leisure.

Delivery of electronic mail is handled by the X.400 messaging standard. The X.500 protocol standardizes the directory service; it views the global directory as a library of telephone books.

E-mail services provide the following forms of electronic communications : intrasystem, intersystem, electronic delivery (fax, Telex, voice-mail retrieval, file-transfer capability), and hard-copy. Only a few communications services offer e-mail links to other services through gateways. For example, MCI Mail links to Dow Jones News/Retrieval. Some offer bulletin boards (BBS) that one can expect to find on consumer services. E-mail communications between services is a trend that is accelerating.

The most popular providers of specialized e-mail services are: AT&T Mail, MCI Mail, DIALCOM, SprintMail. The consumer information service providers offer very good e-mail services. DASnet delivers e-mail between a number of online services. In 1993, this net interconnected 3 million subscribers. Perhaps, it is the future hub of global interconnectivity among all the online services.

The actual cost of sending an e-mail message is surprisingly low — typically less then fifty cents for 150 words. Fax and hard-copy delivery are equally economical. They are usually less than $2 for a domestic fax or a hard copy letter and about $3 or $4 for an international fax (Pournelle and Banks 1992).

MCI Mail also offers electronically transmitted paper mail through printing facilities at various locations around the country and around the world. In such a manner, a letter or file can be sent to the location nearest your recipient. The nearest (local) post office will deliver the mail.

> Unilever e-mail net ties four continents linking local area networks, at 29 locations, on four continents, in a global electronic messaging network. The multinational corporation, with more than 292,000 employees worldwide, signed a multi-year agreement with Sprint that initially links the office automation systems and LANs at 29 sites. Initial implementation involves more than 15,000 users of Unilever's many E-mail systems, among them DEC All-In-1, IBM Profs, Wang Office, Memo and cc:Mail. Unilever users on IBM systems can send messages in their native mode. Other systems link directly to SprintNet via X.400. Besides internal messaging, Unilever users can also communicate with SprintMail users and those on other connected public systems. The company is pursuing a service that integrates existing proprietary mail systems and supports the strategic direction towards open systems. Unilever also put upcoming events, professional training classes, and tips for using e-mail, personal computers and other mainframe systems on the board. The use of the bulletin board feature has really taken off here in the last six months. People are seeing more uses for it and they want to put more and more there.

**Bulletin Board System**

A computer bulletin board system (BBS) is a programmable communications tool for exchanging information among unattending participants where time and space is not a limit. It is a store-and-forward specialized electronic mail shared by all eligible participants. BBS is the place to share your thoughts (items) about a topic which is planned for a given interest group (subject). An item has a limited life span on the bulletin boards, with life spans varying from board to board depending on whether an item has received a reply.

A BBS is organized into:

- conferences
- messages exchange
- item posting
- notes recording
- files handling
- administration services

A *conference* facilitates group discussions. Members of a group read and enter comments into the conference using items and responses. The computer then stores and organizes the comments and presents them to each member upon request. This process facilitates in-depth discussions among people separated by time and space. In reality, conferences can be called boards, forums, and so forth.

For example, PRODIGY has the following boards:

| | |
|---|---|
| ARTS BB | Discuss movies, books, more |
| BULLETIN BOARD CTR | Find all the boards and topics |
| CAREERS BB | Members discuss jobs & careers |
| CLOSE-UP BB | Members discuss current events |
| COMEDY BB | Trade jokes, discuss comedians |
| COMPUTER BB | Members discuss bits & bytes |
| EDUCATION BB | For all concerned with schools |
| ENVIRONMENT BB | Discuss the "green revolution" |
| FOOD & WINE BB | Discuss cooking & fine dining |
| GAMES BB | Discuss games of all sorts |
| GENEALOGY BB | Members research their roots |
| HEALTH & LIFESTYLES | Ideas, issues, and opinions |
| HOBBIES BB | Members discuss their hobbies |
| HOMELIFE BB | Members discuss home & living |
| MONEY TALK BB | Discuss business & finance |
| MUSIC BB | Discuss records & performances |
| PEN PAL BB | Find a pen pal |
| PETS BB | Discussion for pet lovers |
| RELIGION BB | Members discuss religion |
| SCIENCE BB | Science bulletin board |
| SENIOR BB | Seniors' bulletin board |
| SERVICE CLUBS BB | Community organizations |
| SINGLES BB | Discuss common interests |
| SPORTS BB | Members discuss all sports |
| SPORTS PLAY BB | Discuss participatory sports |
| SUPPORT GROUP BB | Share experiences online |
| TEENS BB | Teens discuss common interest |
| THE CLUB BB | Kids meet & talk with friends |

| | |
|---|---|
| TRAVEL BB | Discuss past and future trips |
| TV BB | Discuss shows & personalities |
| VETERANS BB | Info & support from other vets |

An *item* is entered by a participant in the conference to begin discussion on a topic. It includes the initial text and ensuing responses. Items become a permanent part of the conference proceedings; they are immediately visible to anyone who is a member of the conference. Unless an item is deleted, it may be viewed at any time by any participant.

A *response* is a short reply to an item. Every participant is free to respond, and every response is immediately visible to every one who can access the conference. Through items and responses, the conference makes it easy for anyone to share an idea with the group for comment.

An *index* consists of a number of descriptive category names. Associated with each category is a list of the items in the conference which fit into the category. The index is maintained by the conference organizer.

A *message* is a private communication to another participant. Messages are transmitted instantaneously to the designated recipients. A message can be seen only by the person who has the access to the account where the message was sent. It is like an intra-conference electronic mail.

A *bulletin* is a short announcement which is automatically shown once to every conference participant. Any participant may post a bulletin. Bulletins usually announce important events, but they are useful for drawing attention to a particular item or problem.

A *note* is a reminder to yourself; you are the only person who can see your notes. You tell the conference when to show you each note you leave to yourself.

A *file* is a storage place in the computer for text or programs. The conference lets you create, edit, and share files. It can automatically manage three different types of files for you:

- *Participation File* : The conference creates a participation file on your account when you register in a conference. It contains information about your participation in the conference such as items and responses you have seen.
- *Message File* : The conference creates a message file on your account when you register in your first conference. It contains all the messages sent to you which you have not deleted.
- *Temporary Files* : Temporary files, serving as a storage place for text, are created when you use the conference. Normally the files are destroyed when you exit the conference, but their current contents are saved if something unusual happens (such as a computer crash

or the phone being disconnected), and you are notified the next time you access the conference. These files helps prevent the accidental loss of text you are composing [7].

Each implementation of a BBS has its own syntax of commands for steering the communications (e.g., RESPOND, FORGET, PASS, ITEM, RESPONSE, NEW, and so forth). Also, each BBS has its own rules and guidance of expected communications behavior.

Corporate information —such as news releases, insurance and stock plan updates, training classes and department softball games —often get taped up on bulletin boards, distributed in newsletters, or spread by word of mouth. No one is quite certain if all employees get the message. But no longer. Enter electronic bulletin boards, where news items and company facts are displayed for all employees to see. With the touch of a key, mass company-wide messages are no longer a hit and miss gamble.

> An example of how bulletin boards have made a difference can be seen at the Iowa Public Service Company (IPS) and Iowa Power subsidiaries of Midwest Resources Inc., Iowa's largest utility supplier. Prior to e-mail, news releases were sent out to over 3,000 employees through a combination of newsletters and phone calls. The installation of an electronic mail system in 1987/88 opened up a whole new realm of communication. The bulletin boards are used to transmit all kinds of company information from monthly operating letters to daily newsletters. The electronic newsletter serves another function for IPS. Employees are encouraged to ask questions about the company and its policies, anonymously if they choose, and which are then answered on the daily newsletter.

The bulletin boards have gotten so popular, in fact, that Iowa Public Service Company has had to partially restrict employee's ability to add to it. A lot of people use e-forms for phone messages rather that writing down the message and sticking it in a slot, a person simply types it up and sends it electronically. The recipient only has to look at his or her mail to see their phone messages. The scheduling of rooms, equipment, and pool cars is also easier with BBS. At any time, any employee can look at the schedule menu and see which cars, rooms and audio/visual equipment are available. The greatest benefit of e-mail and BBS is the ease in which correspondence is carried over the system. Rather than playing telephone tag with someone, you can send a message on the system and answer it back and forth until your conversation is finished. There are no more loose ends (Cohen 1992).

## Fax Service

A fax machine is a scanner which translates a document into binary signals, sends them via telecommunications lines, and printing it on the other end. Communications via fax is possible in three ways. The first way is via fax machines. The second way is via PCs with a special board and software, perhaps even with a scanner for capturing graphics and pictures. The third way is via consumer information services. The most promising way is sending faxes like e-mail messages. It offers all the advantages of paperless and programmable communications with few disadvantages.

## EDI-Interorganizational Connectivity

Electronic Data Interexchange (EDI) is nothing more than an electronic form of a paper document. EDI is not a fax or e-mail. It is a CPU-to-CPU telecommunications of legally binding machine readable documents. It is the computerized exchange of standard business documents such as purchase orders, invoices, and medical claims. EDI is most frequently used between organization (inter-organizational), but can also be used internally (intra-organizational). Figure 2-4 illustrates a set of typical enterprise users of EDI users.

EDI is application software (e.g., customer orders) integrated with other application software (e.g., inventory management). EDI requires systems to be linked (i.e. networked). Unlike e-mail, EDI is meant to be read by a machine, not by a human being. To give an

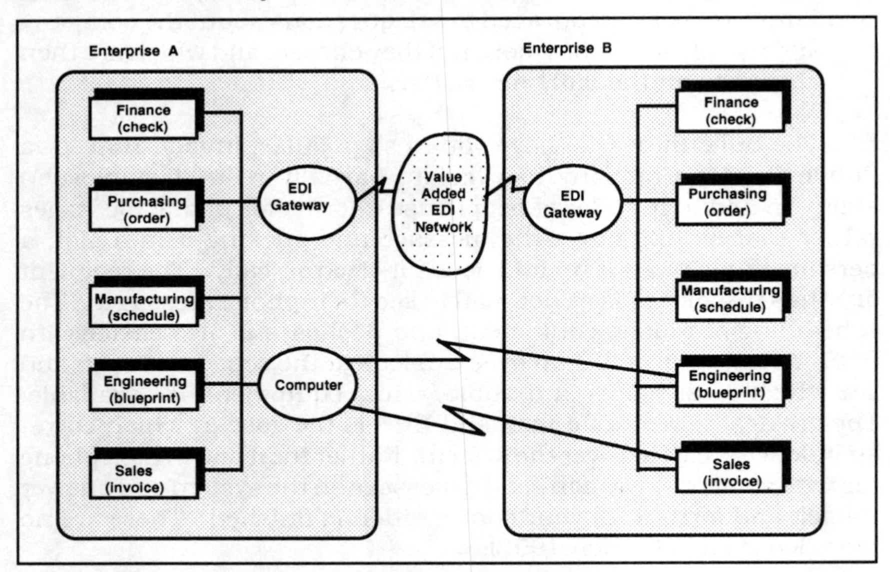


*Figure 2-4: Network-based EDI Application Service*

example, EDI transmission can go straight from a buyer's purchase order application into the seller's order entry application, without human intervention.

EDI speeds up the process by which companies conduct their business. A purchase decision is made and communicated to the supplier at once, instead of waiting for a purchase order to make its way through the typing pool and then the postal system. EDI takes on strategic importance by enhancing relationships between companies and their suppliers on one side and their customers on the other side. EDI helps companies to reduce their inventories and associated carrying costs. In a number of industries, such as the grocery industry, where margins are slim, the cost/benefit that EDI yields can significantly increase profitability.

Some concern can be found in respect to the security of EDI. People worry about unauthorized access to proprietary data, fraudulent collusion between employees of the buyer and seller, and the possibility of mischievous messages being transmitted. Because, however, concern has been so well articulated most EDI system vendors address that issue in a satisfactory way, through callback techniques, multiple password levels, encryption if necessary, authentication if desirable.

The U.S. Commerce Department and the U.S. Customs Service support the evolving international EDI standard called EDIFACT (Electronic Data for Administration, Commerce and Transport). The International Air Transport Association moved its automated systems to EDIFACT as well. The dominant U.S. standard, the American National Standards Institute X12, is used in Canada and has been adopted by the Australians. X12 is significantly different from EDIFACT, the European standard, but the ANSI X12 committee has agreed toward the convergence of the two standards with the goal of a single, international EDI standard.

EDI has a great future as a time-driven, integrative component of the organizational electronic infrastructure. New applications are continually being designed for such industries as aerospace and the ready-to-wear apparel industry where electronic graphics are important in the purchasing process, as blueprints or drawings are required to support the purchase decision.

A new EDI database is evolving in the enterprise. All EDI transactions are stored and retrieved to generate management status reports, sometimes down to the minute. Integrating EDI with other enterprise-wide system makes the information management complex a very time-sensitive and intelligent system. It allows managers to examine what is happening within the company. In some industries, this is very important down to the minute (financial businesses) or at least down to the day. For example, in apparel or retail wear, many products are fads and it is important to know precisely when the fad starts to fade [6].

EDI is a strategic initiative at Texaco — one of just 13 identified by this international petroleum company of some 40,000 employees. James Kinnear, president and chief executive officer said: "EDI will increasingly become a prerequisite for doing business, driven by the market place and sparked by global automation. I encourage you to assess the use of EDI as a resource in your business strategy." Texaco recognizes EDI as a business philosophy which enables the re-engineering of business processes to adapt to changing business paradigms. Texaco's earliest implementations were intra-industry with other petroleum companies in the 1970s. These applications were targeted toward improving the timeliness and accuracy of information. They generate over 80,000 EDI transactions each month with approximately 150 partners. In the mid 1980s Texaco's industrial and retail customers began requesting that the company conduct business with them through EDI. This was the company's entry into standards-based EDI. Texaco exchanges 12,000 transactions a month with 70 trading partners. The company uses five Value Added Networks to transmit EDI documents (Zimmerman 1992).

The proliferation of paperwork in today's business world is problem that just keeps growing and growing. The more file cabinets a company buys, the faster they fill up. EDI is a solution to the problem. It was a problem for Haworth Inc. of Holland, Michigan, a producer of furniture. Prior to EDI, Haworth primarily used the mail to send out orders to suppliers. The company has 350 to 400 total suppliers, so foot-tall stacks of paperwork had to be separated and mailed. It could take suppliers four to five days to receive their mail. By the time they received one week's order, the next week's order was already going out. Since some suppliers wanted to get their orders faster by fax, material planners at Haworth would have to stand at the fax machine and fax a pile of orders. With EDI, suppliers now receive their orders immediately as they are sent, and the information is already on their computers, eliminating the need for any paper transfer. Now orders and order changes are always up to date.

## Global Video Teleconferencing Networks

Any live, point-to-point, electronically aided conversation is a teleconference. Technically, two people talking over an intercom are engaged in a *teleconference.* The word does not become particularly meaningful, however, until you have several people sitting around a speakerphone, talking with one or more people who are hovering about similar setup in one or more geographically discrete areas. At that point, you have a special kind of teleconference called *audioconference.* Add the capability to send some form of graphical display from one of those sites to the other and you have moved into the realm of *audiographic* teleconferencing. Finally if you beam a live, full-motion television picture from point "A" to however many destinations or "receive sites" you wish, and add two-way telephone communications, you have a video conference.

The ultimate video-conferencing "experience" say proponents of the medium is "x" number of groups in "x" different locations seeing and hearing one another live, via satellite or direct-line hookup. It is a synchronous communications which far surpasses electronic bulletin boards, electronic mail and what in the past has been called "computer conferencing." Though several organizations have experimented successfully with two-way audio and video for tele-training and for meetings, "two-way video, two-way audio" is the video-conferencing form of preference for most business applications today.

Video conferencing has been around since the 1960s. And though colleges have long been interested in the potential of the technology, it is the growth in corporate video-conferencing capacity, particularly downlink capacity, that has led to the current boom. The year 1984 was probably the turning point for the coming age of teleconferencing. In that year, American companies purchased approximately 100 teleconferencing rooms, about three times the number of installations made in any previous year at about $250,000 a room. Another 100 rooms were installed in 1985, and for 1986 that figure was considerably higher. In 1984, about 15 corporations were video-conferencing in a serious way. By the end of 1985, that number had doubled (Zemke 1986).

> Meeting tables at Pharmacia-Upjohn can now stretch across oceans. A room with video-conferencing equipment went into service in 1990 in Kalamazoo-Portage, Michigan. Similarly equipped rooms have been put into operations at Upjohn sites in Brussels, Belgium and Crawley, United Kingdom. Video-conferencing involves the use of television equipment to conduct meetings between people at different locations. Television signals are transmitted from location to location using fiber-optic cable and/or satellite transmission systems. "Worldwide Pharmaceutical Marketing will be a primary user of the technology," said Glenn A. Miller, Worldwide Video Communications manager. "We expect our business groups and support staff will use the facilities extensively to meet both domestic and worldwide needs." Reviews of advertising pieces, discussions of marketing plans and exchanges of competitive information number among the video conferencing uses envisioned in marketing. Objectives of Pharmacia-Upjohn global video conferencing system are:

- Improve home office—field communication
- Simultaneous, nationwide delivery of information
- Reduce travel time, expense
- Speed decision-making
- Enhance training, education effort
- Allow reception of other programming
- Provide new audio conferencing capability
- Provide worldwide video communication capability.

Among the planned applications the following ones are the most critical:

- Internal news, information, announcements
- Management meetings
- Briefings
- Conferences
- Focus groups
- Presentations
- Interviews
- National sales conferences
- Legal negotiations
- Training
- Lectures
- Seminars
- Symposiums
- Communication with medical and scientific community
- Communication with customers.

Among the inbound applications the following ones are expected:

- Remote seminars, presentations
- Continuing education courses
- Hospital, Science Network programs
- Commercial programming (CNN, major network)
- Events at the Western Michigan University, Henrietta Complex

In 1993 cost per hour of video teleconferencing was:

- $ 1450 - Continental U.S.
- 1450 - Continental U.S + Canada (major cities)
- 2225 - Continental U.S + Mexico
- 2325 - Continental U.S + Puerto Rico
- 4675 - Europe
- 6100 - Asia

• 14250 - Japan

The worldwide impact of Pharmacia-Upjohn video teleconferencing in 4th Quarter - 1991 was:

| | |
|---|---|
| • Conferences | 100 |
| • Actual Cost | $126,800 |
| • Displaced Travel, People Expense | 568,100 |
| • Savings | $441,300 |
| • Travel Days Eliminated | 271 |

The architecture of Pharmacia-Upjohn global video teleconferencing network is shown on Figure 2-5.

Upjohn can use the private satellite network to transmit programs from one of their studio locations in Kalamazoo to company employees at 17 U.S. field sites, and when needed, to sites around the world. The receivers enable users of the system to more efficiently and effectively reach employees with information on strategic plans, product launches, training and company news. One of the advantages of this system is the ability of Upjohn to respond quickly and comprehensively to changes in the business environment and to opportunities in the field. The Halcion incident, in 1992, was just such an example, when waves of negative press regarding Pharmacia-Upjohn's *Halcion* sleep-aid drug, reinforced the importance of communicating quickly with employees. Pharmacia-Upjohn was able to defend the drug and reject the unjustified accusations.

VideoTelecom Corporation (VTC) of Austin, Texas, unveiled in 1991 its new product for multimedia conferencing called MediaMax. MediaMax incorporates the best capabilities and features of audio, motion video, graphics video and computer conferencing, adding to it what VideoTelecom calls "time conferencing," which is perhaps the most revolutionary aspect of this new product.

In the past, users who wanted all these capabilities had to custom-configure a video codec, audio and video-conferencing equipment, computers and a variety of software, and even then they did not have the time-conferencing feature. VTC has integrated all these capabilities and more into a system that, from outward appearances, looks like a regular video-conferencing roll about—i.e., an integrated system package.

To understand what the MediaMax system does, and to see how

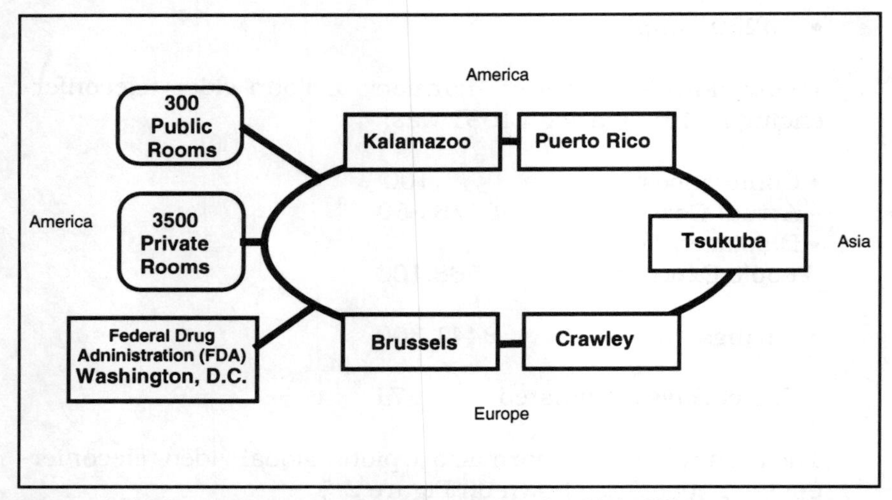


*Figure 2-5: The Architecture of Pharmacia-Upjohn Global Video Teleconferencing*

*Figure 2-5a: A World Wide Videoconferencing Facility at Pharmacia-Upjohn United States headquarters in Kalamazoo, Michigan. Glenn Miller, shown in photo, is a pioneer of global videoconferencing (Courtesy of Pharmacia-Upjohn, photo by Keith Mumma).*

it could change the way business is done, we will use a hypothetical applications example.

> Company X, with headquarters in Seattle, is about to put its new Ultra-Widget into production at its plant in Singapore when Marketing discovers a problem. Using VTC's MediaMax, the Seattle head office immediately holds a multipoint conference

with Marketing in Chicago and Engineering in New York. The video-conferencing component and multipoint control unit together permit people at all their locations to see and hear those at the other locations. In Chicago, Marketing pulls a spreadsheet using the MediaMax's computer and by pushing a stylus on a MediaMax graphics tablet, makes it visible on monitors at all the locations. Participants in all three cities can make changes to the spreadsheet, and then graphs can be generated showing the product changes that are needed.

Marketing then takes a paper drawing of the unmodified product and shows the other locations what changes are needed. Using their graphics tablets, all three locations annotate and revise the drawing until everything is perfect. The engineers load their CAD (computer-aided design) file, make alternations and show it to the other locations.

While this is being discussed, engineers generate a three-dimensional CAD animation showing the effect of the changes, and display it to the other locations. The North American locations have solved the problem, but can the factory on the other side of the world be brought to speed in time to execute the changes?

Enter time-conferencing, VTC's introduction of a form of multimedia messaging/mail. Although it is the middle of the night in Singapore, the Seattle office dials into the Singapore factory's MediaMax to leave a voice and motion-video message about the necessary changes, complete with copies of the spreadsheets and graph, the annotated product drawing, and the animated CAD drawings. Ten hours later, the factory views-hears-reads-processes the message. Using the CAD file, the factory implements the product changes before the staff in the United States is out of bed in the morning (Halhead 1991).

**Business Television Networks**

In the 1980s, the satellites emergence as a channel for video delivery made corporate TV networks possible and influenced the beginning of a new kind of teleconferencing known as business television (BTV). BTV uses telecommunications satellites to broadcast informational television programs from a single location—frequently a corporation's headquarters—to a selected audience, situated at multiple, geographically diffused locations. The programs are usually live, providing the audience the chance to phone in inquires.

Between 1982 and 1988, more than 50 of the largest and most successful corporations in the United States, representing a wide range of industries, installed BTV to enhance their intra-communications ( Irwin 1989). In 1992, there were more than 300 corporations owning BTVs.

BTV is a human communications medium allowing people to see, hear and respond to each other. BTV offers the following benefits:

- Immediacy of delivering time-critical messages and getting instant feedback
- Simultaneous delivery of a message to multiple locations
- Feedback from the field
- Greater access to experts
- Efficiency of training
- Travel reduction
- Motivation of employees by "visiting" them via air
- Increased productivity
- Improved decision-making

The classic applications of BTV are:

- Employee news/information
- Product announcement
- Press conferences
- Special events
- Cross-networking (sharing programs with other companies).

The external programming of BTV services is available in some industries. For example, the following industries can assign to external networking services:

- Health Care: Hospital Satellite Network (HSN), about 1000 hospitals use this service for continuing education and entertainment. The American Rehabilitation Network (AREN) provides an annual series of medical seminars.
- Education: the following networks are the most popular:

  - National Technological University (graduate programs in engineering)
  - University of Phoenix (programs in business)
  - Chico State University (programs in engineering)
  - Institute of Electrical and Electronic Engineering

- Automotive: the most effective network is The Automotive Satellite Television Network (ASTN) providing programs for about 5000 dealers. The programs are designed for sales and service and parts personnel.
- Finance:

-A New York-based Satellite Conference Network provides programs for bankers and accountants
-The Institutional Research Network (IRN) airs programs for money managers.

Corporate networks are not limited by industry. As of February 1989, the industries with the highest number of active networks were those in the finance/banking, automotive, manufacturing, retail and high-tech industries. These are followed by insurance companies, health care and government agencies ( Irwin 1989).

## Global Communications Networks

The **Internet** is *a constellation* of thousands of computer networks used by millions of people located in 100 plus countries. It is being used by academicians, researchers, government workers, military, and business people. The Internet is the best tool of the Electronic Global Village. It connects people from all the continents, 24 hours per day. Internet started in 1969 as a single network called ARPANET (DoD's Advanced Research Project Agency). ARPANET was a packet-switched, store-and-forward, host-to-host digital network of computers. Its goal was to increase research productivity and communications of DoD researchers by interconnecting the super computers involved in defense projects. One of the main objectives was to carry command and control information during a nuclear war. On the other hand, very soon, the computer specialists could develop the network concept and pilot solutions without any special pressure, just for the sake of "networking."

The background for this network was provided by Paul Baran from the Rand Corporation and Leonard Kleinock from MIT who published the idea of computer networking in the early 1960s. Since 1969, the ARPANET has been financed by DoD and implemented by the Bolt Beranek and Newman company (BBN). In October of 1972 the first public demonstration of the ARPANET took place in Washington, DC. In early 1973, the first satellite link of the ARPANET was established between California and Hawaii. The same year a telephone link between distance hosts was also established. In 1975, the network was fully operational based on TCP/IP protocol developed by Richard Kahn (DARPA) and Vinton Cerf (Stanford University).

It is worthy to note that in 1972 a computer value added network —INFOSTRADA was established in Poland. The first three nodes (Singer 10) were installed in Katowice, Warsaw, Gdansk. The network was initiated by Andrew Targowski and later disassembled,

| | |
|---|---|
| 1981 | 213 |
| 1980 | 313,000 |
| 1991 | 376,000 |
| 1992 | 727,000 |

*Table 2-1 The Growth of Computers in the Internet*

| | |
|---|---|
| 1993 | 1,090,500 |
| 1994 | 1,635,750 |
| 1995 | 2,453,625 |
| 1996 | 3,680,437 |
| 1997 | 5,520,656 |
| 1998 | 8,280,984 |
| 1999 | 12,421,476 |
| 2000 | 18,632,214 |

*Table 2-2 The Estimated Growth of Computers in the Internet*

since the Communistic Party did not allow the uncontrollable horizontal transfer of information and communications in the totalitarian regime.

In the early 1980s, the ARPANET was reorganized into two networks; the military went on the MILNET (classified) and unclassified applications stayed on the ARPANET. The connection between these networks was provided by the DARPA Internet. Later the name was shortened to the Internet. In 1985/86, the U.S. National Science Foundation (NSF) had begun to finance the development of the Internet under the name of SNFNET. This inter-network had the goal to interconnect a majority of the research and academic institutions in the United States, not only the best ones such as MIT, Stanford, Berkley, UCLA, Carnegie-Mellon, and so forth. A father of this idea was an Irishman, (not an American) Dennis Jennigs (University College in Dublin), who was working as an exchange scholar at the NSF. Ever since, the NSF has viewed NSFNET—Internet as a backbone of a national wide area network serving different communities of users. In the late 1980s, the Internet expanded into overseas networks, creeping over the globe and reaching all continents (including Antarctica). In 1990, the name ARPANET was dropped altogether.

The growth of the Internet is spectacular. Table 2-1 illustrates the number of attached computers to the network (Lottor 1992). The Internet's administration estimates the yearly growth rate at 120 to 180 percent. Based upon a 150% growth rate, the projected number of host computers connected to the Internet in the coming is illustrated in Table 2-2.

Of course, this exponential growth will sooner or later slow down. Perhaps, instead of host computers, LANs and MANs will be attaching. Imaginably, all the computers in the world will be attached to the Internet. In 1991, about 4500 computer networks were attached to the Internet (Lynch and Rose 1993). In effect, the Internet is the Network of All Networks.

The Internet's major backbone (The U.S. backbone—NSFN) transfers information over digital lines T1 (1.5 Mbps) to D3 (45 Mbps). In September 1991, traffic through the American backbone was a terabyte or a trillion bytes per month, with traffic growing 25 percent per month (Lynch and Rose 1993).

The future speed of the National Research and Education Network (NREN) will be from 150 Mbps to 1 Gbps (even more, 2.4 Gbps for the Optical Carrier OC-48 within SONET/ATM technology). The NREN is planned as a backbone of the Internet. The term *backbone* is being replaced by a new term — *a cloud*, which reflects the fuzziness of the Internet architecture.

The Internet is managed by the following bodies:

- The Federal Networking Council (FNC) formed in 1990 is a liaison unit to the President's Science Advisor who is a head of the Office of Science and Technology (OST). The OST sets the national policy effecting the Internet. The FNC coordinates governmental agencies involved in the planning and operations of the Internet.
- The National Science Foundation (NSF) is the main sponsor of research and education involving the Internet. It operates a backbone network—NSFNET. The Internet itself as a whole does not have one central administrative authority.
- The Internet Society is a nonprofit international organization which collects membership fees from the Internet user organizations to finance the network's development and operations. It incorporates individuals, institutions, and organizations to promote the development of the Internet.
- The Internet Architecture Board (IAB) promotes the technological research and development that supports the Internet. IAB specialists are committed to maintain the smooth functioning and cutting-edge development of the network. The IAB is a part of the Internet Society. The IAB made up of dozens people which supervise about 50 subcommittees within two Task Forces:
  - The Internet Research Task Force (IRTF) for the investigation of new issues involving networking technology.
  - The Internet Engineering Task Force (IETF) for the coordination of the technical operation and evolution of the Internet.
- Network Information Centers (NICs) are created for each network. The NSF created NIC(s) for NSFNET and NREN to consult users in the network's operations.

To understand how the advanced applications work, it is necessary to be familiar with the client/server architecture. In networking terminology, a *client* is an application running on the user's computer, taking chances of its capabilities, and a *server* is a program running on outer-computers that are available through the network. The servers specialize in the organization of data and documents to support client queries.

The following Internet applications are available to users (a short description of their syntax is provided more as an illustration than as an user's guide):

- **Electronic mail** is the most popular application. It gives a real feeling of "traveling" over the globe. It is easy to communicate with someone across 10 time zones or in the same campus or building.
- **Usenet newsgroups** is a Bulletin Board System comprised of numerous computer conferences covering various interest groups. It is ideal for browsing or posting issues or questions, such as "How to fix my laser printer?" or "Where should I stay on my first trip to Graceland?"
- **Finding someone** is an application of a "white pages" of a telephone directory. There is no one directory maintained containing all the addresses of the Internet users.
- **Moving files** is possible by using File Transfer Protocol (ftp). It allows you to move files from one computer to another, electronically.

- **Remote login** to other computers connected to the Internet. It may be used to enter many of public services, including library card catalogs and databases or computing on powerful computers. To do so you need to use the **telenet** command and an address of a given host computer. To enter a computer of Rutgers University, you type in the telenet command an info about your account on this computer (id and password). If you do not have the account, then you may try *anonymous*:
- **Accessing archives** allows one to retrieve any of the indexes of files that are available on public servers connected to the Internet. In the early 1990s, archives indexed approximately 1200 public servers containing roughly 2.1 million files. In those archives can be found scientific and government research (text files) along with software and other data. To find files, you search for filenames containing a certain search string or suggest files whose description contains a certain word. Archie tells you *where* a file is located based upon the provided filename and search criteria. Once you did find the right file at an assumed location, then you can move this file to your host with anonymous FTP. Enter **archie** server which will do searching. To do so, you should use the **telenet** command first to the closest archie server [8] [9]: where *eudora* is a name of an e-mail package for Macintosh which you are looking for at the McGill University in Canada.
- **Searching databases indexes** to find the right document with a command **WAIS** (Wide Area Information Servers). WAIS allows you to find an item in a given library located in a given wide area. WAIS

allows you to search on the basis *what* is in a given database. It retrieves the indexed databases of resources in a given wide-area. The user provides a question in English and receives responses (in text, sound, video, graphics) to this question. At the beginning the user can ask "What library do I look for 'stomach ulcers'?" Later the user can ask "Results of Zantac." The user can mark all documents that interest him/her and WAIS will search for the next documents complying with the question. It is called "relevance feedback" or "knowbot." It allows the user to predefine a query and check for new items every. In the early 1990s there were 400 WAIS libraries on the Internet.

- **Navigating the Internet resources with Gopher.** The Internet Gopher (developed by the University of Minnesota) is a client application which facilitates the navigation through the resources of the network. Gopher provides the user-friendly transfer to other computers, supporting the browsing and searching for documents, logins (teleneting), transferring files, querying databases (archie and WAIS), changing directories, and so forth. It is a menu-driven navigation all over the world.

- **Navigating the Internet resources with the Web.** The World Wide Web (WWW), (W3) is the newest information service of the Internet. Its development started in 1989 by the CERN (Centre European pour la Recherche Nucleaire) in Geneva, Switzerland. It merges the techniques of hypertext (hierarchical documents through the expansion of processable words that link documents) with information retrieval. Its purpose is to provide the user friendly, global access to the web of available information. In the early 1990s, the CERN web of information is the only developed web.
- **Accessing outernets and outerservices.** The Internet can provide an access to many external networks (Bitnet, Compuserve, America Online, AT&T Mail, MCI Mail, AppleLink, FidoNet, UUCP, hundreds of bulletin boards, and so forth) and information services (Dow Jones News/Retrieval, Lexis and Nexis, Mead Data, University Microfilm, and so forth). Altogether, about 40 million people worldwide can communicate electronically among themselves. In addition to a new market created by the Internet, online information providers get an alternate (and often less expensive) way to deliver information to existing millions of customers around the globe. Some universities and research centers now require potential vendors to have access to Internet before they will even consider an information service provider. Why? A direct connection through the Internet can dramatically reduce access costs and improve performance. The Internet is a nonprofit entity with users paying only for access to it, not for service use. Access charges are often several dollars per hour less on the Internet than on commercial services, which can mean a savings of hundreds of dollars each

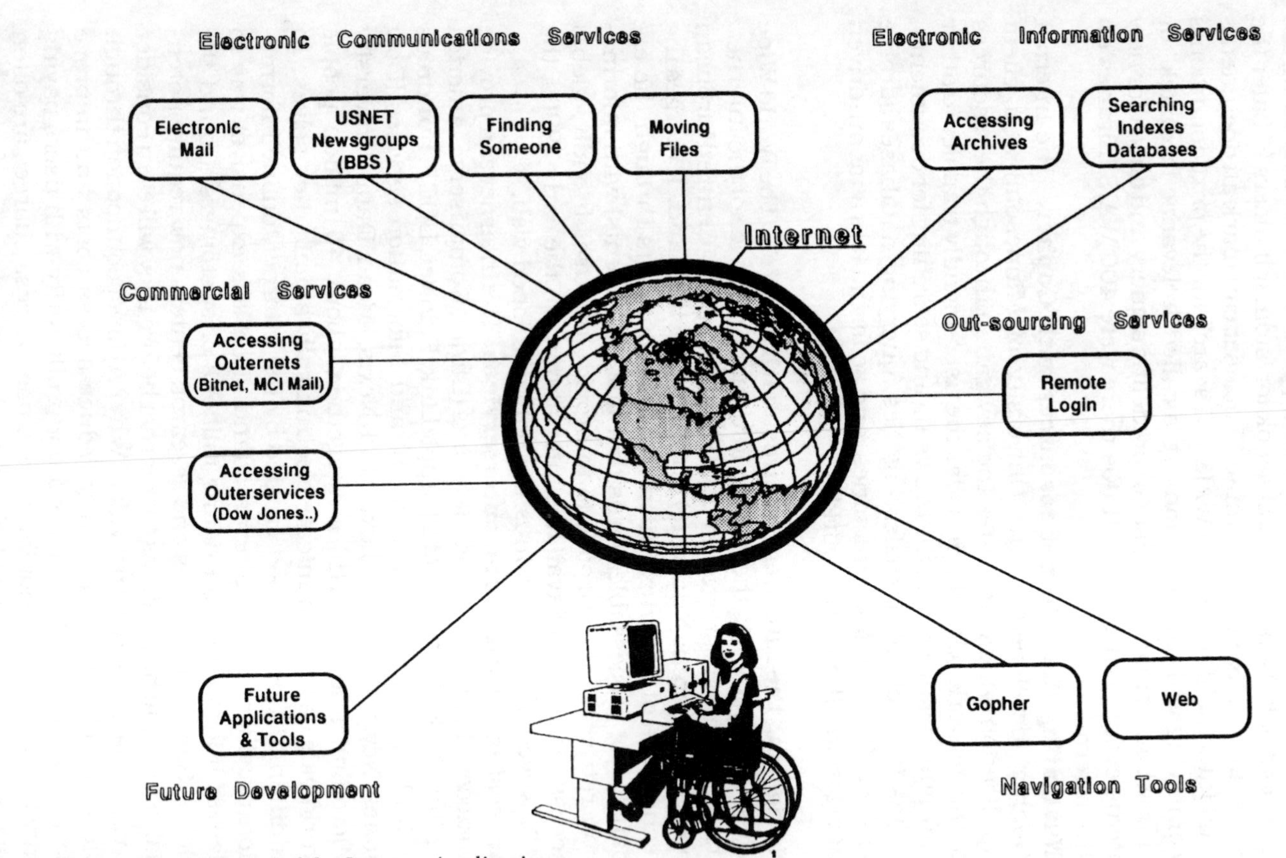


*Figure 2-6: The Architecture of the Internet Applications*

month. One university has recovered the entire cost of its monthly Internet connection fee through reductions in phone costs (Dan 1992).

The architecture of the Internet applications is shown on Figure 2-6.

Today, however, the Internet has grown to encompass a broad variety of services for users in every field. Without slick advertising, software giveaways, or online contests, the Internet has evolved to become the world's largest electronic information exchange.

In practice, the Internet becomes a laboratory for the development of the information infrastructure of the Electronic Global Village in practice. The Internet brings a new level of power to ordinary people using the most basic computers. It makes the Electronic Global Citizen a member of the Global Human Family.

## Local Communications Networks

The Cleveland **Free-Net** (CFN) is the world's first and largest community-wide network . It has 350 applications and tens of interest groups. The CFN, located at the Case Western University in Cleveland, Ohio, [was originated by Thomas Grundner in July, 1986, operating on more than a dozen Unix-based computers. The software is written in object-oriented mode in the C++ language by the Americast Corporation. Based on CFN, other Freenets were implemented in [10]:

- Youngstown Free-net, Youngstown, Ohio
  (Internet: yfn.ysu.edu, Login: visitor)
- Tri-State On-line, Cincinnati, Ohio (tso.uc.edu)
- Heartland Free-Net, Peoria, Illinois
  (Internet: heartland.bradley.edu, Login: bbguest)
- Lorain County Free-Net, Elyria, Ohio
  (Internet: 132.162.32.99, Login:visitor)
- Big Sky Telegraph, Dillion, Montana
  (Internet: 192.231.192.1, Login: bbs)
- Buffalo Free-net, Buffalo, New York
  (Internet: freenet.buffalo.edu, Login: freeport)
- Columbia Online Information Network (COIN), Columbia, Missouri (Internet: bigcat.missouri.edu, Login: guest)
- Denver Free-net, Denver, Colorado
  (Internet: freenet.hsc.colorado.edu, Login: guest)
- National Capitol Free-net, Ottawa, Canada
  (Internet: freenet.carleston.ca, Login: guest)
- Tallahassee Free-net, Tallahassee, Florida
  (Internet: freenet.fsu.edu, Login: visitor)
- Victoria Free-net

• Medinal County Free-net
• Wellington Citynet, Wellington, New Zealand
(Internet: kosmos.wcc.govt.nz)

Based upon the need to regulate Free-nets ,the National Public Telecommunication Network Agency (NPTN) was created to regulate and govern free-nets, and Thomas Grunder was installed as it's first president.

Free-Nets are in operation for 24 hours and are designed for home, school, and business users. There is no fee for the use of Free-Nets. The initial version of the CFN attracted over 7,000 registered users and averaged 500 to 600 calls a day on 10 incoming phone lines. In the mid 1990s, the CFN averages over 5,000 logins a day on 64 telephone lines.

The CFN is implemented on more than a dozen unix machines interconnected (clustered) via Ethernet. The system has 2.3 gigabytes available storage space and has the capacity for 360 simultaneous users.

The CFN displays a list of information categories in the main menu:

1. The Administration Building
2. The Post Office
3. Public Square

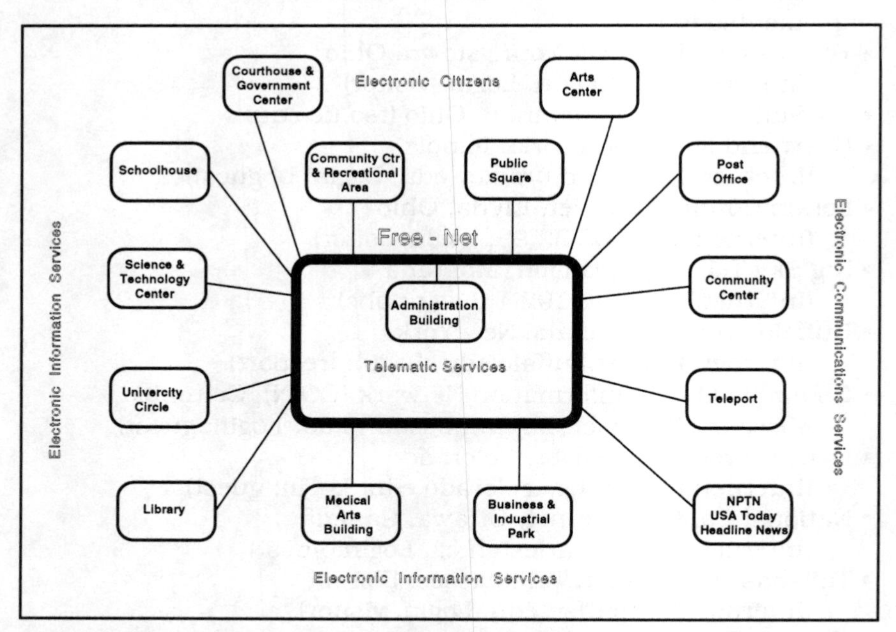


*Figure 2-7: The Functional Architecture of Cleveland Free-Net*

4. The Courthouse and Government Center
5. The Arts Building
6. The Science and Technology Center
7. The Medical Arts Building
8. The Schoolhouse (Academy One)
9. The Community Center and Recreation Area
10. The Business and Industrial Park
11. The Library
12. University Circle
13. The Teleport
14. The Communications Center
15. USA Today

In general, most free-nets are menu driven systems allowing users to access different topic areas by entering their desired choice.

The Figure 2-7 illustrates the functional architecture of the Cleveland Free-Net.

The application architecture of CFN is explained on the following examples:

### The Well

The WELL (The Whole Earth' Electronic Link) network provides local information services for the San Francisco area. It was started in 1985 as a "people's network" (in terms of the 1960s). The WELL is recognized as one of the best conferencing systems, providing an electronic forum for about 100 interest groups (mostly hobbyist, social, technical, and business groups, very vibrant in debating, but not any more arcane than those found on other networks. Each conference has an moderator and the general code of ethics. Other services include: electronic mail, electronic storage, a local monthly online magazine, and access to USENET and UUCP worldwide networks. It can be accessed by direct dial, by TYMNET or from CompuServe services (by command GO PARTI).

## Networking Telecommunications Services

### Teleport, Inc

A new era of telecommunications began in the 1980s. After the sweeping changes of the 80s in the long-distance business (deregulation and a breakup of AT&T), competition is now headed for local phone service. The regional monopolies of New York Telephone, Pacific Bell, Illinois Bell Michigan Bell, and other local phone companies are beginning to be undermined by such technologies as microwaves, fiber optics, and wireless phone systems. Long-distance carriers, cable TV, and new players such as Teleport Communications are trying to grab some of the local communications market.

The biggest telecommunications issue of the 1990s is the competition in local phone service. Competition in local telephone services could have an even greater impact than competition in long distance has had. The local phone industry comprises some 1,300 companies that reach almost every U.S. home and business, employing 650,000 workers, and generating about $100 billion a year in revenue. This is three times as great as all the long-distance companies combined (Lewyn and Coy 1991).

Local alternate-access services are growing rapidly, led by Teleport in New York City and Metropolitan Fiber Systems Inc. in Oak Brook Terrace, Ill. These companies string cables from corporate customers directly to long-distance carriers, bypassing the local phone network. This deprives the local phone companies of the access fees they usually receive for completing long-distance calls. In 1990, New York Telephone collected 2.3 billion is such fees.

Before the newcomers can seriously challenge the monopolies, they need to firm up their foundations. The cost of building alternative phone networks are huge. As a result Teleport is the only bypasser now making a profit.

Teleports are the new gateways for all regional, interstate, and international communications services. Just as seaports and airports are the transfer points for goods and services for a region, so are teleports becoming the transfer points for information. And increasingly, information is the commodity upon which modern economies are being built.

The World Teleport Association (WTA) defines a teleport as "an access facility to a satellite or other long-haul telecommunications medium incorporating a distribution network serving the greater regional community and associated with, including or within a comprehensive related real estate or other economic development."

Teleports serve corporate customers, news organizations and broadcasters as well as long-distance carriers. These companies may use all or only some of the teleport's facilities and services, including its satellite transmission capabilities, its regional distribution network or its real estate offerings. The Staten Island (NY) Teleport, for example, links the private networks of institutional customers like stock exchanges, money-center banks, and large brokerages giving them a cheaper way to reach long-distance carriers than by going through the local company.

Just as a region's competitive position is determined by the quality of its airports and seaports, a region's competitive position in the future will be determined by the quality of its teleport and information infrastructure.

The Teleport Communications Group (TCG) is the nation's largest competitive local telecommunications provider for America's business community. Its network operates in seven major cities collectively reaching over 450 miles.

The services are provided through fiber optic networks for the transmission of voice, video and computer data—monitored 24 hours a day, seven days a week. The following lines/services are available:

- Teleport DS3: (45 Mbps), a cost effective option for high capacity requirements
- Teleport DS2 (6.312 Mbps), a high speed interconnection of Local Area Network
- Teleport DS1 (1.544 Mbps), a standard digital interface
- Teleport DS1E: (2.048 Mbps), the international standard transmission rate
- Teleport DS: The exact bandwidth users need in multiples ranging from 2.4 to 64 Kbps.

- TC Systems, Inc. provides facilities management services under two product lines. Teleport Centrex (sm) delivers telephone service with analog, digital or Integrated Services Digital Network (ISDN) functionality directly to the customer's desktop. TeleXpres Network Services (sm) delivers local and regional calling diversity for PBX users.

Teleport Communications-New York, the founding member of Teleport Communications Group, was formed in 1983 and two years later began providing private line and other dedicated transmission services. TCG was formed in 1987 as the umbrella organization to manage its expansion into new markets, cities and services.

The following cities were the first targets of TCG's expansion:

- Teleport Communications-New York: Began operation in 1985; the network now reaches over 250 miles, serving more than 890 customer locations in more than 350 buildings throughout the metropolitan N.Y./N.J. area.
- Teleport Communications-Boston: Formed in 1988, the TCB fiber optic network totals 35 miles and runs throughout the Financial District, Back bay, Cambridge and Waltham areas.
- Teleport Communications-San Francisco: Formed in 1989, TCSF consist of a 12-mile fiber optic network in the city's financial district to serve the international banking, trade and financial services industries.
- Teleport Communications-Los Angeles: Formed in 1990, TCLA consists of a 10-mile fiber optic system downtown, with an extension of the backbone network proposed to Culver City through the

mid-Wilshire, Miracle Mile and Century City areas.

- Teleport Communications-Houston: Another 1990 creation, TCH began development of a seven-mile system in the heart of the city's Financial District and by year-end 1991 added an additional 80 miles connecting the Brookhollow, Greenspoint, Energy Corridor and Westchase areas.
- Teleport Communications-Chicago: began in 1990, the 10-mile network, initially serving the downtown "Loop," TCC is TCG's first SONET-capable (Synchronous Optical Network) operation. TC-Chicago is the first non-telco local carrier in Illinois to receive permission to operate by the Illinois Commerce Commission.
- Teleport Communications-Dallas: Operations began in early 1991 with a fiber optic network centered in downtown Dallas. Since that time, the network has been expanded throughout the greater metropolitan Dallas area.

Teleport Communications-New York or the Teleport, located on Staten Island (the only location available in New York City with the open sky required for satellite communications), is a venture of the Port Authority of New York and New Jersey, Merrill Lynch, and the City of New York. The components of the Teleport (NY) are:

- A communication-enhanced office park for lease by communications-intensive customers. Among them there are such as Merrill Lynch (transmits information to 29 countries from the Data Center located on 200,000 square foot of space), Nomura Securities International, and others.
- A satellite communications center consisting of an 11-acre radio-frequency shielded satellite communications center which can communicate with any domestic or Atlantic Intelsat Satellite. Satellite transmission is managed by IDB Communications—the leading provider of satellite transmission services for radio, television and data/voice communications in the United States. There are 12 earth stations operating at the site, with IDB's earth stations serving top corporations, major broadcast networks and many others.
- A regional fiber optic network which spans Manhattan's business district, and extends into Brooklyn, Queens, Staten Island, and to Jersey City, Newark, North Brunswick and Princeton in New Jersey.

Teleport has chosen to compete on the basis of service and flexibility rather than on price. In general, it has pegged its price close to those of New York Telephone— sometimes higher. But that could change. Teleport has shown a striking ability to use the state-of-the-art equipment to keep costs down, presumably giving it leeway to cut prices. While Nynex and other regional Bell companies employ an average of 42 workers for every 10,000 telephone lines, Teleport employs fewer than two (Andrews 1991).

## U.S. Signal

U.S. Signal Inc., is a Grand Rapids based company which in 1992 completed a 100-mile fiber optic ring in the greater Grand Rapids (Michigan) area. The fiber network delivers high technology voice, data, and video services to area organizations.

State-of-the-art technology allows U.S. Signal to provide the most advanced two-way interactive video and audio networking capability in the country. Broadcast quality multimedia signals are delivered directly to a company's conference room or site, providing the customer with the ability to communicate with:

- Voice: providing by-passing linkage to ATT, MCI, Sprint and others
- Video:
  - Other local organizations (local video conferencing)
  - Area educational facilities for distance learning
  - Satellite teleconferencing services (conferences and seminars)
- Data:
  - LANs local interconnection
  - CAD/CAM connection
  - Digital radiology and telemedicine (medical networking)
  - Host-to-Host connectivity
  - Multiple terminals
  - Computer resource sharing

These services are characterized by:

# fiber optic high-speed transmission of multi-media
# linkage of business and education
# linkage of high schools and colleges
# broadcast quality
# integration of the metropolitan area with satellite cities and towns
# reliability of service (alternate path)
# up and down link-nationwide and worldwide access

The main source of U.S. Signal's income is in voice local-to-long distance connectivity. A customer must pay $3000 per month for long distance calls to expect savings from U.S. Signals services (fee

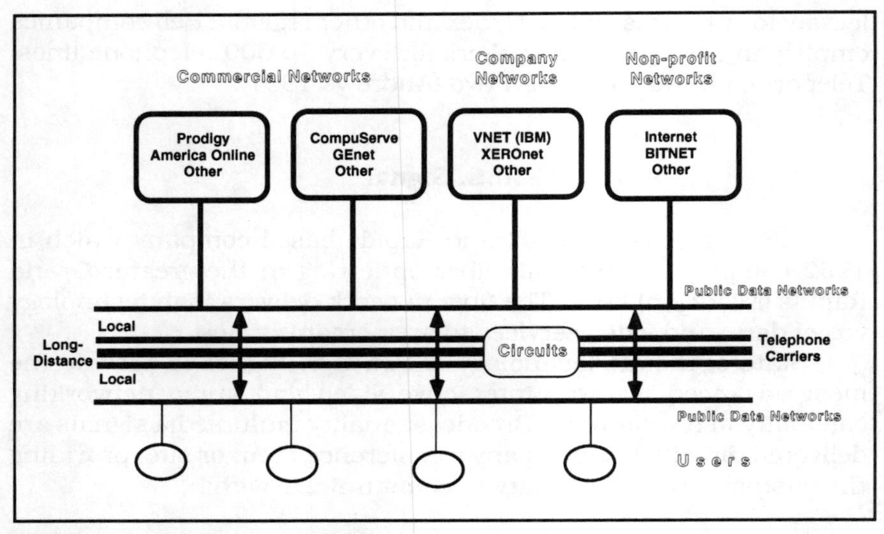


*Figure 2-8: Public Data Networks*

$500 per month). Usually, such a customer will make about $200 to $2000 of savings. The cost of setting a fiber optic ring up varies from a city to city, depending on the downtown construction. It cost the city of Grand Rapids $20,000 a mile to lay their fiber optic line. The total cost to wire a city, varies from 2 to 10 million dollars.

U.S. Signal is a company specializing in servicing mid-size cities, however, it also serves such large cities as Detroit. U.S. Signal and Teleport companies are examples of the future competition in local telecommunications services. It is expected that the market for such services can reach $10 billion a year, that is 10% of the total local telecommunications market in the United States.

## Public Data Networks

A public data network is an information infrastructure that is publicly accessible for a fee from remote locations. In the United States they resale circuits from the telephone companies to users of commercial, company, and non-profit networks. In Europe and some Asian countries, public networks are owned by the governmental agencies (Poste, Telephone, et Telegraph—PTTs).

Since the public networks provide a variety of telecommunications services (transmission, computing power, storage, and software rental, private networks management, trouble shooting, and so forth) they are also called value added networks (VANs).

The architecture of the public data networks interaction with private networks is shown on Figure 2-8. Among the most popular public data networks are in (Quarterman 1990):

- United States: Telenet, TYMNET, Accunet
- Canada: Datapac
- Europe: EURONET
- Australia: AUSTPAC
- New Zealand: PACNET
- Japan: DDX-P, VENUS-P, NIS/TYMNET
- Malaysia: MAYPAC, MAYCIS
- India: VIKRAM
- Latin America: TELEPAC, Morelos, RACSAPAC
- Middle East: several VANs

All these networks are interconnected globally, providing the second layer of the information infrastructure of the Electronic Global Village. The first layer is provided by the telecommunication utility companies.

## Cable and Broadcast TV Information Services

In the early 1990s, the major cable operators and telephone companies are competing and collaborating to bring the communicopia to the residential neighborhood, while the Clinton Administration is scrambling to see how the government can join in the development. Driving the explosive merger of video, telephones, and computers is the following technological advances:

- The ability to translate all audio and video communications into digital information.
- New methods of storing this digitized data and compressing them so they can travel through existing phone and cable lines.
- Fiber-optic wiring that provides a virtually limitless transmission pipeline.
- New switching techniques and other breakthroughs that make it possible to bring all this to neighborhoods without necessarily rewiring every home.

Suddenly ,the brave new world of video phones and smart TVs that futurists have been predicting is not decades , years, or even months away. In 1994, Hughes Communications offered DirTv, a satellite system that delivers 150 channels of television through a $700 rooftop dish the size of a large pizza pie. At about the same time, Tele-Communications, Inc (TCI), the world biggest cable-TV operator, began marketing of a new cable decoder that can deliver 540 channels. It also started programming services, based on this decoder, to 100 cities. Time Warner is running for years a 150-channel system in Queens, New York, and in 1994 launched an interactive service that will provide video and information on demand to 4,000 subscribers in Orlando, Florida.

When the information highways come to town, channels and nightly schedules will begin to fade away and could eventually disappear. In this post-channel world, more and more of what one wants to see will be delivered on demand by a local supplier (cable system, telephone company or a joint venture) from giant computer discs called servers. These will store thousands of movies, the current week's broadcast programming and all manner of video publications, catalogs, data files, and interactive entertainment. Remote facilities, located close to the entertainment industry centers, will provide additional offerings from HBO, Showtime and other former Channels.

All the major cable companies, such as TCI, Time Warner and Cablevision, as well as telephone companies plan to deliver that type of service in the mid 1990s. In the 1994, U.S. West started the delivery of "video dial tone" to 13 million subscribers across 14 states.

"Make no mistakes about it," said Vice President Al Gore, who was talking about information highways long before they were fashionable. "This is by all odds the most important and lucrative marketplace of 21st century." If Gore is right, the new technology will force the merger of television, telecommunications, and computers, consumer electronics, publishing, and information services. Apple Computer chairman, John Scully, estimates that the revenue generated by this megaindustry could reach $3.5 trillion worldwide by the year 2001. In 1992, the entire U.S. gross national product was $5.9 trillion (Elmer-Dewitt 1993).

The applications of cable and broadcast TV information services can be classified as follows:

- Interactive shopping channels
- News & information services
- Classified adds
- Libraries & data banks
- Financial services
- Video catalogs
- Television archives
- Long-distance phone services
- Other

Some familiar components of the TV landscape will disappear. Local affiliate stations, which have the exclusive right to pick up network shows and distribute them to viewers in their localities, would seem to have no function—except as suppliers of local news and other community-based programming. The video rental stores will disappear also.

Some futurists look forward to this new world, forecasting a burst of creative programming for niche audience and the withering of mass-audience pap.

# Conclusions

A new world of networking telematic services is coming to your home, business, and institutions — sooner than you think. A new information civilization, however, should support information ethics allowing the promotion of well informed citizenry, improving liberal democracy and making humankind more wise in living on the Earth.

### Endnotes

[1]. The image of advanced home health care presented in this section is based principally on forecasts by the Institute for Alternative Futures. It also reflects interviews with Michael McDonald, Windom Health Systems; Dr. Barry Zallen, Harvard Community Health Plan; Bob G. Thompson, University of North Carolina Medical School; and Peter Tolos, The Center for Medical Informatics. Other sources include the testimony of Dr. Barry Gilbert, of the Mayo Clinic Foundation, to FCC hearing on the future of telecommunications: "Applications of Broadband Switched Digital Networks to the Practice of Medicine and Delivery of Health Care;" a summary of new products from Interpractice Enterprises; issues of the Medical Documentation Update published by the Medical Records Institute; and "Healthcare for Aging America: The Role of Telecommunications," remarks by Mary Gardiner Jones to the National Engineering Consortium.

[2] Following Quotron's acquisition by Citicorp in 1968 for $680 million, two major firms—Merrill Lynch & Co., Inc. and Shearson Lehman Brothers, Inc., now know as Shearson Lehman Hutton Inc. —announced they would not renew their contracts with Quatron because they consider Citicorp a competitor. ADP has recently begun installing a personal computer-based stock quotation system for registered representatives at Shearson and Merrill. If these installations are completed, and ADP achieves a one-for-one replacement of the terminals at both Merrill and Shearson, Quotron's network of approximately 100,000 terminals could be reduced by up to 30 percent and ADP could surpass Quatron as the leading stock quotation provider in the U.S. (Water Information Systems, Transcript of Quotron-Reuters-Telerate Conference, New York, NY, November 1988, p.19). To date, ADP's conversion of terminals at Merrill and Shearson is running behind

schedule, and Quotron has added more terminals than it has lost (Roxanne Taylor, Quatron, Los Angeles, CA, personal communication, August 1989).

[3] Speech by Robert Ethrington, international marketing manager for transaction products, Reuters Holdings, PLC, New York, NY, July 1988.

[4] Waters information Services, data Broadcasting marketplace, New York, NY: 1989.

[5] The whole section on Financial Information Services is based on a source: U.S. Congress, Office of Technology Assessment, Electronic Bulls & Bears: U.S. Securities Markets & Information Technology, OTA-CIT-469, Washington, DC: U.S. Government Printing Office, September 1990, pp.132-136.

[6] "EDI: Extending the Enterprise. From an interview with Victor S. Wheatman of Input. The Consultant Forum, vol. 5, no. 3.

[7] Confer V™ Beginer's Guide, Kalamazoo: Academic Computer Center of the Western Michigan University, 1993.

[8] The syntax of commands can vary for each host. For more detail information on "How to Use the Internet" look at books: E.Krol, The Whole INTERNET, User's Guide & Catalogue, Sebastopol, CA.: O'Reilly & Associates, Inc. 1992, T. LaQueuey, The INTERNET Companion, Reading, MA.: Addison-Wesley, 1993.

[9] Archie was developed by researchers from McGill University.

[10] This section is based on the individual study project "Kalamazoo Free-Net" by Scot Albright, under the author supervision, at Western Michigan University, Business Information Systems Department, Winter 1993.

## References


Andrews, E.L. (1991), That Local Call Goes Up for Grabs," *The New York Times*. December 29, section 3.

Cohen, Ch.,(1992), "E-mail's Bulletin Board Gets the Word out," *Communication News*, August 1992, p. 22.

Delivers, Y., and P-A. Pays. (1988), Personal communications to J. S. Quarterman, The Matrix, Bedford, MA.: Digital Press, 1990, p. 511.

Dern, D. (1992), "Internet System Experiencing Meteoric Growth," *InfoWorld*, September, p.56.

Elmer-Dewitt , P. (1993), "Take a Trip into the Future on the Electronic Superhighway," *Time*, April 12, pp. 50-55.

Halhead, B., (1991), "Video Telcom's MediaMax," *Business Communications Review*, July 1991, p.63.

Herman, T. (1989), "Big Dealers Keep Monopoly on Bond Data," *Wall Street Journal*, April 11.

Irwin, S. (1989), "Introduction To Business Television," in *The Teleconferencing Manager's Guide*, ed. Kathleen J. Hansell, White Plains, NY.:

Knowledge Industry Publications, p. 146-166.

Lewyn, M. and P. Coy (1991), "The Baby Bells Learn A Nasty New Word: Competition," *Business Week*, March 25, pp. 96-101.

Lottor, M. (1992) "Internet Growth (1981-1991); RFC 1296, Working Group Request for Comments, Network Information Systems Center, Menlo Park, CA.: SRI International, January.

Lynch, D.C. and M.T. Rose (1992) *Internet System Handbook*, Reading, MA.: Addison-Wesley Publishing, Inc.

Moore, A.M., and R.M. Savey, (1992) *BITNET for VMS Users*, Burlington, MA.: Digital Press.

Olson, R., M. G.Jones, C. Bezold (1992), 21st Century Learning and Health Care in the Home: Creating a National Telecommunications Network, Washington: The Consumer Interest Research Institute.

Peterson, T. A. (1992), *The Future of U.S. Retailing*, New York: Quarum Books.

Philo, E. and Kenneth Ng (1989), *Reuters Holdings PLC*, Goldman, Sachs & Co, New York, NY, February p.5.

Pournelle, J. and M. Banks (1992) *Pournell's PC Communications Bible*, Redmond, WA.: Mocrosoft Press, p.281.

Rugge, S. and A. Glossbrenner (1992), *The Information Broker's Handbook*, Blue Ridge Summit, PA.: Windcrest/McGraw-Hill, p.15.

Quarterman, J.S. and J.H. Hoskins (1986), "Notable Computer Networks," *Communications of ACM*, vol.29, no. 10 (October), pp. 932-971.

Quarterman, J.S (1990), *The Matrix, Computer Networks and Conferencing Systems Worldwide*, Bedford, MA., Digital Press.

Snider, J., T. Ziporyn (1991) *Future Shop*, New York: St. Martin's Press.

Zimmerman, B. (1992) "How Texaco Views Corporate EDI Initiative," *Communications News*, September 1992, p. 24.

Zemke, R. (1986), "The Rediscovery of Video Teleconferencing," *Training*, 28, September, pp. 28-43.

# CHAPTER 3

# Electronic Money

## History of Money

Adam Smith (1950) regarded a "propensity to truck, barter, and exchange one thing for another" as one of the basic ingredients of human nature." It is fact that a man has been occupied in the process of "truck, barter, and exchange" from very early times. The medium in which prices (value) are expressed and exchanged, debts discharged, goods and services paid for, and bank reserves held, is called money. Money is the medium of exchange, whereby goods and services are paid for. It is also a measure of wealth.

Scores, perhaps hundreds, of different objects have served as money at one time or another, including such things as slaves, gunpowder, skulls, the jawbones of pigs, gold, and silver. The origins of money, like those of trade, lie far beyond the earliest written history. The function of money as a medium of exchange is closely related to that of standard value. For example, a certain cattle could serve as a measure in Homeric Greece. It is a matter of fact that the word "pecuniary" is derived from *pecunia*, the Latin word for money, which in turn, come from *pecus*, cattle. The standard value led to the birth (about 750 BC) of coins with an information written on its surface—"how much value does it represent"? The Romans were perhaps the first proponents of the state budget deficit, since they not only manipulated the appearance of their coinage to suit political ends, but also manipulated its value to suit the financial needs of the state (Morgan 1965).

In 17th century Europe (Sweden, England, Venice, Amsterdam), the banks begun to circulate paper money as bank-notes. Notes were being passed from hand to hand as substitutes for coin. When the Bank of England was founded in 1694, the issue of notes was one of its most important functions. British bank-notes were claims to payment in gold on demand. The circulation of bank-notes was, however, smaller than that of coins and continued to be so until the First World War.

After 1914, and particularly after 1945, money as inconvertible bank-notes and deposits repayable in such notes became the main mean of wealth exchange in all highly developed economies of the world. Today money is a way of counting costs and benefits, storing wealth, paying for work, goods, and services, measuring capital, supporting international trade and so forth.

During the war period of 1939-44, however, gold had more confidence than paper money that was not backed by European governments. In 1956-65, *Pax Americana* rules (created at Bretton Woods in 1944, and through the International Monetary Fund) established the dollar (paper money) as the world currency that was used to oil the development of Western Europe and *eo ipso* the U.S. 1965 began the years of the triumph of paper money. The intensive development of Atlantic and Pacific economies caused a bigger supply of money and inflation as well as an international monetary crisis. In 1963-1974, an abstract standard of money was invented and implemented by the Group of Ten, the largest western and Asiatic democratic countries. A Composite Reserve Unit became a measuring unit of a common abstractive currency backed by reserves in gold and applied in international exchange. In 1971, gold disappeared as the standard of paper money (a U.S. decision under the Nixon administration) and paper money values began to float (Wiseley 1977).

In the 1980s, paper money was contested by plastic money that could warm up the American economy via intensive consumer purchases on credit. In the 1990s, electronic money is becoming a new medium of exchange in transaction processing, particularly among financial institutions in the money and capital markets.

The history of money is one of money technology. From goods, coins, and paper to plastic and electronic forms, it is in fact a conversion from material to information money in order to make its application easier and particularly to build financial computer-based information systems. A consumer or investor applying electronic money gets a competitive advantage because their organization can immediately access computerized information on a businesses state of affairs and develop decision options. Electronic money's by-product is the opportunity to make more money or spend it more wisely.

Electronic money has an intrinsic added value that can be generated if computer supported decisions are right and made properly. All old types of money were rather passive, just a medium. While electronic money is still a medium, but with a message. In addition, electronic money is less expensive since its management is provided through computers and networks (financial information systems) not by labor and a mail intensive environment. Also, that type of media accelerates the circulation of money and makes it more productive and less costly. The following formula defines electronic money potential ($P_e$):

$$P_e = V_f + V_o - (C_i + C_m)$$

Where:

$V_f$ —face value of money
$V_o$ — opportunity value of money
$C_i$ — cost of information
$C_m$ —cost of money

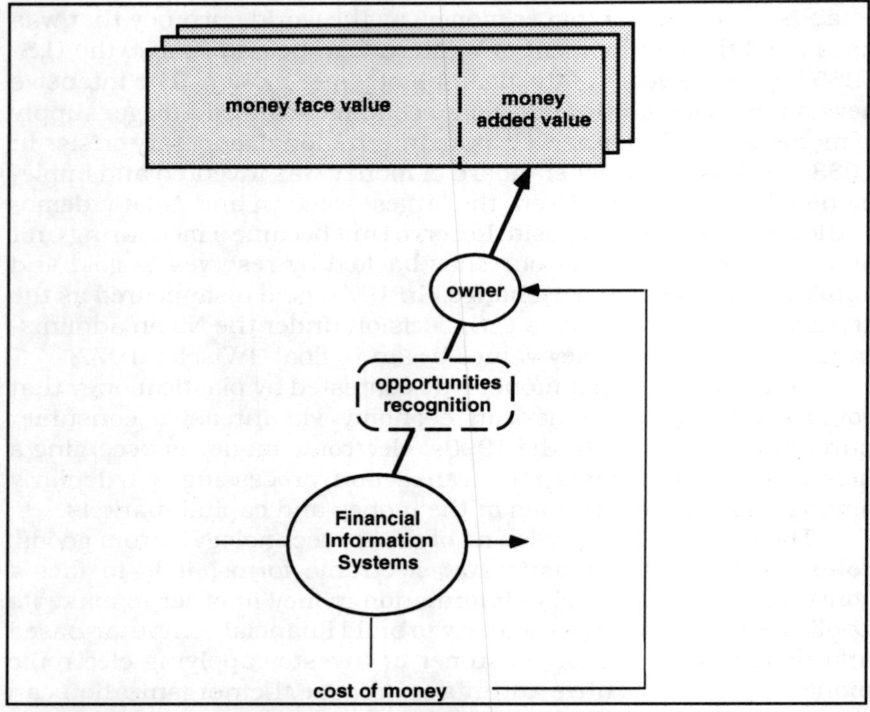


*Figure 3-1: The Potential of Electronic Money*

## Wealth Exchange Alternatives

The history of social evolution is to a certain degree the history of wealth exchange system's evolution. In early, primitive societies, *barter exchange* was in use; it was based on a medium that was a good ("cattle"). In further development of the primitive-agrarian society, indirect barter was based on a specific commodity as a medium of exchange.

*Money exchange* based on gold or silver coins was used to facilitate payments for business transactions. To store an increasing volume of coins (currency), warehouses were introduced. They took coins on deposit and provided receipts for stored "wealth." Soon, the warehouse owners noticed that it was improbable that all owners would withdraw all coins at same time. They began issuing notes with partial backing in money.

This gave birth to the *banking system* which could "create" money. Now the money supply could oil economic activities and create more wealth. Consequently, the banking system began to issue saving books, Certificates of deposit (CD), checks, loan lines, credit cards, and so-called financial products.

By the end of the 18th century, trade of securities (*stock exchange*) took place in London (1773) and later in New York (1792). By the end of the 19th century, the New York Stock Exchange and London Stock Exchange were effectively trading stock and bond shares among their owners through the floor brokers and their parent firms. Stock exchanges, also known as bourses, play important roles in the economic life of all major developed nations. Their primary mission is to provide a convenient means for persons to invest money in business enterprises and to liquefy their investments at any time by selling their securities. A by product of stock exchanges is the generation of information on economic activity in the nation. In return, it provides a new opportunity for investors, since they can be aware of the economic climate in the country by analyzing capital market indexes such as Dow Jones, Standard and Poor (S&P 500), and so forth.

In order to take full advantage of the banking and stock exchange systems, the *Electronic Fund Transfer Systems* (EFTS) was invented. It replaces paper money with electronic money processed by computers and their networks. EFTS is a tool to communicate, transport, integrate, and share information among financial institutions and their customers. On October 28, 1974, the Congress of the United States provided for the creation of a National Commission on Electronic Fund Transfers. It is a new information infrastructure for the facilitation of payment mechanisms. But not only is it an electronic tool for payments, it is also a tool for the generation of new financial and information services.

Once EFTS is in operation, *on-line exchanges* can take place. Not only users from the financial institutions, but also consumers and private investors can fix deals electronically via Automated Teller Machines (ATM), Point-of-Sales (POS), credit and debit cards, smart cards (with a chip on it to instantly compute a new balance after a transaction took place), and information kiosks located in public places as in malls, libraries, hospitals, airports, bus and railroad stations, and so forth.

With the proliferation of home computers with modems, a consumer can shop and pay from home. It gives birth to a *virtual exchange* through home banking and trading. It not only facilitates the transaction, but expands the choice of financial services throughout the world. Table 3 - 1 compares different alternatives of wealth exchange systems.

In the following sections the modern exchange systems alternatives will be analyzed.

# Banking Exchange

### The Rise and Decline of Modern Banking

Commercial banks play an important role in facilitating economic activities. On a macroeconomic level, they represent a primary conduit of Federal Reserve monetary policy. Efforts to

| Alternative | Goal | Medium | Opportunity | Gateways |
|---|---|---|---|---|
| Barter Exchange | Fulfillment of basic needs | Goods | None | None |
| Money Exchange | Improved payments | Coins | Wealth accumulation | To other cities and countries |
| Banking Exchange | Smoothed payments | Paper and plastic money | Business activity increased | International and national trade |
| Stock Exchange | Liquidity of ownership | Shares | Information and new investments | To brokers and firms |
| EFTs | Instant payments | Electronic money | Integrated and shared information, new deals | To all financial institutions |
| On-Line Exchange | Instant payments | Electronic money | Time saving, cashless society, crime-free society | To all financial instituions |
| Virtual Exchange | Instant deals and better choice | Electronic money | Quick decisions, money game | To whole financial world |

*Table 3- 1 A Comparison of Wealth Exchange Systems Alternatives*

control a nation's money supply initially effect the level of aggregate economic activity by changing the availability of credit at banks. On a micro-economics level, commercial banks are the primary source of credit to most small businesses and many individuals. A community's vitality is typically reflected in the strength of its major financial institutions and the innovative character of its business leaders.

The etymology of the word "bank" can be traced to the French word "banque" (meaning chest) and the Italian word "banca" (meaning bench). "Chest" suggests the safekeeping function, while "bench" refers to the table, counter, or a place of business for a money changer. The latter suggests the transaction function. The two basic functions that commercial banks perform are (Hempel, Coleman, and Simonson 1983):

(1) The safekeeping function (providing a safe place to store savings, it was the main mission of first warehouses with coins); and
(2) The transaction function (furnishing a means of payment for buying goods and services).

Banks also perform a lending function, supplying liquidity to the economy. In general, banks make commercial loans and accept demand deposits.

The first bank in the United States, the Bank of North America, was established in Philadelphia in 1782. It could issue bank notes that were exchangeable for metallic coins. In 1863, the National Bank Act was created by the Congress in order to make the national currency system uniform (the U.S. Treasury Department began to print nationally standardized bank notes), to regulate the organization of privately held banks, and to provide a new source of loans to finance the Civil War. In 1913, in response to the financial panics and economic depressions of 1873, 1884, 1893, and 1907, Congress established the Federal Reserve System. Prior to December 31, 1980 (earlier for the New England states), commercial banks had a monopoly of the transaction function as they were the only financial institutions to have demand deposit or checking powers. Currently, commercial banks share this power with savings banks, savings and loan associations, credit unions, money market funds, mutual funds, and brokerage firms that offer check-writing services (Gart 1989).

As a result, commercial banks declined in importance. Numerous less-regulated non-bank competitors such as General Motors, Sears Roebuck, and American Express now offer a wide array of bank-like services. Corporations, aided by Wall Street, are selling ever greater volumes of commercial paper, which has cut into banks' short-term corporate borrowing business. Commercial credit firms and insurance companies are actively providing long-term financing. And money market funds, which offer better yields than certificates of deposit, have lured away a large share of the banks' deposit base. There is an opinion that banks in their traditional form are already obsolete.

## Technology-driven Changes in Banking

The development of electronic payment mechanisms is the answer to the banking decline. Its influence, however, cannot be overestimated.. The appropriate federal regulations should accompany the technological race. This race can be analyzed through the growth of consumer banking facilities (Table 3 - 2).

The application of data transmission technology (*remote jub entry* to the central computing facility) doubled the number of banking facilities for customers in the 1960s. In 1975, the introduction of plastic money issued by banks (the banks immediately lined up behind Visa and MasterCard against competitors such as American Express, which dominated the credit card business) increased the number of payment facilities by 3714 percent. Again, the application of ATM and POS machines through the EFTS expanded the number of payment facilities by 154 percent. The predicted number of customer payment facilities will be increased by 3000 percent with the introduction of home banking (*virtual exchange*) technology by the year 2000.

The fight for customers through the application of information technology among financial institutions is apparently taking place. Who will win? So far, in the 1990s, the banks have the technological edge over other financial institutions. However, the banks should not take that for granted since all financial institutions sooner or later will be at the same level of technological might.

## System Architecture of a Bank

The system architecture of a bank is illustrated on Figure 3 - 2. There are six federations of banking information systems:

| Year | Technology | Number of Banking Locations |
|---|---|---|
| 1945 | Home offices and facilities | 18,000 |
| 1965 | Branch banks (data transmission) | 35,000 |
| 1975 | Bank card outlets (plastic money) | 1,300,000 |
| 1980 | EFTS-type facilities | 2,000,000 |
| 2000 | Home banking (Interactive ) | 60,000,000 |

*Table 3- 2 The Growth of Banking Facilities in Relation to Technology [7]*

- Management Information Systems which include: Executive Information System, Personnel Information System, Accounting Information System, Purchasing Information System, Facilities Information Systems, and others for the support of intra-bank management.
- Financial Product Information Systems which support the product planning, designing, and marketing functions of a bank.
- Operations Information Systems which support a bank's cash and funds management at the link with a customer.
- Inter-organizational Information Systems which manage the electronic transfer of a bank's funds via local, regional, and national computer networks. This federation is integrated with the federation of Operations Information Systems.
- International Information Systems which manage the electronic transfer of a bank's funds to/from international financial partners and customers.
- Office Information Systems which automates office communications, time management and archiving.

These banking information systems are integrated and shared via a bank's infostrada, which is an interconnected set of private local, regional, and national networks and gateways

In a bank's systems architecture of the 1990s, there is a shift of emphasis from data to time management. The progress of information technology until the 1990s satisfied bank users' needs for electronic number-crunching computing. In the 1990s and beyond, a bank's staff and executives want to communicate among themselves and with customers. In the past, the effort to develop a banking information system was concentrated on database loading in 95 percent and on providing access to it in 5 percent [8]. In the 1990s, the effort is being reversed. The access to stored information via fuzzy queries (supported by artificial intelligence) is a sought priority by the end users and customers staying either at ATMs, POS, or at home.

## The Future of Banking

The decline in banks' dominance of finance has been stunning. In 1874, 37 percent of all financial institution's assets were held by banks. By 1989, they had lost a quarter of their share as their slice of the market plummeted to 27 percent. Even in their core business —short-term corporate lending — banks' share slumped from nearly 80 percent in 1975 to about 55 percent in 1989 (Yang et al 1991).

Banks may hold their influence in a few niches such as small-business lending, check clearing, and *relationship banking* (person-

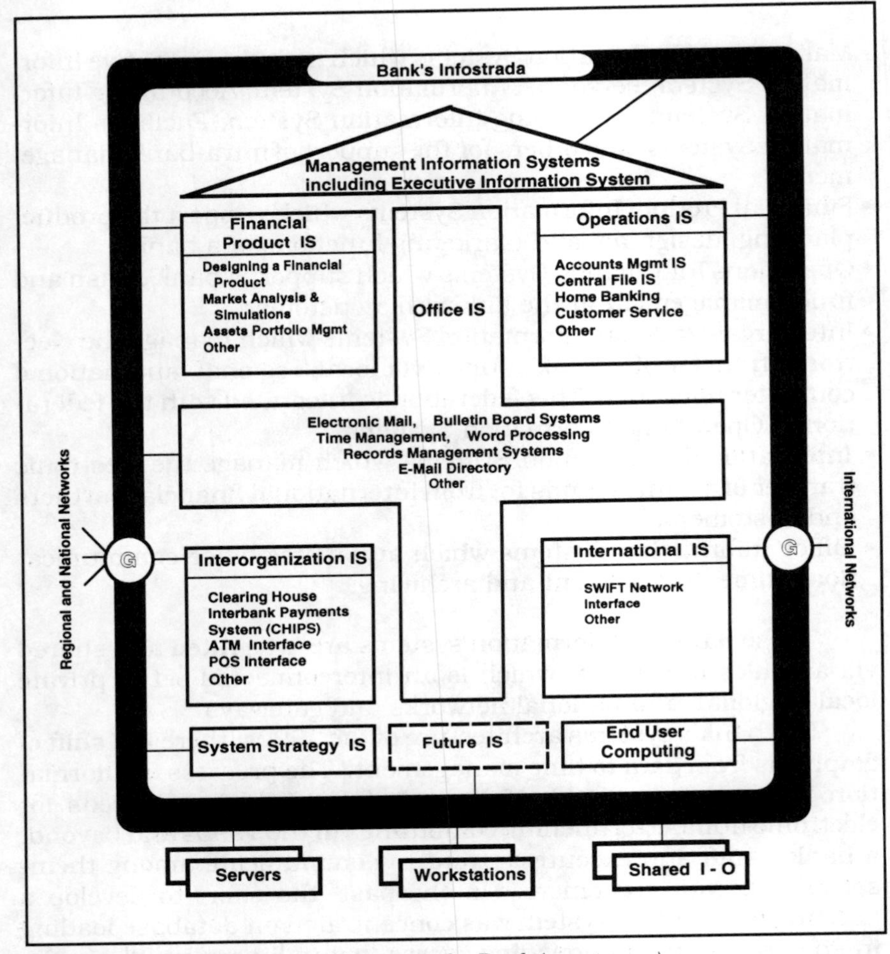


*Figure 3-2: The System Architecture of a Bank (a gateway)*

alized services). But eventually, most of today's banks will be just another set of actors in the financial services free-for-all where everyone will be able to invade rivals territories. The leading figures will likely not be financial department stores but highly successful niche virtuosos. The chief beneficiaries will be borrowers, savers, and investors who will have virtual access to a wide range of services (*virtual exchange*) at lower prices.

# Stock Exchange

### Securities Information Systems

At the beginning of 19th century, delivery of a message (or a market quote) from New Orleans to New York took from 4 to 7 days.

The telegraph was first demonstrated in 1844. By January 27, 1846, telegraphic communication linked New York and Philadelphia, via Network. Until direct lines were installed a few months later, messengers ran between the telegraph office and Wall Street. It was two more years before the New York and New Orleans foreign exchange markets could directly communicate, but by then message time was nearly instantaneous (Garbade and Silber, 1978). Financial markets were quick to realize the possibilities. The *New York Herald* of March 3, 1846, mentioned that "certain parties in New York and Philadelphia were employing the telegraph for speculating in stocks." The use of the telegraph greatly reduced price differences between the participating markets.

A successful trans-Atlantic cable was completed on July 27, 1866. Four days later The New York Evening Post published price quotations from the London exchange. The first cable transfers occurred about 1870 and arbitrage between the London and New York exchanges began immediately. This led to further reductions in price differences between markets.

The third invention that revolutionized the exchanges was the stock ticker introduced in 1867. Before that, reports of transactions were recorded by "pad shovers" —boys who ran between the trading floor and the brokers' offices with messages. Several ticker companies had men on the trading floor to type results directly into the ticker machine. These reports went to the ticker companies' headquarters and were retyped to activate indicator wheels at local tickers which then printed results on paper tape.

In 1878, the telephone, successfully tested 2 years earlier by inventor Alexander Graham Bell, reached Wall Street. Until then, a messenger carrying a customer's order could take 15 minutes to get to the floor; with the telephone, it took 60 seconds. By 1880, most brokers had telephones linked directly to trading floors, and in the next few years, telephones were installed by the thousands. Finally, in 1882, the Edison Electric Illuminating Co. gave Wall Street electric lights (Gardner 1982).

By 1880, there were over a thousand tickers in the offices of New York banks and brokers. In 1885, the New York Stock Exchange (NYSE) began to assemble the information for ticker company reporters to ensure consistency. The New York Quotation Co. was created by NYSE members in 1890 to consolidate existing ticker companies and integrate the information distribution. This did not eliminate "bucket shops," where the ticker tape output was rigged to swindle investors [4].

The development of Securities Information Systems in the 20th century took place in the scope of the following systems [4]:

- Trading Support Systems

- Market Surveillance Systems
- Clearing and Settlement Systems
- 24 - Hour Global Trading Systems

*Trading Support Systems* under the form of fully electronic transmission and storage of trading information began in the 1960s. Quotation devices were first attached to ticker circuits to provide bid and ask quotations and prices. An improved stock ticker was introduced in 1964 that could print 900 characters per minute and report transactions without delay of up to 10 million shares per day. The pneumatic tube carried information to the ticker and quotation system, until it was replaced with computer readable cards in 1966. Reporters on the floor recorded the transaction on a card and put it into an optical scanner. The scanner read the information into a computer where it entered the ticker system. About this time the Central Certificate Service was created as an exchange subsidiary to computerize the transfer of security ownership and reduce the movement of paper. In 1973, this became the depository Trust Company. The computer display of dealers' bids and offers called NASDAQ (National Association of Securities Dealers Automated Quotations) began to operate in 1971.

During the 1960s, the security industry had a back-office crisis, since it could not keep up with the paper-work for high transaction volume. In 1972, the Securities Industry Automation Corporation (SIAC) was established by the New York Stock Exchange (NYSE) and the American Stock Exchange (AMEX) to coordinate the development of data processing systems. Three systems were implemented by SIAC in the 1970s:

- Market Data Systems (MDS) for the processing of the last sale information,
- Designated Order Turnaround System (DOT) —automation of delivery of small orders [fewer than 199 shares] from member-firm offices to exchange floors,
- Common Message Switch (CMS) to let member-firms communicate with the other SIAC systems.

The regional stock exchanges were developing the same systems. In 1969, the Pacific Stock Exchange (PSE) automated some trade executions through the COMEX system. It allowed traders to complete a trade without any human intervention. In the 1970s, this system was redesigned as SCOREX to allow a specialist to offer within 15 seconds a better price.

Information systems reengineered the entire exchange structure. The Cincinnati Stock Exchange and the London International Stock Exchange (ISE) do not use physical trading floors but implement trading through the computer. Other exchanges in Toronto,

Madrid, Brussels, Copenhagen, Zurich, and Frankfurt are "floorless" too. The American NASDAQ and British ISE integrate screen-driven quotations with negotiations by telephone. Other American stock exchanges choose not to automate floor operations and prefer to keep floor specialist supporting them with computerized trading systems.

*Market Surveillance Systems* Computers are applied to trace illegal activities during mergers, acquisitions and other major corporate security-oriented transactions. Systems are designed to monitor market data and detect unusual patterns. They are later analyzed to define the cause of the fluctuation. The expert systems will be used in the future analysis.

*Clearing and Settlement Systems* are applied to support the exchange process of security ownership from the seller to the buyer. The Pacific Stock Exchange developed a Continuous Net Settlement (CNS) system in the 1960s to process high volume transactions.

*24 - Hour Global Trading Systems* are applied to facilitate trades 24-hours around the clock. In 1987, the Chicago Mercantile Exchange (CME) together with Reuters developed the Post (Pre) Market Trade System called GLOBEX for "global exchange" in off-hours to automatically match orders. In 1989, The Chicago Board of Trade (CBOT) announced a plan for the AURORA system which can select the counter party for their trade. In 1990 GLOBEX and AURORA were integrated [4]. In the 1980s, all major international stock exchanges began to establish links with each other. In 1984, the Montreal Exchange linked with the Boston Exchange; in 1985, the Toronto Stock Exchange linked with AMEX and later with the Chicago Midwest Exchange. By the end of the 1980s, the New York Stock Exchange and London Stock Exchange hooked up their operations. In the 1990s, exchanges in New York, London, Tokyo, and Hong Kong extended their trading hours to accommodate the 24-hours cycle. Global trading provides new investment opportunities for traders and investors who can integrate information globally and allocate money where the best deals can be fixed. As a result of global trading, the total international equity business almost doubled from $400 billion in 1985 to $740 billion in 1986. In European markets, international business has become a driving force. Foreigners accounted for 40% of total equity turnover in West Germany, 35% in France, and 36% in Britain. Foreign investors were most active in the United States and least active in Japan (Estabrook 1988).

## Securities Information Services

The Reuter Company is a pioneer in providing securities information services world wide. As early as 1850 ,Paul Julius Reuter first used pigeons to fly stock market quotations between Brussels and

Aachen, Germany. Once an underwater telegraph line was implemented between Dover and Calais, Reuter was applying this line to deliver financial news from London to Continental Europe. Today, Reuters is still a leader in selling financial information services.

The customers of financial information services can be divided into the following categories:

- Financial news,
- Stock quotations,
- Value-added products,
- Foreign exchange data,
- U.S. Government bond data.

*Financial news* providers are Dow Jones (leading vendor of stocks news in the U.S.), Reuter (leading international leader), Knight-Ridder (commodity news leading vendor) and on-line news vendors such as Associated Press, McGraw-Hill, Financial News Network, and Market News Service.

*Stock quotations* in the U.S. are dominated by five companies; Reuters Holdings PLC, Quotron Systems Inc., Automatic Data Processing Inc. (ADP), Telerate Inc. (owned by Dow Jones), and Knight-Ridder Inc. These companies operated about 426,000 terminals as of February, 1989 (Philo and Ng 1989). There are about 200,000 terminals receiving real-time and on-line data from U.S. stock exchanges [4].

*Value-added products* contain historical data, analysis, research, and niche information which is disseminated through existing on-line channels of such companies as Quatron, Reuters, and Bridge Brokerage System. The successful companies which provide enhanced information are: MMS International, First Call, and so forth.

*Foreign exchange* data is dominated by Reuter which introduced on-line service in 1971 and in 1981 implemented the Monitoring Dealing System replacing a telephone as the medium to negotiate between two parties transactions with a terminal. Its strong competitor, Telerate, provides similar services and a system, The Trading Service, for multi-party conversations over on-line monitors. As a result of this competition, Reuter implemented the Dealing 2000 system which supports auction-like market where bids and offers from both parties are exposed and aggregated per price.

*U.S. Government bond data* are distributed by the only vendor — Telerate, which has the exclusive rights till the year 2005 [4].

## The Future Trends

The solutions that should improve the integration and sharing of security information include the following:

- Enterprise-wide systems integrating and displaying on the same screen data and information from different sources and vendors
- Self-service computing by end-users who should develop a custom-made solution with such tools as Computer Aided Software Engineering (CASE),
- Conversion of all records into electronic standard format for better retrieval among different service providers,
- Expansion of the customer reach through advanced multimedia (ISDN) telecommunication technology, which should promote *on-line exchange* through the desk-top video conferencing among traders and customers,
- Connectivity (gateways and protocols) to National Information Infrastructure to develop the networked mass market for the financial services and support *virtual exchange* of wealth.

The future of Wall Street depends on information technology penetration of customers, collaborating financial firms, and on the global economy. Is Wall Street becoming irrelevant? Big institutional investors are using its services less and less. Many corporations are handling their own investment banking. New electronic trading technologies could wipe out the jobs of thousands of brokers, traders, and salespeople. Emerging securities markets with global scope are winning business from U.S. exchanges; the American financial system will never be the same.

The proliferation of advanced information technology in the form of cheap but powerful user-friendly software and workstations is doing violence to the Street's competitive advantages. Technology is making once exclusive market information instantaneously accessible to anyone with a terminal. It is fostering new automated trading systems open to all players — not just security firms. It is destroying the old-boy network and replacing it with dealings based on price and service. Technology is also transforming once intricate, labor intensive activities such as trade execution into a simple mechanical function. This is bad news for much of the Street's work force, especially many brokers, sales people, traders, and back office people. If these workers will not "add value" to their operations, they will be eliminated. Information technology, however, is both a destroyer and a creator. It also opens new prospects and opportunities (Welles and Roman 1990).

**Public Policy**

The world is moving toward electronic around-the-clock and around-the-globe securities trading [5]. The U.S. regulatory structure will have to support the U.S. markets competitive advantage over foreign security markets and protect the objectives of domestic policy in an industry driven by change. At the same time, it has to protect the public interest in operating a "clean" investment environment. From 1955 to 1982, there were only two occasions when stock market prices fell more than 4 percent in 1 day. From 1982 to mid 1990, there have been 10 such episodes. The crash of October 19 and 20, 1987 disclosed three serious problems that should be solved:

- The limits of computerized systems when trading volume spikes (over 600,000 shares exchanged per day as a result of programmed trading),
- Limits of human traders under the pressure of market crash,
- Recurring excessive short-term volatility that may promote further crashes.

Steps have been taken in all of the markets to correct such problems and increase the capacity of the information infrastructure [4]. During this computer-driven crash of 1987, investors lost about $1 trillion in wealth. The public policy should include a premise that security markets should provide to every citizens an equal access (through law and technology) to the wealth-producing industry.

## Electronic Fund Transfers Exchange

The Electronic Fund Transfers System (EFTS) is a payment system in which the processing and communications necessary to effect economic exchanges and the processing and communications necessary for the production and distribution of services incidental, and/or related, to economic exchanges are dependent wholly or in large part on the use of electronics [6]. EFTS contains a cluster of related practices and information technology that employ electronic impulses generated and interpreted by computers to debit and credit financial accounts. Each such debit and credit transaction is termed an electronic fund transfer. Electronic impulses, rather than paper, are applied to change an economic transaction.

EFTS is described by Bequai (1981) as a growing array of financial services. Among the services are: wire transfer of funds, direct deposit of income checks, periodic or authorized payments, check verification, and credit card authorizations. Point-of-sale (POS) systems, automated teller machines (ATM), and automated clearing houses (ACH) represent more advanced form of EFTS. These financial services relay on computers and have the potential to operate locally, regionally, nationally, or internationally.

EFTS employs electronic money as a tool to provide a superior way to communicate and transport data electronically and allows for an access to financial electronic records. One expert, Sprague (1974) has defined EFTS in its broadest sense as embracing all of the following functions:

1. The electronic transfer of funds between corporations and banks nationally either on a direct basis or through the Federal Reserve System.
2. The automation of preauthorized debit or credit payments for individuals and enterprises in which automated clearing houses would be focal point for handling, clearing, and settlement functions among banks and the Federal Reserve.
3. The authorization for an execution of both cash and credit exchanges of value on either an immediate or a delayed basis as determined at the point of sale. Both thrift and demand deposit accounts would be used to third party transfers.
4. The provision of a comprehensive range of financial services for both individuals and organizations at points of convenient locations. Such services would include: cash deposit and withdrawal, the transfer of funds between accounts, budget management services, and so on.
5. The establishment of local, regional, and national electronic clearing houses for effecting all of the above.

These functions can be reduced to the generic functions:

- Clearing transfers through data networks (ACH),
- On-line remote services (ATM, POS),
- Pre authorized debit/credit payments.

The heart of EFTS is the automated clearing house. The ACH movement originated in California in 1968 by ten California banks who formed a Special Committee on Paperless Entries, known as SCOPE. Its mission was to develop and implement a system of "preauthorized paperless entries." Based on SCOPE, the Atlanta Committee on Paperless Entries (COPE) was formed. The first ACH modeled on SCOPE was in operation in December of 1974 in San Francisco. Soon, other cities such as Los Angeles, Boston, Minneapolis/St.Paul and others were implementing ACHs. In 1974, the National Automated Clearing House Association was formed to facilitate the application of ACH in all 12 Federal Reserve Districts, representing population centers.

The locally oriented ACHs are integrating with the nationally oriented Fed Wire and Bank Wire and world-oriented SWIFT network which was developed by the Society for Worldwide Interbank Financial Telecommunications. These networks create an Electronic Global Village of financial operations.

The flow of electronic money through EFTS is illustrated on

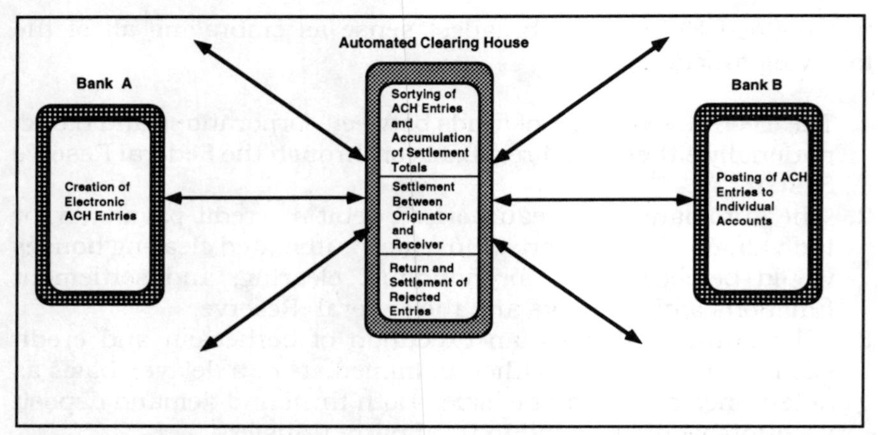


*Figure 3-3: The ACH Processes*

Figure 3 - 3.

The EFTS environment allowed for the development of other services such as electronic mail for member banks. Among these systems one can mention the Electronic Message System (EMS), the Electronic Computer Originated Mail (ECOM), Electronic Message Services (EMSS), and Intelpost. Promoters of these systems view EFTS as an environment which will replace present postal, telex, and cable communication systems in message and funds transfers.

EFTS represents an array of new opportunities for merchants to promote new services and to the computer and telecommunications industries to promote new technologies. For the public, EFTS provides an environment for the applications of on-line and virtual exchanges that should make customers, consumers, and investors' life easier with a tool for new opportunities of increasing wealth. EFTS is a financial information infrastructure of the Electronic Global Village.

## On-line Exchange

In the 1970s, the commercial banking community was putting into place automated teller machines (ATM) as a free-standing and remotely-located combination of a bank office and cash dispensing machines. It was the beginning of on-line exchange of wealth by the customer. The comprehensive ATM is capable of providing a bank customer practically any service available from a standard teller operated office including cash withdrawals from savings or checking accounts, cash deposits; inter-account funds transfer, withdrawals charged against credit cards;, and bill payments by account-debiting. And all of the latter can be executed at less cost per transaction and greatly increased customer convenience in terms of time and location [1].

The widespread public acceptance of automated banking facilities is evidenced by the rapid growth of ATM installations. For example, in 1969 there were two ATMs installed in the U.S. In 1974, 2,200 and in 1980, 35,000 (estimate) ATMs were in operation in the U.S. (Skoba 1974). According to an estimate by *Bank Network News*, about 500 million transactions per month were made through ATMs in 1990. That is nearly 700,000 transactions per hour every day. A cost per transaction via ATM was about 68 cents in 1988 which is relatively high (Egner 1991). To offset this cost, the economy of scale was promoted at the level of ATM networks. Inter-state networks are consolidating and serving multiple banks. For example, in 1988 a super-regional Southeastern ATM network was created. It consolidates such networks as Honor in Florida, Avail in Georgia, MOST in Virginia, and Relay in the Carolinas. Another measure to lower a cost per transaction is the application of satellites. In 1990, the First Union 700 branches in the Carolinas, Georgia and Florida were networked by GSTAR II, owned by GTE Spacenet Corporation.

Generally, users visit five times per month an ATM station. In 1987, 35% of U.S. households used ATMs monthly, in 1989 this percentage climbed to 44%, and is still growing (Egner 1991). In the 1990s, ATMs are a permanent component of on-line banking, almost as the commodity that has become a competitive necessity.

The banking industry is split into two camps. One is struggling with ways to keep customers out of branches so they can lower their operating costs, and the other is figuring out how to leverage the branches and control costs by introducing on-line exchange.

Both strategies require significant investment in technology. The first involves investing in home-banking systems, off-site ATMs, or similar out-of-branch ventures. Banks in the second camp are opening automated on-line centers within existing branches in order to give customers the option of performing transactions and obtaining information from automated devices or working with a teller or platform officer in more traditional branch settings.

Such centers typically include ATMs, sel-service terminals, and personal computers (PC) for non-cash transactions, account inquiries, and obtaining other information.

> For example, Hibernia National Bank in New Orleans is rolling out a new feature. A PC in a private booth the bank calls a "Personal Banking Center." The PCs are loaded with special software for customers to use. The need to deliver banking services more efficiently was the original motivation behind this solution. Using the PCs, customers can obtain information on retail checking accounts, certificates of deposit, and loans. Instructions on the screen are easy to follow, and a bank

employee is stationed in the area to answer questions about using the PC. The booths also have telephones, and telephone extensions for specific customer service personnel appear on the screen.

Citibank's Financial Center ,at Grand Central in New York City, houses 20 ATMs that perform 65 types of transactions. About 40,000 transactions are made every week. Even opening new accounts and applying for loans can be accomplished on the go. Commuters can pick up a Loan Express Kit or a new account application on the way home, drop it off the next morning, and have a loan application request answered that day. The bank also provides an ATM card, or CitiCard, the day a new account is opened (Arend 1992).

Another on-line exchange application concerns a smart card which may replace several plastic cards. A scenario of your wallet in the year 2000 is presented below :

Remember how you wallet bulged back in 1993? A phone card, three credit cards, a bunch of receipts, ATM card, frequent-flier cards, company ID, Blue Cross, pictures of the kids, and of course, cash. It's slimmer now. Let's take a look. First, the company ID. It lets you into your office network from any computer anywhere. This one is made by Security Dynamics in Cambridge, Massachusetts, and costs about $10 a year. It has a microchip and three-year battery. Say you are on the road and need to check a file; turn on the laptop, dial up the Internet, and connect to the local network at your office. To get in, type your password; then give the numberthat appears in the window of the card. It's a random number that changes each minute, in sync with a number in the network. That makes a double lock: If someone steals your card, he can't get in without the password; if somebody guesses the password, he can't get it without your card.

You have only one credit card; every merchant takes all the major ones. Yours gives mileage credit on AmeriDelt; in a store, it works like old ones. When using it to shop on-line, you include a digital signature with your order to prevent fraud. It's done with "public-key cryptography," invented in California by Whitfield Diffie and Martin Hellman. Every person has two keys— formulas as computer uses to code messages. Encryptions with one key can be unscrambled only with the other key. One key is public, listed in directories with your name and address, the other is your deepest secret.

You send an order across the network to L.L. Bean. It consists of your name and is encrypted with your private key, the list of items you want and your credit card number. When Bean receives it, they look up your public key. If it unlocks the message, only you could have sent it. The process takes no longer than giving your card number over the phone used to, and it's safer. The driver's license and medical card both incorporate microchips. The license remembers your car registration and insurance data—and the speeding ticket you got last week. The health card holds your medical history. When you see a doctor, she or he updates it and gets instant reimbursement from your managed-care network.

There is no currency in your wallet, only a cash card, which you use more often than any other. You stick it in a slot on your computer or a cash machine, enter your password, and load up the microchip from your bank account. Bingo—an electronic $100 bill! Get coffee at the 7-Eleven, put the card in the cash register slot, and it deducts 75 cents. Cash cards evolved from the smart cards used by European phone and transportation systems. Back in 1993, GemPlus, a French outfit, formed a venture with VeriFone, a U.S. maker of credit card scanning systems. They called it Veri Gem. The next year VeriGem rolled out a cash card in a small Northeastern state by working with supermarkets, fast-food joints, convenience stores, gas stations, and at least one big bank.

What is nice about this cash is that it is anonymous, like the old kind. You can buy cigarettes on the sly, pay your housekeeper, and pick up a copy of FORTUNE without getting a subscription come-on in the next day's E-mail. You can even use digital cash on-line. Lots of people do, including crooks. The global cash economy is growing so fast that President-elect Clinton says she is worried about government's ability to regulate and tax commerce. The Treasury may soon be hurting for revenue[2].

Point-of-Sales (POS) is on-line retail device to exchange information money at the moment of purchasing or selling using a debit card. POS directs debit transactions replace checks, cash, and credit cards as a medium of payment. The entering of a four-digit Personal Identification Number (PIN) is a faster process than identifying a name, address, and phone number. The grocery and oil industry lead in the applications of on-line debit payments. Since the cost of one such transaction is about 35 cents, there is a question who does benefit from the POS payment? A merchant or a bank? Another question is should it be an on-line or batch clearing house operation? ATM cards are the most obvious and popular POS direct debit vehicle for many debit programs.

There were about 50,000 POS terminals installed around the U.S in 1988-89. Most of these devices are linked to regional ATM networks, although ACH POS devices represent a growing segment. POS debit transaction volume has soared during the same period from 3 million transactions a month to more than 13 million transactions a months (Egner 1991). Consumers perceive the debit payment system as the opportunity to stop less frequently at ATM machines to get cash, since payment through this system is a "cash payment" via electronic money.

## Virtual Exchange

Home banking and trading are precursors of virtual exchange by the customer who does not need a broker or window banker as an intermediary anymore. The customer can choose through the computer any electronically capable financial institution and make a transaction. It gives the customer control over his/her deal and teaches/informs him/her about the state of affairs and opportunities. The latter option is more promising that just the "control." Since it leads to the generation of additional value of resources that are in the possession of the customer.

Virtual exchange in the 1990s is just in *status nascendi*. It is at the stage of empowering the customer with the electronic tools for remote access and execution. The level of financial institutions electronic integration is still very limited, but steadily upgrading by those firms that are competitive and are looking for better customer service. In the following early examples, trends of virtual exchange will be analyzed.

Home banking started its first operations in the 1980s and was executed via home computers or interactive television. The function of home banking was not yet fully understood, and therefore, the entire concept was in question. In the 1990s, along with the development of more user friendly software and a better understanding of task functionality, many banks will return to delivering financial services via home banking.

Home banking services can be successful only if a bank's operations are fully automated and information is in an electronic format transparent for inter-organizational systems and the customer. Then, the bank can deliver financial services to the customer at home with the capability to generate opportunities. If such additional value does not exist for the customer, just the automation of transactions processing from the home can be appreciated only by the computer hobbyists.

By extending electronic access to account balance information directly to the customer at home, is the first step in home banking. After that, the focus should be on the selection of suitable technology to provide that access, and on the scope of additional financial services that will empower the home customer in tools to generate more information, more opportunities, and test the concept of money allocation, and support the decision-making on future money.

The selection of access technology cannot be conditional to the personal computer. It is still a device with limited market penetration. A telephone set with a special keyboard, like that in the French Minitel or the Citybank's Enhanced Telephone with automated voice, can be considered as a medium that could bring profit to the home banking providers. Another solution can be a smart card that inserted to the telephone set could provide informative connectivity to the customer's financial resources.

Once the customer has a right device to access the banking environment, he/she is looking for the basic information of each account and integration of information from multiple accounts (Central Information File—CIF). As a result of this consolidated information, a synergetic opportunity may appear for the customer. The first consequence of it will be the capability of making transfers among accounts. The following capabilities could concern the opening of a new account or purchasing some financial products as CDs. Even so, the customer may wish to get additional information or application via mail. The home banking should deliver hybrid services, including connectivity to a live bank worker.

From the bank's point of view, the analysis of home banking integrated transactions should lead to the better understanding of the market needs and design of more market sensitive products mix. Intelligent home banking should bring new opportunities not only for the customers but also for the financial services provider. To keep the customer loyal to the bank's credit card, home banking can promote payment for transactions by credit cards and their indirect link to the checking account. In such a way, the bank could limit the costly processing of check and afford some discount for merchants for accepting credit cards.

The future of home banking is in the limitation of 14.3 billion payments by checks annually (94 million households paying between 10 to 12 bills per month) and involving the third-party transfer of funds, once the customer prepaid the repetitive bills. Criteria that will decide the success of home banking should include: convenience, easy of use, accuracy, reliability (Egner 1991) and new opportunity .

Another avenue of virtual exchange is home trading. In 1990, of 40 million individual investors in the United States, an estimated

2 million use PCs, and the securities industry claims that perhaps 100,000 are using them to manage portfolios (Siegfried 1989). By the end of the 1990s, individuals should have the technology available on home workstations to incorporate on-line trading, real-time quotes, graphics, portfolio management, on-line news, reports on investment activity, and historical data. Some of these services are now available but not readily accessible; "windowing" software to split the screen and merge other services may be expensive and difficult to operate.

Largely in 1984-1990, individual investors have begun to use at home trading systems based on a personal computer. Many of them have been quoted as saying that these systems give them a feeling of being "in control" (although none of the systems provides automated execution) and better equipped to compete with the institutional funds' professional investment managers. This perception is provided by brokers who provide the systems, and who have been alarmed by the perceived "flight of the small investor "[3].
The industry estimated that 400,000 individual investors will be using trading systems by 1992 [4]. Such estimates sometimes display more enthusiasm than analysis, but it appears that the number of users have tripled in 1987-1990 [4].

The most popular home trading system, provided by the largest discount broker, claims approximately 50,000 users. Several similar systems such as those of Charles Schwab, Inc., Fidelity Investments, and Quick & Reilly claim about 10,000 to 12,000 customers each [4]. They allow the investor, at his home computer to:

- Access research databases,
- Receive real-time quotes,
- Place orders and receive confirmations,
- Track the progress of a portfolio, and
- Set up dummy portfolio and track their progress.

Trades ordered through one of these systems go to a broker who routes the order to an exchange. The Fidelity Express Service claims that trades are checked within the system without human intervention and go directly to the exchange floor. The customer usually gets immediate confirmation of a trade, or if there is to be a delay of a minute or longer a confirmation is left in a "mailbox" in the system. The advantages to the investor are access to information before the trade, the ability to place orders 24 hours a day (but they can be executed when the change is open), and a slight reduction in transaction time, chiefly because there is no wait on the telephone for a broker. (Traders are said to take 15 to 20 seconds, in most cases). The feeling of "greater control," although it may exist, is not highly justified [4].

A do-it-yourself trading firm was founded in 1988 —All-Tech. It generates an astonishing total of nearly 2% of the national over-the-counter market's $5.2 billion daily volume. The firm has branches in Dallas, Minneapolis, and New York City. The trading is going through the National Association of Securities Dealers Automated Quotation System (NASDAQ), the world's largest stock market after the New York Stock Exchange. All-Tech and dozens of similar firms have become Wall Street's version of off-track betting parlors. Visitors to the Suffern office in New York find plumbers, bartenders, retirees and out-of-work lawyers crammed elbow to elbow in nine rooms filled with clerks, computer screens and high hopes. The screens show various "bid" and "asked" prices offered by dozens of market makers— firms such Goldman Sachs and Morgan Stanley that deal in specific stocks. All-Tech is merely using technology to give the little guy an even break. Under the small-order system, each market maker must trade at the displayed quote price with no chance to adjust to whatever other firms are doing. The system simply assigns the transaction to the market maker with the highest bid or lowest offer at the time. Result: if the Merrill Lynch specialist happens to be away from his desk when a stock starts moving, a savvy bartender could swiftly pinch $250 from his hide for every quarter-point change in price. Without the automated system, All-Tech's clients would have to call brokers and have them place orders with a market maker. By the time all that was done, the price gap would probably have vanished. The trading is restricted to 1000 shares per transaction and the number of transactions for each account is restricted to just 10 a day (Behar 1993).

In 1789, Wall Street got its start as a financial capital when the first Congress met there and authorized the issue of $80 million in war bonds. Ever since, in the last 200 years, Wall Street was the place where the majority of all capital-oriented transactions took place. But due to virtual exchange, the longtime gap between Wall Street and Main Street is closing. What used to be done in New York can now be done almost anywhere. Thanks to the reach of computers and their networks, an options trader can drop work when the market closes, step out his back door, and hike in the Rockies a minute later. Information technology is driving this transition along with real-estate costs and a new generation's insistence on the kind of lifestyle hard to pull off in New York. These forces are reshaping the geography of the securities industry, relocating thousands of jobs, billions of dollars and the power that goes with them. Along the way, attitudes are changing, as Wall Street learns what all American business is learning: it does not matter much where you call home.

New York's share of securities-industry employment in the United States dropped to 30 percent in 1992 from almost 50 percent in 1970 . Meanwhile, much of the industry's growth has happened elsewhere. Mutual funds, an industry born and bread in Boston with major firms like Fidelity spread all over, was once dwarfed by the brokerage giants based on Wall Street. Perhaps because that spread, mutual fund companies business has boomed, with equity volume of 475.4 billion in 1992 (Slomon 1993). Also, NASDAQ, the computerized market that operates from any dealer's desk, has taken a big chunk of business away from the New York Stock Exchange, where market makers must be on the floor. Companies that are Wall Street customers also have moved themselves outside of Manhattan.

No matter how far they are out of New York City, securities people can now assume that Wall Street will come calling. In 1992, Intel, a giant in microprocessor production, eliminated in-person quarterly briefings it had held for 22 years. Now, this discussion is conducted via international video conference with 180 participants.

Information is power in business, particularly in the capital market. It is virtually everything on Wall Street, and the information infrastructure is providing a radical solution of virtual exchange of wealth. There is nothing on Wall Street that is proprietary or protected. There are about 150 stocks at NYSE, which are in continuous 24-hour demand worldwide and which account for half the exchange's volume. In the 1993 NYSE launched an automated system to cross institutional trades without any involvement by specialists. It will compete with the Crossing Network and POSIT, which is expanding overseas. This type of system devastates much of the rest of Wall Street's work force. Driven largely by firms trying to remain competitive, the virtual exchange is transforming many activities once performed by skilled individuals into no-brainers—simple, clerical functions more effectively handled by customers with an access to computers and their networks.

Virtual exchange transforms trading into a remote terminal job. All stock exchanges have installed systems to handle small orders more or less automatically. The Big Board's DOT system, which routes orders under 4,000 shares directly to a specialists' post, eliminated a lot of workers from back offices and floor posts. Such facilities as Crossing Network and POSIT let investors cut out brokers and dealers. Some trades are done entirely electronically, from initiation by computers programmed to spot trading opportunities to automated clearance and settlement. Computers and networks are taking over the capital market. Technology is eroding Wall Street's once central role as an adviser to clients. Customers now have access to the same news, prices, quotes, and other market data that Street firms have. Also, many customers have the same, if not better, number- crunching capability. Today, the advisees sometimes know more than advisers (Welles and Roman 1990).

## A New Electronic Financial Order

For generations the international financial system, as well as the national financial systems of the United States, Japan, and the United Kingdom, were subject to a blizzard of government regulations that restricted transactions and limited product innovation. As a result, a world emerged in which financial strategy played only a minor role for corporations and business strategy only a limited role for financial services firms. In this environment, financial firms and their clients relied on products and transactions that were based on relationships as substitutes for those that compared to price performance.

This simple world of regulation and relationships became a complex one of choice and freedom in the 1970s and 1980s. The deregulation, securitization, internationalization, and electronization that took place in 1980s and 1990s triggered a fundamental change in the financial world. The new financial order breaks with the Bretton Woods system of fixed exchange rates and learns from the inflation of the 1970s, the U.S. budget deficit, and the Japanese accumulation of international assets. All these factors played important roles in the transition from regulated simplicity to the complexity of choice and are essential in understanding today's financial environment.

One of the most important contributors to the complexity of the financial world is the spontaneous advance of information and communications technologies. The new electronic systems facilitate the transactions of the new financial order. Meerschwam (1991) argues, however, that the new opportunities for product complexity they offer are thoroughly in the spirit of the price-driven environment. He agrees, although, with the argument that information differences between issuers and borrowers are the *raison d'etre* of intermediaries, and that information & communications technology have made information much more widely available, and financial disintermediation is taking place. Furthermore, he says that electronization and product complexity have also introduced new risks. Program trading and portfolio insurance— sophisticated strategies dependent on price-sensitive products—offer good examples. There is no hesitation that their use triggered the stock market collapse on October 19, 1987. The program trading system was not designed to function in an environment of very large (high volumes trading) and rapid changes, where imbalances and prices did not carry information about the market conditions.

Figure 3 - 4 illustrates the architecture of the National Exchange

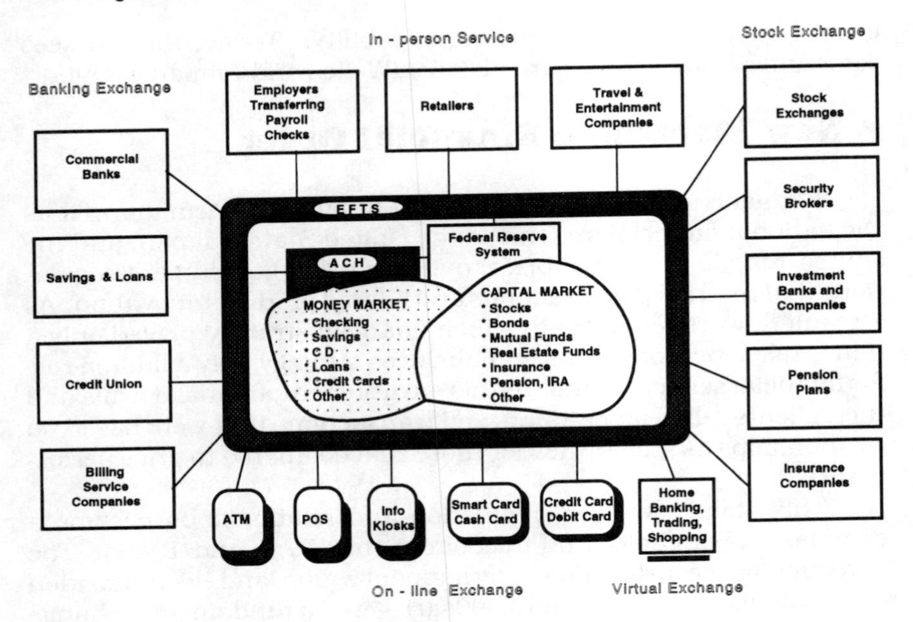


*Figure 3-4: The Architecture of the National Wealth Exchange System*

System. This system introduces *de facto*—a new electronic financial order. Its solution is based in information technology which integrates and smoothes a flow of electronic money among the customers and financial institutions. The EFT System is the core financial information highway which provides convenience, control, and connections. Through this information highway all phases of financial institutions development (from banking exchange to virtual exchange) can be achieved. Only one condition has to be met, the participating institutions should have a gateway to EFTS.

Financial institutions of a new order are becoming an information-processing service industry. Banks are shifting from depository institutions to information-processing-handling centers which can add value to electronic money. Electronic Data Interexchange (EDI) is replacing paper handling. The sophisticated image technology can generate data from checks, and financial networks (such as CIRRUS, PLUS, CHIPS, SWIFT) can transfer electronic money and their added value information around the nation and globe. The new order means that the distinction among depository institutions and other financial institutions becomes a blur. The geography of making financial business, particularly in state restricted banking, will gradually disappear since the information highway makes these restrictions irrelevant. Perhaps, even the legal limits on interest rates will be eliminated by the markets' "invisible hand." As a result of this strategy, the multiplicity of regional, national, and international networks has invaded the financial territory. It is a temporary and even healthy stage of the growth process. The mature level of networking will require the consolidation and sharing of common resources, namely the financial information highway. The new emerging information highway has to provide reliability and throughput to process 1.5 trillion transactions only in the United States, and perhaps, about 5 trillion transactions world-wide on a daily basis.

There are two factors which contributed to the emergence of a new electronic financial order: the deregulation of the banking industry with the passage of the Depository Institution Deregulation and Monetary Control Act (MCA) in 1980, also called the Big-Bang of finance, and the deregulation of the telecommunications industry, which caused the break up of AT&T in 1984. The former energized the financial community, the latter supported this new business energy through technology. Both factors are interdependent and equally important.

The new electronic financial order is just in the making. It will require further system-related regulations, such as common solutions in the scope of reliability (by-passing paths), standardization, interconnectivity protocols, security, and privacy (possible computer surveillance).

# Conclusions

The emergence of a new electronic financial order provides a strong foundation for the operations of an information civilization. It is a component of the information infrastructure, which acts as a "gas and sugar" to energizes the individuals and organizations' performance. Electronic money not only supports existing operations but also generates new added value through new information and opportunities.

Businesses, institutions, the economy, and society at large will change or even shift modes of acting. The five-century old print civilization will be replaced by a new information civilization. New problems, opportunities, and solutions will appear as the new order matures. For some time, however, both civilizations will act concurrently and intensify management, political, and social issues.

## End Notes

1. United States Investor/Eastern Banker, January 13, 1975, p.27.

2. "Your Wallet in The Year 2000," by T.A.S. FORTUNE Special Issue, Autumn 1993, p. 162.

3. The "small investors" do 18.2 percent of trading in 1990, down from 19.7 percent in 1987, according to a study by the Securities Industries Associa-

tion. This has been decreasing for years.

4. U.S. Congress, Office of Technology Assessment, Electronic Bulls & Bears: U.S. Securities Markets & Information Technology, OTA-CIT-469, Washington, DC: U.S. Government Printing Office, September 1990.

5. U.S. Congress, Office of Technology Assessment, Trading Around the Clock: Global Securities Markets and Information Technology, OTA-BP-CIT-66, Washington, DC: U.S. Government Printing Office, July 1990.

6. A definition offered by the National Commission on Electronic Fund Transfers (NCEFT) established in 1974 by the U.S. Congress (Public Law 93-495).

7. Source: American Bankers Association, Research and Planning Group, also in Colton and Kraemer (1980).

8. Author's estimation.

## References


Arend, M. (1992), "High-tech Banking Centers Add Value to Branches," *ABA Banking Journal*, November, pp. 3941.

Behar, R., (1993), "Bypassing The Brokers," *TIME*, June, pp.42-43.

Bequai, A., (1981), *The Cashless Society: EFTS at the Crossroads*, New York: John Wiley & Sons.

Colton K.W., and K. L. Kraemer (1980) *Computers and Banking*, New York: Plenum Press, p.32.

Eastbrook, M., (1988), *Programmed Capitalism*, Armonk: NY: M.E. Sharpe, Inc.

Egner, F.E. (1991), *The Electronic Future of Banking*, Naperville, Ill: Financial Source books.

Garbade, K.D., and W. L. Silber, (1978), "Technology Communication and the Performance of Financial Markets," *The Journal of Finance*, vol. XXXIII, no.3, June, pp.819-832.

Gardner, D.S., (1982), "Marketplace: A Brief History of the New York Stock Exchange," for the New York Stock Exchange, Office of the Secretart.

Gart, A., (1989), *An Analysis of the New Financial Institutions*, New York: Quorum Books.

Hempel, G., A. Coleman, and D.D. Simonson, (1983), *Bank Management: Text and Cases*, New York: John Wiley & Sons.

Lipis, H. A., Th. R. Marshal, J.H. Linker (1985), New York: John Wiley & Sons.

Meerschwam, D.M., (1991), *Breaking Financial Boundaries*, Boston, MA: Harvard Business School Press, p.203.

Morgan , E.V., (1965), *A History of Money*, Baltimore, MD: Penguin Books, p.16.

Philo, E., and K., Ng, (1989), *Reuters Holdings PCL*, Goldman, Sachs & Co., New York, NY: February, p.5

Skoba, R.A., (1974), "The Impact of Marketing on Electronic Funds Transfer," *United States Investor/Eastern Broker*, September 23, pp.17-18.

Smith, A., (1950), The Wealth of the Nation, ed. E. Cannan, 6th ed., Methuen, vol.1, p.15.

Sprague, R. E. (1974), "Electronic Funds Transfer Systems: The Status in Mid-1974," *Part I, Computers and People*, August, 1974, p. 34.

Solomon, E. (1991), *Electronic Money Flows*, Boston: Kluwer Academic Publishers.

Welles, Ch. and M. Roman, (1990), "The Future of Wall Street," *Business Week*, November 5, pp. 119-124.

Wiseley, W., (1977) *A Tool of Power, The Political History of Money*, New York: John Willey and Sons.

Yang, C., H. Gleckman, M. McNamee, Ch.Hawkins, P. Coy, (1991), The Future of Banking, *Business Week*, April, pp. 72-76.

CHAPTER 4

# Electronic Knowledge

## A Pathway for Literacy, Productivity, Innovations, and Democracy

The explosion of innovative technologies in 1980s and 1990s, initiated by the rapid electronic information creation and utilization, has been dramatic. The now identifiable developing trend challenges the former national economic cornerstone of reliance on manufacturing productivity with the need to enhance literacy, increase productivity of services and production, accelerate innovations, and strengthen democracy to meet the requirements of an information-dependent world. As a result, the pivotal need for libraries has to be defined in order to store and disseminate public information and knowledge, which are instrumental in improving literacy, productivity, innovations, and democracy. These qualities are critically based on the storage and dissimulation of information and knowledge [1].

### Literacy and Libraries

The effects of illiteracy permeate the fiber of the nation, undermining the ability of its citizens to live and work in the world of today and to meet the challenges of a rapidly changing world of tomorrow.

The globalization of the world marketplace and its information resources dictates what, where, and how we educate our citizens. These citizens must compete in the world arena and develop literacy in all forms to effectively absorb information in many new forms and formats.

Though the United States boasts one of the highest standards of living, literacy remains a vexing problem: The U.S. ranks 49th in literacy among the 158 member countries of the United Nations. In real terms, Project Literacy U.S. estimates that as many as 23 million adult Americans are functionally illiterate, lacking skills beyond the fourth-grade level, with another 35 million being semi-literate, lacking skills beyond the eight-grade level. One state, Texas, estimates in 1988 that illiteracy cost that state $17.2 billion yearly through lost productivity, unrealized tax revenue, welfare and crime-related costs. With literacy's integral role in an individual's self-image and as a common denominator that brings people together, the toll in human terms is evident.

Trends show that diversity will increase during the 21st century. Fully 30 percent of U.S. school children are from racial or cultural minority families. Early in the next century that percentage is projected to increase to 35 percent. Some states are expected to have no "majority" group by the Year 2000. Coupled with this trend is the representation of every known religious denomination, more than 100 different languages, and the spectrum of special-needs Americans from the gifted and talented to those with learning difficulties and physical limitations. The need and scope of the task become apparent.

In addition, some trend data project an increase in the number of children living below the poverty line, which has been shown to create further challenges to learning and literacy. The current national high school dropout rate of over 30 percent is another measure of literacy.

Literacy provides leverage for responding to the needs of the increasingly diverse population in a fast-paced, competitive world. Literacy is, in fact, the fulcrum for increasing productivity, creating innovations, and strengthening democracy. Libraries and information services are at the heart of the nation's education program which fights illiteracy.

## Productivity and Libraries

Old definitions give way to new. Productivity, the measure of a worker's output in relation to resources, most often has been associated with raw materials and tangible resources. But a labor-intensive economic system is being supplemented by an information-based economy. That old definition now has expanded to include information and knowledge as a resource—and involves reliance on judgments about source credibility, timeliness, format, and utility for application to the end product. These factors are not easily measured by traditional productivity standards, but are critical in an Information Age which can cloud a worker's sense of productive contribution to society.

The abundance of technologies and associated information places new demands on people in the work force who must adapt to these changes. The velocity and rapid turnover of information has created today's "knowledge worker" who must be prepared with lifelong learning habits, access to relevant information, and analytical skills to remain productive in his/her chosen field. Some estimates indicate that today's worker will have to update skills every three years.

Collection, preservation, and retrieval of information and knowledge in a timely and useful form for the end user is a major goal if we are to build and maintain a productive, competitive work force in an interconnected global market. The nation which moves to an information-based economy, harnessing knowledge through technology, and applying it through an educated work force will assure its people economic independence and the standard of living they desire.

The emergence of the "knowledge worker" requires us to recognize libraries and information services as educational institutions for lifelong learning of "knowledge skills." The Information Superhighway built by the National Research and Education Network should encourage the creation of cooperative information and technology partners at all levels of the economy. The sharing of national intellectual resources should lead to higher productivity.

## Innovations and Libraries

The competition among nations and companies in developed environments is more and more relying on innovations. Citizens, workers, firms, and nations should actively seek out pressure and challenge, not try to avoid them. The task is to take advantage of sources of information and knowledge and thus create the impetus for innovation.

Innovations are generated by imagination and knowledge. Imagination alone is not good enough to create worthy innovation. Knowledge is a key ingredient of innovating power and a library and information services is a tool to disseminate that knowledge.

Beyond the pressure to innovate, one of the most important advantages a society, individual or firm can have is early insight into needs, environmental forces and trends that others have not noticed, but will be important in the future. This can be done through the intensive application of library and information services.

### Democracy and Libraries

As dependence of information grows, the potential increases for the emergence of an Information Elite —the possibility of a widening gap between those who possess facility with information resources, and those denied the tools to access, understand, and use information.

This dichotomy could threaten to send fissures into the democratic base of the nation. Thomas Jefferson's warning that the success of a democratic society depends upon an "informed and educated" populace could well have been proclaimed today. The intellectual freedom to access information and pursue truths, make judgments, and achieve goals as full participants in society is the bedrock of a strong democracy.

Today, more than ever, information and knowledge is power. Access to it — and the skill to understand and apply it — increasingly is the way power is exercised. Information has become so essential that a large and growing part of federal, state, and local government, academic institutions, and the private sector work force is engaged in information-related activities. Tens of thousands of organizations, from small businesses, publishers, and associations to global industries, work in the trade of information distillation and delivery.

In the U.S., information delivery systems include more than 30,000 public, academic, and special libraries, and an estimated 74,000 school libraries and media centers. It is a telling testimony to the insight of Benjamin Franklin that, in 1731, he established the nation's first library, the Philadelphia Library Company. In the world, there are an estimated 750,000 public, academic and special libraries containing about 100 billion volumes.

As literacy is a key to productivity, innovations, and democracy, so literacy, productivity, and innovations are essential to a strong "educated" democracy. All are intertwined, interconnected, interdependent, and inseparable. Thus, the role of libraries as "schoolrooms for lifetime education" is central to the nation's long-term political stability and prosperity. The democratic nation should afford equal information opportunity to the broadest number of citizens for their full and informed participation in all aspects of the nation's economic, political, cultural, and intellectual life. The support of a strong democracy is the intellectual freedom to inquire, discover, question, validate, and create.

The challenge remains to provide integrated, cohesive, cooperative national policies and programs to crystallize the continued educational contribution of libraries to enhance literacy, productivity, innovations, and democracy [1].

## Role of Information Technology in Libraries

All libraries employ a variety of information technologies in support of their mission of "allowing people to utilize information," (Turoff and Spector 1976). A library is an institution that acquires, manages, and disseminates information. Moreover, "a library is a bibliographic system regardless of the situation in which is placed, and the task of the librarian is to bring people and graphic records together in a meaningful relationship that will be beneficial to the user," (Wilson 1977).

Information technologies are not "new" to libraries. A broad scope of IT have been applied by libraries for years and have effected all aspects of library operations and services. In fact, almost every library function has been altered to some extend by electronics, computers, and telecommunications.

Libraries may employ one or more of the following technologies and/or technological applications: microcomputers, on-line data services (bibliographic search), networks such as OCLIN (On-line College Library Center) and RLIN (Research Libraries Information Network (RLIN), automated information systems, electronic bulletin boards, optical disk technologies such as videodisk and CD-ROM, facsimile, and microfiche and related equipment.

### On-line Database Services

On-line database services, such as DIALOG, BRS, and other computerized retrieval systems, cover a wide array of continually expanding subject areas. Each database is a compilation of textual, statistical, and/or bibliographic information. Bibliographic and referral databases are sometime called reference databases, whereas numeric and textual-numeric databases are called source databases. In 1979-80, there were 400 databases, 221 database producers, and 59 on-line service available. By 1987, there were 3,169 databases, 1,494 database producers, and 486 on-line services [2].

These services allow rapid access to information sources, integrate information for the user, permit libraries greater flexibility in a choice of format, and provide access to previously unavailable information. Use of these services also allows the library to be less dependent on paper or hard-copy indexing materials.

### Library Communications Networks

Two or more libraries may form communication networks utilizing information technologies to enhance the exchange of materials, information, or other services. The formation of local, state, regional, and national networks has significantly altered the operation of libraries. There are several types of networks —bibliographic utility, regional service organizations, and others. WESTNET, SOLINET, AMOGOS, CLASS, and the like are regional service networks that facilitate the expansion of the bibliographic utility. Although bibliographic utilities began as a means for libraries to reduce costs of cataloging, their primary function today is for sharing of resources. One example of bibliographic utility is OCLC, a major computer-based cooperative network with over 8,000 members and employed by all types of libraries nationally and internationally. The OCLC network assists librarians in:

- acquiring materials
- cataloging materials
- ordering custom-printed catalog cards
- initiating interlibrary loans
- locating materials in member libraries
- accessing other databases
- searching for government documents.

Many depository libraries use OCLC with management information systems [3].

## Electronic Bulletin Boards

Libraries are applying electronic bulletin boards in support of library operations such as:

- interlibrary loan (ILL)
- resource-sharing functions
- access to current information located elsewhere
- professional dialogue on library development and operations
- managerial communication

The Wisconsin Interlibrary Service (WILS) network is one example of the growing use of bulletin boards in libraries. The WILS network is used by 30 member libraries and handle over 90,000 requests a year [3].

Libraries are subscribing to governmental bulletin boards. For example, the Remote Bulletin Board System (RBBS) of the National Science Foundation contains useful information for research activities.

## Optical Disks

Optical disks technologies include videodisks, compact audio disks, CD-ROMs, optical digital disks, and others. Despite a lack of common information access and retrieval standards, an increasing number of vendors are introducing database services on optical disks (they use the differential reflection of light from a mirror disk surface as a means of reading information). Databases are popular on this type of a medium, among many, one can list: the Wilson business database, Books in Print, Ulrich's Periodicals Directory, Forthcoming Books, AGRICOLA on agriculture, and so forth. The Library Corporation sells the Library of Congress (LOC) MARC catalogue records on four optical disks.

Since CD-ROM disks cannot be updated (they are replaced), therefore they are not practical for time-sensitive data.

### Facsimile

Facsimile machines are a very quick method of transmitting printed information between libraries. For example, the National Library of Medicine will send up to 20 pages of professional literature to a member in support of emergency patient care.

## Alternative Futures for Libraries

With the increasing number of information technology-driven projects in libraries, the impetus to automate and to include electronic information in depositories and disseminate it is strong. IT products improve access to library information and negligence to include these products/systems could limit the public access to information and knowledge. The following alternatives of libraries futures are offered:

Alternative I: *Status Quo,* where information is stored and disseminated in paper and microfiche formats with a few CD-ROMs and a few on-line files.

Alternative II: *Automated Library,* where information is stored in a paper format, however, cataloging, retrieval and circulation of information is automated by computer information systems and services.

Alternative III: *Electronic Library,* where information is stored and disseminated in the electronic format via telecommunications and computer networks,however, there is a depository (Information Museum) of books, journals, and documents, and only certain type of information and services are provided locally in a timely manner.

Alternative IV: *Virtual Library,* where information is stored and disseminated in the electronic format to the end users at workplaces, schools and colleges, as well as at homes. Furthermore, a depository type of a library is eliminated.

| Alternative | Goal | Information | Maintain info. | Gateways |
|---|---|---|---|---|
| Status Quo | Growing Collection | Paper, Microfiche, CD-ROM | Catalog Cards | Very Limited |
| Automated Library | Improved Service | Paper, Microfiche, CD-ROM | Automated | To Several On-Line DB and Info Services |
| Electronic Library | Improved Access | Electronic Format, Local Services of Scope | Automated and Networked | To Many On-Line DB and Info Services |
| Virtual Library | Improved Choice | Electronic Format | Automated and Networked | To All Available On-Line DB and Info Services |

*Table 4-1 A Comparison of Library Alternatives*

The last two alternatives of the future libraries implicate the loss of many patrons to private sector information providers. Libraries, particularly public ones, will lose business, and more important, will lose the financial and political support of middle-class citizens. This group of citizens constitute the major group of home computer systems users with access to electronic or virtual libraries dispersed in the world. They will support scope (niche) electronic information providers rather than near-by libraries of scale.

Table 4-1 compares goals and solutions of the four library alternatives.

This comparison implicates a radical change in the library organization. Large collections of paper information are becoming more a matter of history than a goal of improving the access and choice. Those libraries with more gateways to electronic information and services will achieve the higher status than those with large old library buildings.

## Automated Library

The Automated Library Alternative dominates the present level of academic and research libraries' development paradigm. Its architecture is depicted on Figure 4-1, applying the Targowski model (Targowski 1990).

The Automated Library aims to improve services to the patrons either through widely available services (OCLC) or via a so-called the library integrated system. The integrated systems are usually installed by software providers, such as Library Corporation, Information Dimension, multiLIS, Data Trek, CARL, NOTIS, CoBIT,

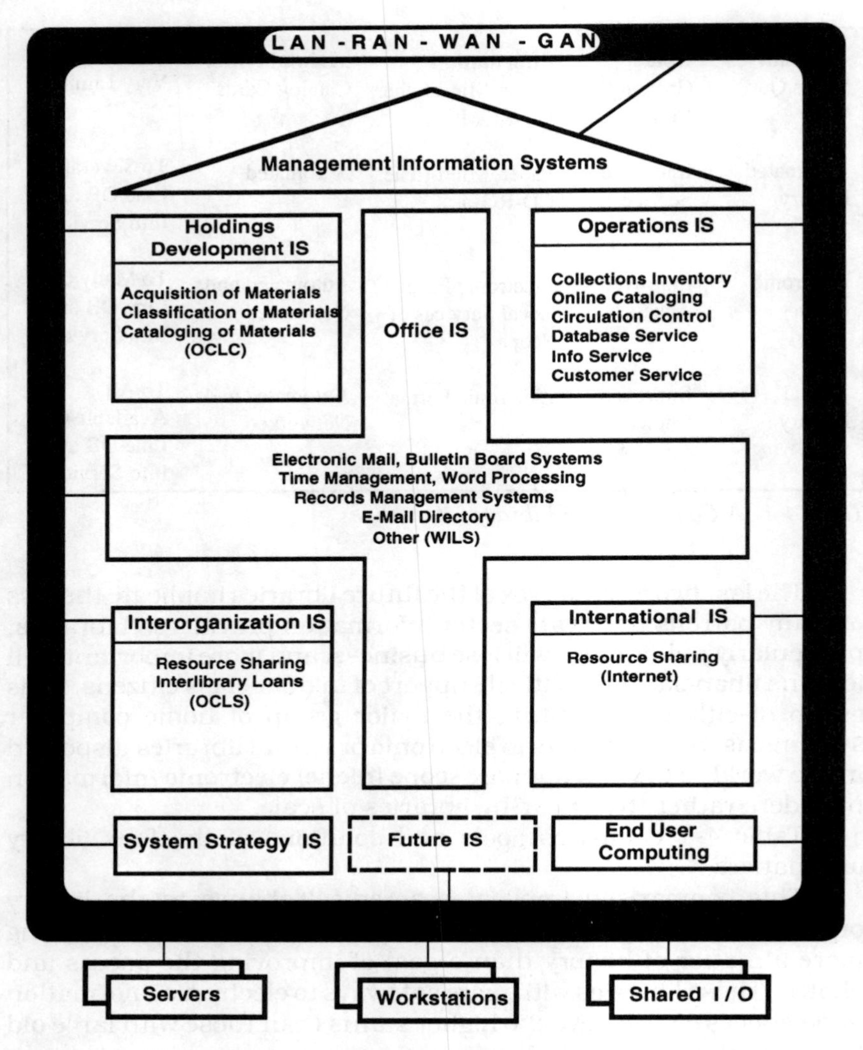


*Figure 4-1: The Architecture of Automated Library*

CARL, and others.

A library at this stage of development is at a cross-roads. It is a "prisoner" of its own old holdings and wants to improve its general library services via automation. It maintains its presence in two opposite environments. By tradition, it wants to expand its own holdings, but being under pressure for cutting edge solutions, it wants to apply as many resources as possible for the automated services. Of course, such libraries face a lack of funds and staff support. These libraries may be in a state of confusion with a lack of vision (unfulfilled promises and expectations).

## Electronic Library

The Electronic Library Alternative is a solution of a new quality. It is still a "library" which provides the automated services and electronic easy access to holdings stored electronically. While these holdings may be stored locally, it will be more probable that more and more holdings will be stored elsewhere. A library gateway will transmit patron requests through LANs, MAN, WANs, and GANs to other electronic libraries. The old type of a status quo library will disappear. The electronic library will become a place where a certain type of information or service is available, perhaps, locally oriented. The architecture of this library is illustrated on Figure 4-2.

This type of a library will prevail in academic and research institutions. Public libraries can survive only if they will develop locally-oriented holdings and services. The public libraries cannot be so general anymore. They will find that library services have to be provided to some people in specialized areas only. The public libraries may change from a storehouse of books and other materials to clearinghouses or that of a repository of programmed experiences.

Shuman (1989) names such a library an "Experience Parlour." According to the quoted author, the staff now call themselves *experience facilitators* or *neotravel agents*. People are not looking any more for passive usage of two-dimensional films and pages. They

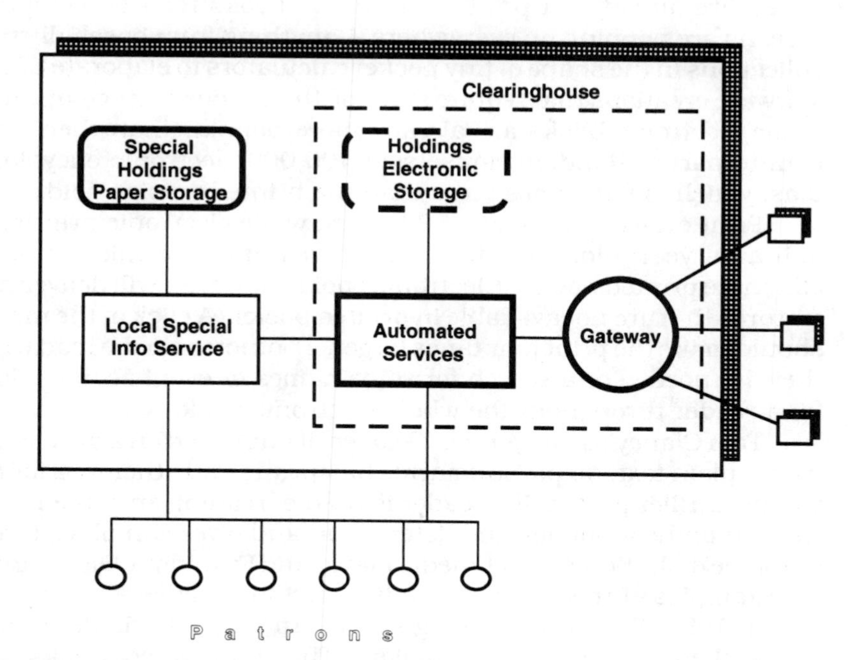


*Figure 4-2: The Architecture of Electronic Library*

seek "all-at-onceness" and actual participation in the experiences they witness. People don't just want entertainment anymore, they want roles. Users of the experience parlour visit them for the following reasons: (1) to escape routine, (2) to test a new lifestyle before abandoning their own, (3) to be entertained in novel, involving ways, (4) to be challenged by different or exciting circumstances, (5) to learn about something new, and (6) to realize fantasies which, for a multiplicity of reasons, will have to remain fantasies.

Such television shows as "Fantasy Island," "Lifestyles of the Rich and Famous," "The Millionaire," "Dynasty," and "Dallas," motivated people to become these "fortunate" people from the shows. Therefore, if they cannot immigrate to the United States or Western Europe or other developed countries to fulfill their dreams, they would like at least to experience dreams under conditions as realistic as possible. Needless to say, the majority of North Americans, West Europeans, and so forth would like to implement their dreams too.

The successful development of the electronic library depends on the development and cooperation of other electronic libraries. This process will be incremental, mostly driven by private information vendors and the climate created by government for the application of the electronic infrastructure. Libraries, however, cannot wait for a "supply of ready solutions." They have to define a vision of their own environment and experience pilot and demo systems.

The electronic library is not far away, since the first electronic books are already "in print." Electronic books read on computer screens are popping up everywhere —anything from baseball-trivia collections in the shape of tiny pocket calculators to elaborate $2000 software creations that require state-of-the-art desk top-computers. Some electronic books are already more popular than their print counterparts. Random House sold 400,000 electronic encyclopedias, which is four times more than the printed version sold.

Print evolved in the last 500 years, while electronic printing is only a few years old. It will take some time until electronic books will eliminate printed books. Electronic books, however, will deliver new features that are not available in printed books. A click of the mouse should make the print four times larger. Footnotes can be read when they are needed or a search for some names or events can be done for a reader throughout the whole electronic book.

Tom Clancy's *Hunt for Red October*, if published on a disk, could have a plain text, maps, submarine blueprints and other toys for the techno-thriller junkie. If a reader is tired of reading on screen, they can put on headphones and listen to an audio version played from the same disk. Sony's multimedia Player and Franklin's Digital Book are examples of this trend which is tested in practice.

In 1992, IBM begun selling schools the tools students need to create their own electronic books. The Illuminated Books and

Manuscripts system is pricey —$2000 for the software and about $8000 for the adapted PC. It comes with five documents already "illuminated" —each one appears on the computer screen as a printed page, and the students can easily order up sound and video presentations to accompany the text. Actors, for example, provide various readings of *Hamlet*; politicians, including Joseph P. Kennedy II and Daniel K. Inouye, discuss the Declaration of Independence. The IBM package also includes an additional 100 books on disk— from the collected Shakespeare to the Bible — ready to be illuminated by the students by attaching their own text, audio and images, linking them all together through the technique called hypertext.

In the 1990s, some publishers of academic textbooks, such as McGraw Hill, offered instructors a custom designed textbook for a given class. Publishers are beginning to store electronically written textbooks and later can assemble needed chapters into one specific textbook.

William Gibson a prophet of Cyberspace, who created much of computer-hacker mythology with cyberpunk novels like "Neuromancer" and "Mona Lisa Overdrive" produced a limited edition of "Agrippa." It is an electronic book that one can read on a computer only once. After that it turns into nonsense. The book costs $1500 a copy. It is rather an investment than fun to read (Rogers 1992).

Paper and electronic books will coexist for decades to come. In the near future, however, books will be delivered electronically to bookstores, then printed on demand, one copy at time or down loaded to a customer's computer on location or through a network. Today, this type of a readership requires computers that are still expensive.

In the coming future, a public library will able to provide easy and inexpensive access to many electronic works of artists, scientists, and professionals. This will be the beginning of the knowledge democratization process.

## Virtual Library

The Virtual Library Alternative is the implementation of a new way of accessing information. So far people have been reaching out for information, visiting libraries, news stands, and so forth. In this alternative, information will come to people, even "automatically." An user not only will look for a bigger choice of information via global networks and electronic holdings but will be receiving information via smart "info absorbers." The user will program an info absorber that will look for electronic information of interest anywhere on the globe. Even without patron intervention, such information may soon

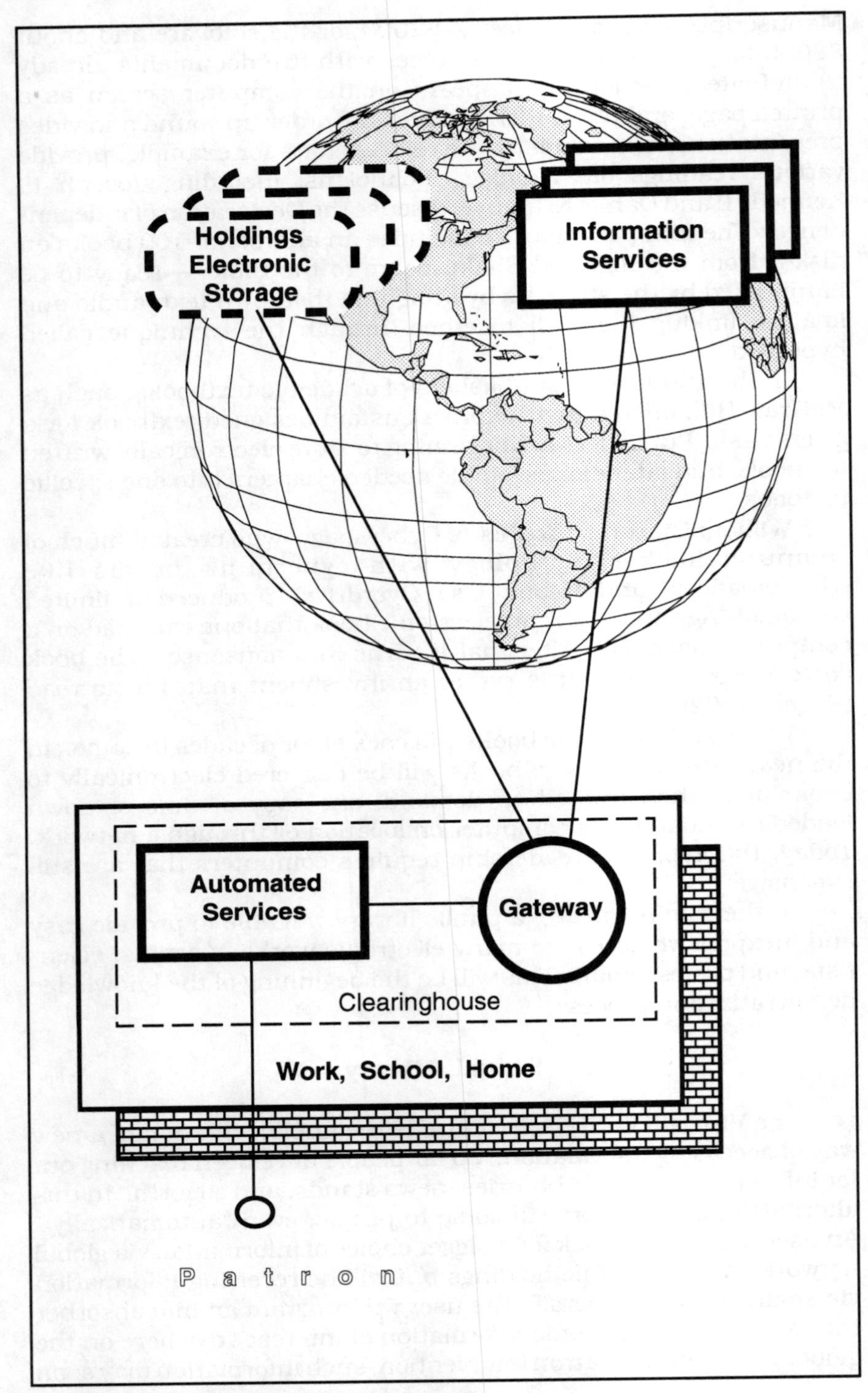


*Figure 4-3: The Architecture of Virtual Library*

flow to his/her work-school-home computer.

The architecture of the virtual library is shown on Figure 4-3.

Readers will find it more convenient to search for information from their own place of work rather than visiting libraries or even using/buying libraries intermediate services. A library's classic functions are downloaded and carried-out from the user's location. A virtual library, however, provides not only convenience, but more important, it provides a much larger choice of information. At this stage of information infrastructure development, private and governmental information providers will take over library services. A patron will be able to access electronic sources of information directly and also get direct information services.

## Informing the Nation

The libraries, regardless of their alternatives are depositories of human culture. They preserve and disseminate information, concepts, knowledge, and wisdom. They are tools of human cognition which make us wiser and more aware of our needs to survive and blossom. An informed man is a free man who knows how to sustain healthy environments in economy, education, and politics. " If a Nation expects to be ignorant and free in a state of civilization, it expects what never was and never will be...if we are to guard against ignorance and remain free, it is the responsibility of every American to be informed." (Thomas Jefferson, July 6, 1816).

In the 1990s, the U.S. is "A Nation at Risk" which, with the help of better information, concepts, knowledge and wisdom should become "A Learning Nation" and eventually "A Nation of the Informed." Libraries and information services are essential to a learning and informed nation.

To be an informed nation, some solutions for the information infrastructure have to be provided. The nation has to be led, focused, and facilitated in order to implement these solutions. The executive and legislative branches of the federal (national) government should serve as a catalyst to energize the local, state, federal, and private sector in information areas.

The focus on ways of achieving the informed nation can be explained by the list of following recommendations groups (that include 95 specific recommendations), by the 1991 White House Conference on Library and Information Services (with 984 participants, this conference was a summary of 100,000 Americans participating in local, state conference forums who provided 2,500 policy proposals) [4]:

- Availability and access to information
- National information policies

- Information networks through technology
- Structure and governance
- Service for diverse needs
- Training to reach end users
- Personnel and staff development
- Preservation of information
- Marketing to Communities

To be informed, a person has to have information rights, something like *habeus corpus* or rather *habeus scriptus*. The mentioned conference offered the following People's Information Bill of Rights [4]:

- All people are entitled to free access to the information and services offered by libraries, clearinghouses, and information centers
- All people are entitled to obtain current and accurate information on any topic
- All people are entitled to courteous, efficient, and timely service
- All people are entitled to assistance by qualified library and information services personnel
- All people are entitled to the right of confidentiality in all of their dealings with libraries, clearinghouses, information centers, and their staff
- All people are entitled to full access and services from library and information networks on local, state, regional, and national levels
- All people are entitled to the use of a library facility or information center that is accessible, functional, and comfortable
- All people are entitled to be provided with a statement of policies governing the use and services of the library, clearinghouse, or information center
- All people are entitled to library and information services that reflects the interests and needs of the community.

The attention has to be brought to the need for appropriate reading materials and programming for children, youth, and adults with assorted disabilities: perceptual, neurological, binocularity, ocular motility, dyslexia, etc., as well as blindness, and to increase the education of librarians adequately to meet the demands of handicapped young and old people.

The growth and augmentation of the nation's libraries and information services are essential if all citizens, without regard to race, ethnic background, or graphic location are to have reasonable access to adequate information and lifelong learning.

## Informing the World

The access to information and ideas is indispensable to the

development of human potential, the advancement of civilization, and the continuance of enlightened self-government. The emerging satellite communication networks and other technologies offer unparalleled opportunity for access to education opportunities in all parts of the world. The emergence of the Electronic Global Village (Targowski 1991) offers the opportunity that the learning potential inherent in all children and youth will be reinforced throughout the globe, especially in the scope of literacy, reading, research, and retrieving skills.

Libraries and information services available electronically at all corners of the world should limit existing information ignorance among the world's almost 4.5 billion people. These people do not necessarily have access to good information sources. With the help of their electronic public libraries and Global Area Networks, these libraries should provide information to uninformed people. The uncensored information and knowledge is the first prerequisite to abolish "information slavery." in the world.

"Information slavery" is one of the main sources of the growing spread of fundamentalism, racism, and nationalism in the Second and Third World countries. Conflicts in the 1990s, such as in former Yugoslavia, the former Soviet Union, the Middle East, or some African countries are the result of a lack of education and still existing mentality of the 18th and 19th centuries. In times when Western Europe eliminates its own borders, in Southern Europe, former Yugoslavia, thousands of people are killed just in order to keep strong ethnic lines.

The people from all countries need to be informed and better educated through their whole life to know of all their options and be wise to solve their own problems in a rational manner. At the stake is "to be or not to be" of human kind and..... each of us.

## Conclusions

The development of libraries and information services is driven by accelerated information technology progress and a national strategy of using information as a key resource in supporting the economy and democracy. The eclectic influx of automated and electronic systems causes some confusion among library management, sponsors, and patrons. This needs to be replaced with a vision of ultimate solutions and needs support for given library services. On the one hand, the strong pressure by private information vendors will move libraries on to higher levels of solutions. On the other hand, the patrons will search for better information services and choices. In effect, it may lead to the death of the classic library. Certainly, if the virtual library becomes a reality, then libraries will evolve into clearinghouses and distributors of information services.

## End Notes

1. This overview of challenges facing libraries draws from remarks of the 1991 White House Conference on Libraries and Information Services by President George Bush and keynote speakers Congressman Major R. Owens (D, N.Y.); William T. Esrey, Chairman and Chief Executive Officer, United Telecommunications, Inc.; and Mary Hatwood Futrell, Senior Fellow and Associate Director of the Center for the Study of Education and National Development, George Washington University.

2. Cuadra Associations, Directory of On-line Databases, New York, NY: 1986, vol. 7, No. 3, p.v.

3. U.S. Congress, Office of Technology Assessment, Informing the Nation: Federal Information Dissemination in an Electronic Age, OTA-CIT-396, Washington, DC: U.S. Government Printing Office, October 1988.

4. Information 2000, Library and Information Services for the 21st Century., Washington, DC: the U.S. Government Printing Office, ISBN 0-16-035978-3, 1991

## References

Rogers, M. (1992), "The Literary Circuitry," *Newsweek*, June 29, pp.66-67

Shuman, B, A. (1989), The Libraries of the Future, Englewood, CO: Libraries Unlimited, Inc. p.89.

Targowski, A. (1990), *The Architecture of Enterprise-wide Information Management Systems*, Harrisburg, PA: Idea Group Publishing.

Targowski, A. (1991), "Strategies and Architecture of the Electronic Global Village," *The Information Society*, vol. 7, No. 3. pp. 187-202.

Turoff, M and M. Spector (1976), "Libraries and the Implications of Computer Technology," *Proceedings of the AFIPS National Computer Conference*, vol. 45.

Wilson, P. (1977), *A Community Elite and the Public Library: The Uses of Information in Leadership*, Westport, CT. p.xii.

# Part II

# Enterprise Information Infrastructure

# CHAPTER 5
# Virtual Business

## Enterprise-wide Computing

Enterprise-wide computing is a new challenge for technology to find solutions for business information problems. Questions senior executives are asking more frequently include:

•How do we change our position in the market?
•How can we make more money in less time?
•How can we spend time better?
•How do we reduce inventories?

Quite often, enterprise-wide computing is a part of the solution. Enterprise-wide computing contributed to the creation of a "global marketplace." The ability to link different facilities at different locations around the world offers the promise of really extending an organization's reach beyond the traditional boundaries of an organization, not just dispersing "pockets of business." At the same time, the competitive pressure of these emerging international markets are largely responsible for creating increased demand for more sophisticated technological advancements. The same companies — once satisfied with the efficiencies created by linking computers together — are now looking for "global solutions" that allow them to go beyond these initial efficiencies to develop competitive barriers.

Enterprise-wide computing means linking computer applications from different departments, different locations, and even different companies to achieve some common purpose. Enterprise-wide computing has the potential to help a company be more *effective* and *do things right.* Large, multinational and global companies can use enterprise-wide computing to capture the synergy among business units and at the same time keep corporate overhead costs down. For example, a company that buys raw materials from many different countries might put in a sophisticated, networked purchasing system that not only locates the best prices but also factors in exchange rate and other variables. A company with similarly configurated manufacturing facilities in different countries might use a global information system to help make decisions about where to manufacture certain products. If things are slow in one plant but backed-up in another, it may be possible to shift manufacturing to the slower plant rather than pay overtime at the busy plant. It is very hard to be *flexible* without an enterprise-wide information system to evaluate the trade-offs quickly [1].

Enterprise-wide computing is the next step in the evolution of computing. Over time, computing has extended beyond a highly centralized data processing function. In the 1990s, they want to focus on the glue, and networking, that will put this whole array of computing devices together.

A key success factor to enterprise-wide computing is the architectural planning of information technology components as tools supporting a business strategy. An architecture providing for integrated applications across multiple tiers of the enterprise is vital. The architecture should provide information across the enterprise, much like a utility provides electricity. When discussing architectures, it is important to understand the various components comprising a business information architecture:

- Business aims architecture, such as a configuration of creed, mission, goals, strategy, objectives, policies, tasks, resources, processes, organizational structure, management control, culture and so forth that describe a class of a business (e.g.: world class manufacturing aim).
- Information architecture describing the integration of applications and information technology systems.
- Communications architecture provides a vision for the ideal barrier-free communication inside and outside of the enterprise. It connects internal "island of automations" into one "enterprise continent" within the electronic global village. It allows for local action and global thinking.

If the information architecture leads only to information transport, it is a computer networking. However, if that network is based on an multimedia communications architecture to support the integration of enterprise utility, including telepower applications (such as for example: telepresense in the virtual mode) throughout the enterprise, than it is enterprise-wide computing. The information architecture and communications architecture determine the corporate electronic infrastructure as a evolving solution, or intermediary solution (corporate network computing), or as an ultimate solution (enterprise networking).

## Information Architecture

Information systems integration and connectivity can be achieved through architectural information strategic planning. The architecture of information systems brings harmony between the means of information technology and the business environment by establishing systems relationships based upon the intended system's functions in the organization. Synchronized with business strategic planning, enterprise-wide systems architecture becomes a tool of information strategic planning. It offers a long-term vision of solutions to corporate information-communications problems. With an architectural design, systems are zoned much like industrial and residential areas zoned by municipal governments (Targowski 1990 p.119).

The five blocks of system architectures proposed are :

- Federated Information Systems
- Data/Information/Knowledge
- Computer Software
- Computer Systems
- Communications Networks

Enterprise-wide Information Management Architecture development is defined by the Targowski/Rienzo model on Figure 8-1 (Targowski and Rienzo 1990). Architectural planning is carried out in a series of five steps:

*Step 1:* Translate an organization's mission, goals, and strategies into information mission, goals, and strategies through business planning.

*Step 2:* Identify the information needs of business functions through the Enterprise Processive Model.

*Step 3:* Create Federated Information Systems architecture of the information Management Complex, utilizing the Bill of Systems

Processor technique, incorporating functional information needs of the business as well as information mission and goals. Systems planning should recognize independent computerized applications already in place, represented in Figure 5-1 by the Information Archipelago.

*Step 4*: Develop architectures of Data/Information/Knowledge, Company Software, Computer Systems, and Communications Networks.

*Step 5:* Design architecture of the overall Information Management Environment showing system connectivity and integration through graphic design.

The Federated Information Systems Architecture of the Information management Complex (Step 3) supplies a logistical plan for the information flow between data sources, data files, knowledge

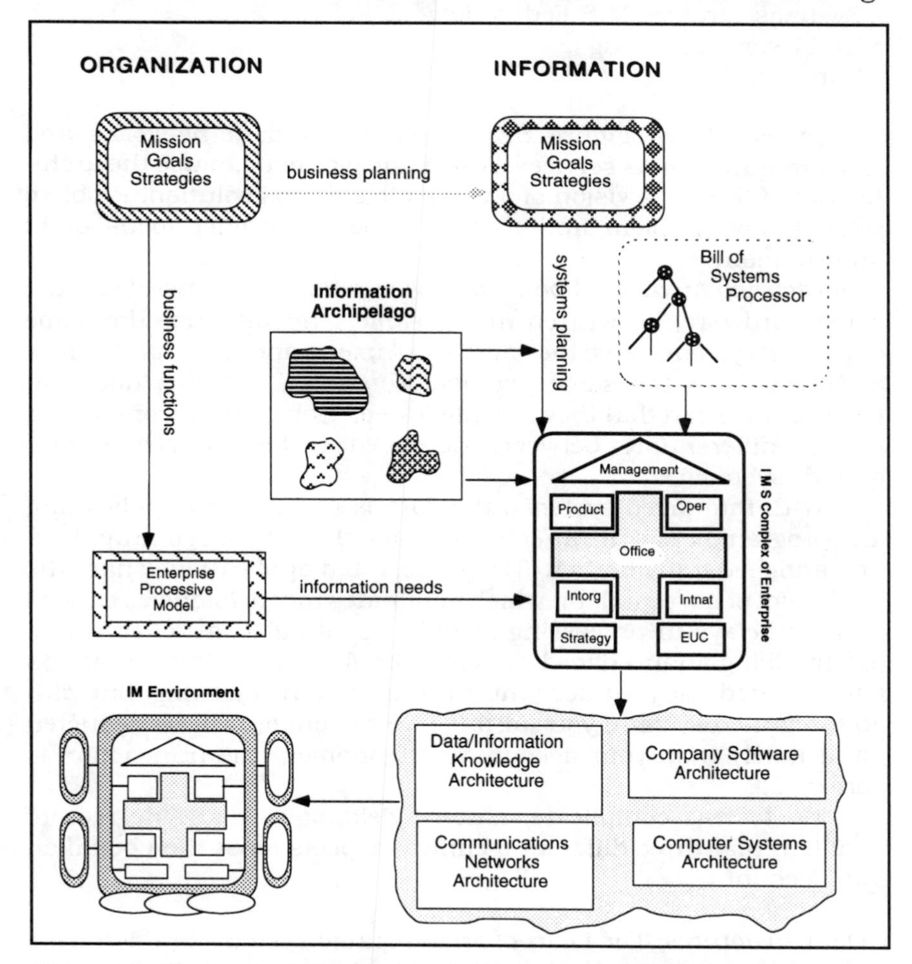


*Figure 5-1: The Targowski/Rienzo Model of the Development of Enterprise-Wide Information Management Architectures*

files and reports for users. It furnishes conceptual integration of exclusively recognized units of information systems.

The dynamic architecture of the generic organization Information Management Environment is shown in Figure 5-2. A generic set of federated application systems is shown inside the Infostrada polygon. The Information Management Complex consists of the following information systems federations (ISF) (Targowski 1990a):

- Management IS (MISF), including Executive IS (EIS)
- Office ISF (OISF), e.g., office automation
- Product ISF (PISF), e.g., Computer Aided Design
- Operations ISF (OISF), e.g., Computer Aided Manufacturing
- Interorganizational ISF (IoISF), e.g., Electronic Data Interexchange
- International ISF (InISF), e.g., Transborder Data Flow
- Systems Strategy MIS Federation
- End User Computing (EUC)
- Future ISF

Infostrada in Figure 8-2 encompasses data networks and telecommunications services. A systems planner can use the architecture of IME as a vision of a complete systems solution. Problem solving, not problem finding should be the proper focus of IS improvements.

*Clearing Away the Confusion.* Two systems may operate on the same hardware, be written in the same language, use the same operating system, serve the same set of users, and they may both be accessed. from the same personal computer—yet this does not necessarily mean that they are integrated. Bob Curtice from Arthur D. Little differentiates between "interfaced" systems and truly "integrated" systems.

With interfaced systems, the data is tied to each application, and programmers write interfaces to take the output of one application and feed it to another. Truly integrated applications share the same pool of current data, which eliminates inconsistencies caused by time delays in processing or data updating. That difference is major. Using an interfaced system, a bank may not know that you have cleaned out your account until early in the morning, but with an integrated system, your airline fare to London will be deducted immediately from your account at the moment the transaction is performed.

Most firms complicate matters yielding so many degrees of variation that three different technical aspects must then be taken into account:

- Data structures: The heart of an integrated systems is a database that contains current information available to any application that

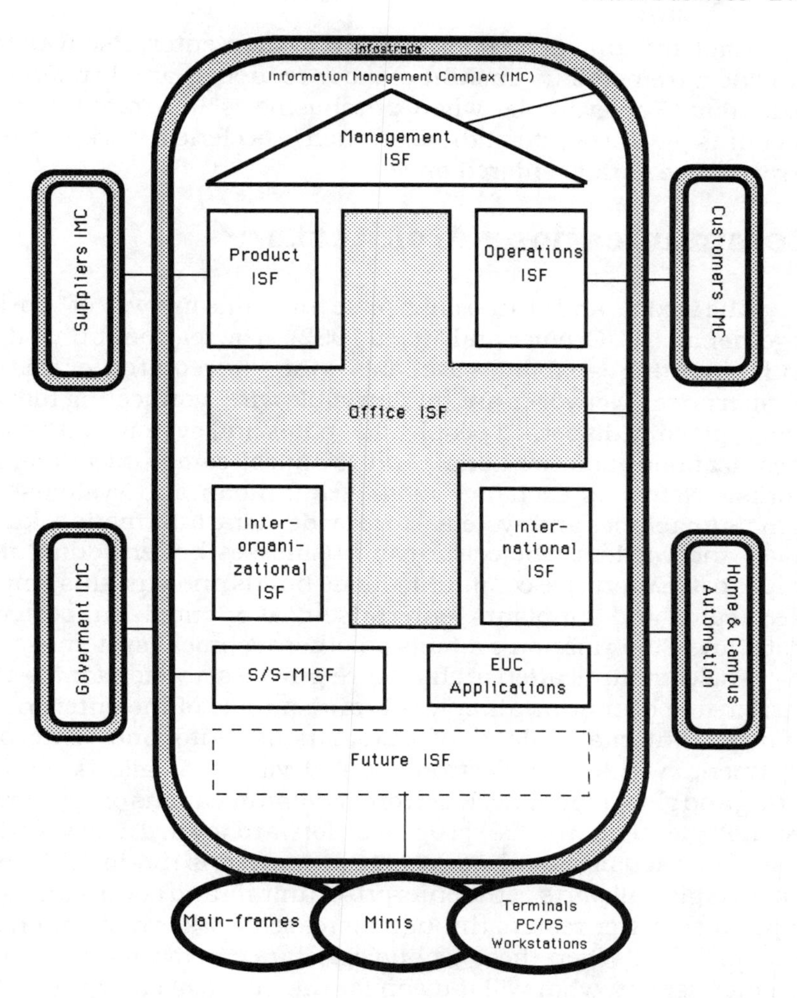


*Figure 5-2: The Dynamic Architecture of the Generic IM Environment (IME)*

seeks it. Often, existing databases may have to be completely restructured or re-entered in a different format to meet the needs of an integrated computing environment.

- Applications: Integrating applications requires changing or rewriting each application allowing integration with other applications and data.
- User interface. Applications can be integrated or seem to be integrated to the user, when they employ a common user interface, such as consistent menus, commands, keyboard functions, or a windowing user interface. Although applications requiring modification to use different user interfaces can increase effort dramatically.

Implementing a single integrated system enterprise wide may provide a tremendous amount of labor reduction and human error reduction. For example, when purchasing is integrated with the accounts payable system, there is much less clerical work and fewer errors since data is entered once.

## Communications Architecture

The flexible and virtual enterprise functions in a way of "working together apart" (Grenier and Metes 1992). The telecommunications and computing capabilities available is the glue required for this type of enterprise. Networks are the "invisible wire" connecting together geographically dispersed people, machines and systems into social, organizational, informational, and technical communications networks. Networks exchange messages (human and systems) and signals (machines and systems) to provide: data, information, knowledge, and wisdom (choice) to run businesses better adding more value to it. Enterprise communications builds and sustains empowered teams and subplants, enterprise-goal oriented supported by distributed information systems as other technical systems.

The communication architecture provides a vision for the ideal barrier-free communication inside and outside of the enterprise. It connects internal "island of automations" into one "enterprise continent" within the electronic global village. It allows for local acting and global thinking. Electronic communications or "telepower" technologies (Pelton 1989) propels us forward toward a new world of global consciousness and interaction, an "enterprise brain." Telepower is converging all parts of the enterprise, uniting and controlling them to produce better results through increased worker motivation and responsibility. When these technologies are empowered with artificial intelligence, what will it mean for the future of enterprise? Will they control a CEO on a ranch in Wyoming or in a summer house in Maine? Or, vice versa, as we still believe in it. The new telepower technologies are creating additional telepower corporations. Such Fortune 500 corporations as General Motors, IBM and TRW are diversifying into other lines of businesses to include telepower products and services. For example, General Motors acquired Electronic Data Systems (with 60,000 employees) to promote telepower inside of GM plants and offices.

Telepower is not only increasing enterprise performance but developing a whole new distinct enterprise, perhaps more efficient, but also more socially problematic. This new enterprise will be based on the co-evolution of machine and mind, or stated differently, hardware design and management problem solving if you will, possibly developing a new conscious technology. Machines will incorporate a greater capacity for intelligence allowing for the enterprise and world-wide communication of thoughts and action. On the other hand, management-yet human-based systems will embrace more technology into their entities that will become more, and more intelligent. Glenn (1989) is positive about this technology. The merger of two world views: the one emphasizing shared consciousness and growth (mystic) with the other emphasizing a technology-method for organizing the world (technocrat)—is positive. Since without the mystic, we end up with a robot like civilization and without the technocrat our civilization will lack the organization to absorb the pace of change and will crumble into chaos.

If we push technology to its ultimate limit, we can predict that in the 21st century, multipurpose intelligent (mystic) robots will replace virtually the entire U.S. work force, and people will receive a salary to "enjoy life." A household robot could serve as chef, maid, and family doctor, with other robots designing and managing them. However, a lack of work ethos will not cause life enjoyment but life emptiness and boredom, like today, having an unpredictable impact on human behavior.

The communications architecture of the enterprise is shown in Figure 5-3. The enterprise is analyzed in five views, with each characterized by its capability to develop networking techniques or technologies.

*The Organizational View.* This view is identified with an organizational chart and management structure. The organizational lines of communications are intensified by telephone, voice mail, electronic mail, computer teleconferencing, time management, electronic filing and word processing with its "sending" function. These technologies communicate via: voice, text, video, data, and image.

In the mid-80's, many people realized the value of answering machines and demanded access to this communications feature not only at home but at work too. Unfortunately, each answering machine requires its own port on telephone switches—at a cost of $800 per port.

> For example, MCA Universal Studios in Southern California (420 acres, 176 buildings, and numerous sound stages serving thousands of people each day) began looking for alternatives. Its business depends on people and materials being in the right place at the right time. In 1989, MCA Universal Studios began using AT&T's Audix Voice Messaging System to replace answering machines. Instead of "standing in" for answering machines, the system has vastly improved communications throughout the lot. Not only have customers given it rave reviews, but it has turned out to be a sound financial investment. Telephone costs have been greatly reduced as the studio eliminated many

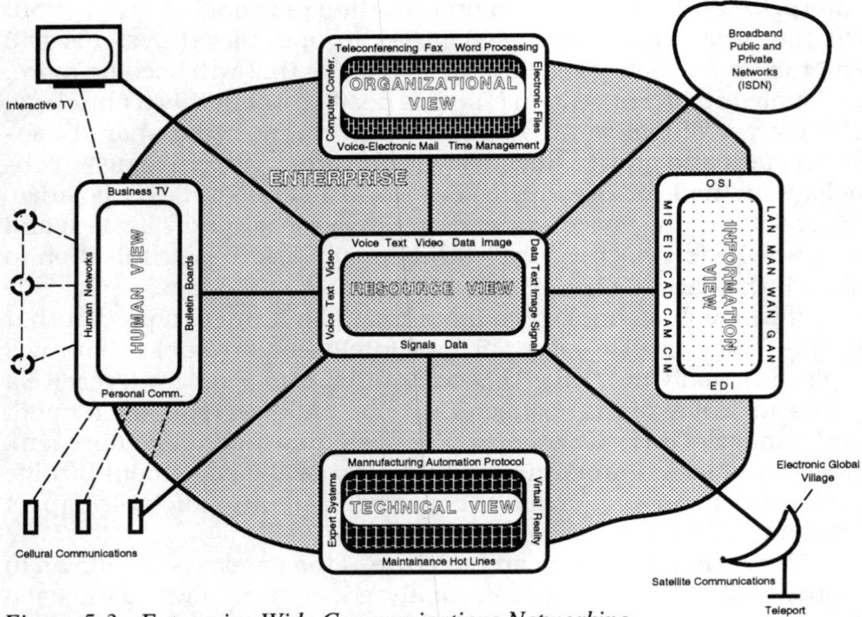


*Figure 5-3: Enterprise-Wide Communications Networking*

answering machine ports supported by its telephone switch. In fact, the system paid for itself in the first year. Studio productivity has also improved. For example, the studio eliminated many missed calls, reducing telephone tag, enabling directors, producers, and other to schedule more uninterrupted time encouraging greater creativity, reduced costs, and improved the utilization of human resources.

Universal City receives more than 12,000 calls per month requesting routine information about public tours, directions, special events and filming schedules. Before Audix, it took three people to respond to these calls. Today, this information is provided in both English and Spanish using the automated attendant feature. Callers can select information using the telephone touch-pad, or they can hold the line for the operator. As a result, the studio is better able to utilize the employees' skills answering non-routine calls requiring human judgement. Calls can be sent to the Audix System or forwarded to another department. The system even allows for multiple outgoing "one-to-many" broadcast messages. Managers at MCA Universal's record sales division use the broadcast feature to distribute information simultaneously to more than 200 people coast-to-coast. Sales representative on the road can make one phone call to access their messages. Field offices use the system to provide information to remote locations and to keep in close contacts with customers, headquarters, artists and many others. In 1992, MCA used 23 automated attendants in the Audix System to handle caller requests for information from the mailroom, payroll/tax department, restaurant, and human resource department. Anticipated demands will require the addition of several more automated attendants. The restaurant/commissary uses the system to inform callers about daily menus, which frees the phone lines for business purposes. The mailroom has improved productivity by using the Audix System to direct incoming rush deliveries and to handle general information inquiries regarding services. Human resources uses the system to list job openings and provide information on insurance benefits. The next step in the voice communications improvements will be the addition of an AT&T Conversant Voice Information System in 1992. The Conversant will be programmed to receive job applications over the phone system, which will then prompt callers for responses to be recorded on the Audix. The MCA's ability to receive and process job applications will be greatly increased (McNeil 1992).

Research has shown that up to 83 percent of all business telephone calls do not reach the called party on the first attempt (Rowe 1988). Some other telephone services include:

- Automatic Calls Distribution (ACD). Automates calls to people taking orders, handling reservations, answering questions.
- Caller ID. Over 90% of all 800 lines in North America can now give you a number of the person calling as the call is coming in. With it, the computer attached to the telephone switches can route the call to your LAN.
- Contact Management. When did you last speak with your customer? What did he/she last buy last? Remind you of client birthdays, anniversarys,etc. Make sure to call him on his/her birthday. What did he/she say during your last contact; did they say to call back in 10 days for a decision? These question can all be answered easily, improving relations with your customers and clients.
- Dial-by-Name. Dial TARG for Andrew Targowski. Dial SEBA for Walt Sebastian. So much easier than remembering numbers.
- E-911. Emergency number which retrieves database stuff. This technology is a gigantic life saver.
- Text Messaging. "Johnson! your biggest client is fuming on line 1." You would pick that message up fast—no matter to whom else you were talking. That is the point here. Fast, short, accurate messages

when time is short and customers are uptight.

- Time Tracking. Time is money. Here is a simple way to use the system to track billable hours—whether you on the phone or not.
- Voice mail. No better way to: get message-meanings,leave detailed messages; grab a message, add comments and pass it on and avoid time-zones snags.
- Call Accounting. Detailed reports on all calls you made and received. Useful for cost allocation, checking phone bills, catching abusers, etc.
- Other services can include: transfer calls, set up conference calls, hold a call, or pick up a held call, call transfer from the PC keyboard, instead of the phone's touch pad. Bring up menu of choices, then a list of possible users to transfer to, and it is done.

Electronic mail is one-to-one and one-to-many text communication with automatic answering and transferal features in a mode of store-and-forward by the computer network (ALL-In-One on Digital or Office Vision on IBM networking computers). It avoids writing memoranda and speeds up reading them as well as answering and transferring them. In the City Hall of Kalamazoo, within one year from the introduction of E-mail in 1991, the number of send messages exceeded 24,000 among 150 subscribers. It is like sending 160 memos per person per year or a message every second day. It means increased productivity by 3 times in the first year of applying a new communications tool. The communications ladder crosses the organizational boundaries, since many messages are sent across horizontal lines of interest than the vertical ones through the organizational chart. It is against the old paradigm of management hierarchy, but it is up with the new paradigm of democratic and short sleeves practice—result-drive.

Computer conferencing is text communications on a networked computer, it is a form of enhanced electronic mail. Analogous to a public bulletin board but only available for viewing by a predefined group of management and staff users. True conferencing systems are used for detailed discussions within continuous topics by participants usually known to each other. Automatic separation of messages into categories is usually supported. Sophisticated user interfaces are often provided. These can display lists of subjects of messages per category, and the user can select messages (either to display or to avoid) by subject, sender, who has read the message and logical combinations of these and other attributes. Even larger groups can be accommodated simultaneously on conferencing systems involving groups thought impractical over the telephone. This is because computer mediated systems can arrange that only one participant can hold the floor at a time. Computer conferencing provides both clarity and immediacy almost at the level of traditional face-to-face communications but at reduced cost.

Teleconferencing or video-conferencing simulates (motion video) the real business environment. Some things never change. Teleconferencing provides the digital handshake, a human touch-voice, gesture, and body language; that extra dimension to judgment that comes only with eye-to-eye contact. It gives also speed, convenience, and cost-control unequalled by any airline, providing video-retrieval and computer power-sharing capability in real time. In general, it is a journey from voice to video.

> At The Upjohn (pharmaceutical) Company in Portage, Michigan one can see that video-conferencing is an alternative to travel. When W. E. Upjohn founded the Upjohn Pill and Granule Company in 1886, the task of communicating with his three brothers and 12 employees in one location was a fairly simple matter. Today, about 18,000 Upjohn workers in 200 world-wide locations are invited to discover the advantage of an interactive meeting with colleagues in another city without traveling.

Businesses use video-conferencing for more effective communication and cost savings. With the use of voice and visual communication, it is possible to present reports, documents, charts, slides and graphs in an efficient manner to individuals in different locations. In 1992, over 8000 public and private sites around the world were equated with videoconferencing facilities. Such a facility features:

- Incoming, outgoing large screen TV monitors
- Multiple cameras (with zoom, focus, pan capability)
- Projection systems for documents, charts, slides, diagrams
- Video, audio tape recording
- Encryption, privacy
- Facsimile copiers
- Simple controls

Information can be transmitted directly from computer to be displayed over a television screen at different locations using a combination of fiber-optic cables and satellite transmissions. Participants can then jointly view and revise documents with other participants at dispersed geographical locations. Once a document is revised, the system is then capable of instantly generating hard copies—essentially video snapshots—of the transmitted documents. Confidentiality is insured by scrambling the video and audio signals at their orgin point and unscrambling them at their final destination. This will be highly beneficial in drawing up contracts and resolving regulatory affairs on sensitive issues. Previously, in Figure 2-5a, a teleconferencing facility at the Upjohn Company is shown.

> Video-conferencing equipment at Upjohn placed into service in the spring of 1990, now extends meeting tables across oceans. Similarly equipped rooms have been opened at Upjohn site's in Brussels, Belgium, and Crawley, United Kingdom. "Worldwide Pharmaceutical Marketing will be a primary user of technology," said Glenn A. Miller, Telecommunications Planning Manager. "We expect our business groups and support staff will use the facilities extensively to meet both domestic and worldwide needs." These facilities can be used to diagnose problems with manufacturing or laboratory equipment. A videotape of malfunctioning equipment can be transmitted to engineers at different sites. (INTERCOMM 1990).

> Reviewing of advertising pieces, discussing marketing plans, following-up on marketing meetings and exchanging of competitive information number among the video-conferencing uses envisioned by Upjohn for marketing. Upjohn's top executives applied this medium for crisis management in 1992, when one of their drugs (*Halcion*) needed support in order to counter attack unfounded accusations. In October, 1992, Dr. Theodor Cooper, CEO, recorded a videotaped message to employees about the company's position on the controversy and its plans for defending *Halcion*. Video-conferencing permitted company executives in the U.S. and Europe to coordinate communication strategies for Halcion with regulatory agencies, employees and journalists.

The Upjohn Company together with other pharmaceutical companies founded a teleconferencing facility at the US Federal Drug Administration office in Washington. Doing so reduced traveling costs incurred by the pharmaceutical companies and expedites drug approval by the FDA.

Another communications service is Electronic Data Interexchange (EDI). EDI is the computerized exchange of standard business documents such as purchase orders, invoices, and medical claims. The exchange must often take place between companies, although it is also correct to speak of intra-company electronic data interchange.

Looked at from another perspective, EDI is a software application integrated with other software applications. It is important to note that EDI refers to the transfer of legally binding documents. Unlike electronic mail, EDI is meant to be read by a machine, not by a human being. An EDI transmission, to give one example, can go straight from a buyer's purchase order application into the seller's order entry application without human intervention.

## Enterprise Networking

Enterprise networking tends to grow larger and more complex as the enterprise learns to exploit distributed computing, client/server computing, and integrates its smaller applications-specific or localized networks into a single enterprise network. Problems, although, can occur when two enterprises merge. The dissimilar computing networks, with potentially dissimilar characteristics, have to be interconnected in order to support daily business operations. The dissimilar networks need to interconnect so that any terminal or application program in one network can access any terminal or application program in the other network.

A shift in the way users and developers think about information systems is needed, away from concerns only about how each individual new application is to be implemented, and toward how such an application can be designed to fit into the corporate electronic infrastructure. The new "infrastructure thinking" is based on the concept that an information infrastructure is essential to:

- The rapid introduction of new products and services to the marketplace,
- The instant communications between customers, suppliers, and business partners,
- The instant links between networking units of a corporation.

The corporate information infrastructure can be seen as the Enterprise Network Architecture (ENA) evolving from four distinctive environments:

A. Group Net:
- Departmental or team environment
- Includes self-contained applications such as in engineering (CAD), business administration (personal productivity)
- Includes: workstations, PCs and applications servers (LANs)
- Small company of knowledge workers

B. Location Net:
- Building or campus environment
- Includes intra-location applications and communications
- Includes: backbone network, bridges/routers, wires, gateway
- Hospital, college, university, city hall

C. Enterprise Net:
- Hierarchical transactions and information processing

- Connected to a large number of remote users
- Includes: proprietary host-distributed computers
- Insurance company, airline, chain retailer, banking

D. Enterprise Utility:

- Complex of information systems federations
- Intra- and inter-organizational information/communication systems and services
- Multivendor telecomputing platform
- End-to-end telenetwork (LAN-MAN-WAN-GAN-WAN-MAN-LAN) managed from the central location.

The migration paths from the group net to the enterprise utility are depicted in Figure 5-5.

The migration paths of the evolving corporate electronic infrastructure is determined by the following steps:

- LAN of the 1980s— supports the sharing of files that are loaded to a PC within a framework of local/departmental islands of non-critical computing
- Client/Server of the 1990s —supports the transfer queries rather than entire files and develops corporate connectivity utility
- Enterprise Network of the 2000s—supports the integration of applications across multiple network platforms, with the transparent user access. The network computing coherent platform leads to the enterprise utility.

The Enterprise Network of the next century, now only a few years away, is composed of various data communication, processing and information management technologies (e.g., WAN, LAN, video, voice, image) which form an integrated electronic infrastructure, an Enterprise Network linking corporate departments with one another to facilitate the efficient exchange of information. An Enterprise Network is a valuable corporate resource which provides timely and complete access to information in order to supply a competitive advantage to that enterprise in the particular market(s) served.

An Enterprise Network might encompass a single floor (LAN), several buildings or campuses covering many blocks (MAN), cities (WAN), or continents (GAN). Figure 5-6 illustrates an Enterprise Network, and Figure 5-7 depicts an integrated Enterprise Network.

*The Enterprise Network Architecture b*uilds on the distributed intelligence, self-management features, and communications capabilities of peer-to-peer communications. More specifically, ENA should support:

- Management of all devices, applications, and data on the network

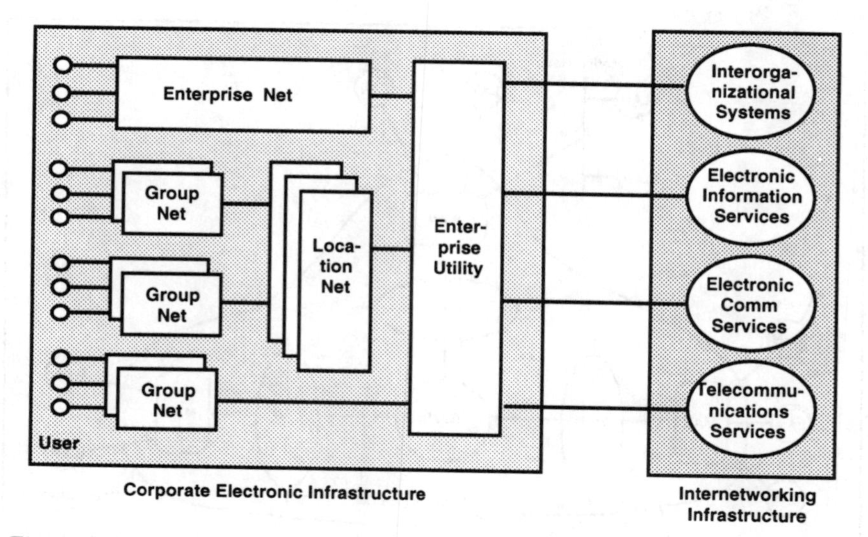


*Figure 5-5: The Migration Paths in the Network Computing Environment of EGV*

- Open system with published interfaces
- Multivendor commitment
- Standard-base approach to enterprise information management, with a strong commitment to OSI standards.

*The open system* strategy should be based on two premises:

1) Open computer systems are based on international standard that specify six characteristics:

- interoperability: applications can work together in a heterogeneous environment,
- portability: application can be moved easily between platforms,
- scalability: application can be downscaled or upscaled,
- heterogeneity: applications are supported by different platforms,
- distribution: resources and processes can be used on a local system or another system on the network,
- vendor neutrality.

2) These standards must be set by an open process in which anyone may participate; the results of which are available on equal terms to all.

When people today talk about open systems, they do so in terms of "open environments" that actually allow users to mix and match different standards-based components — including installed proprietary systems —and still work effectively together as one system.

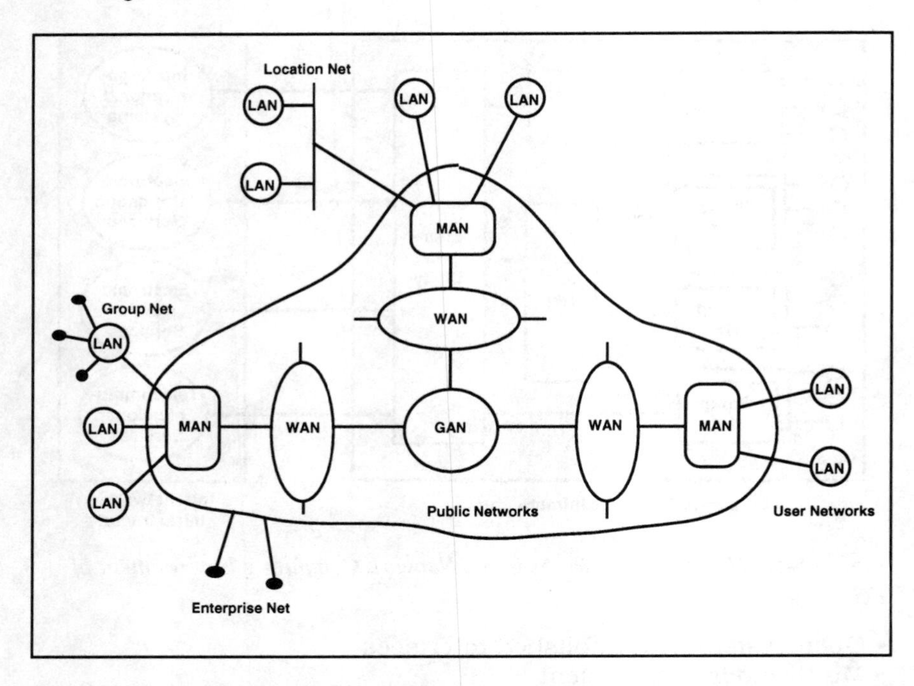


*Figure 5-6: The Enterprise Network Architecture*

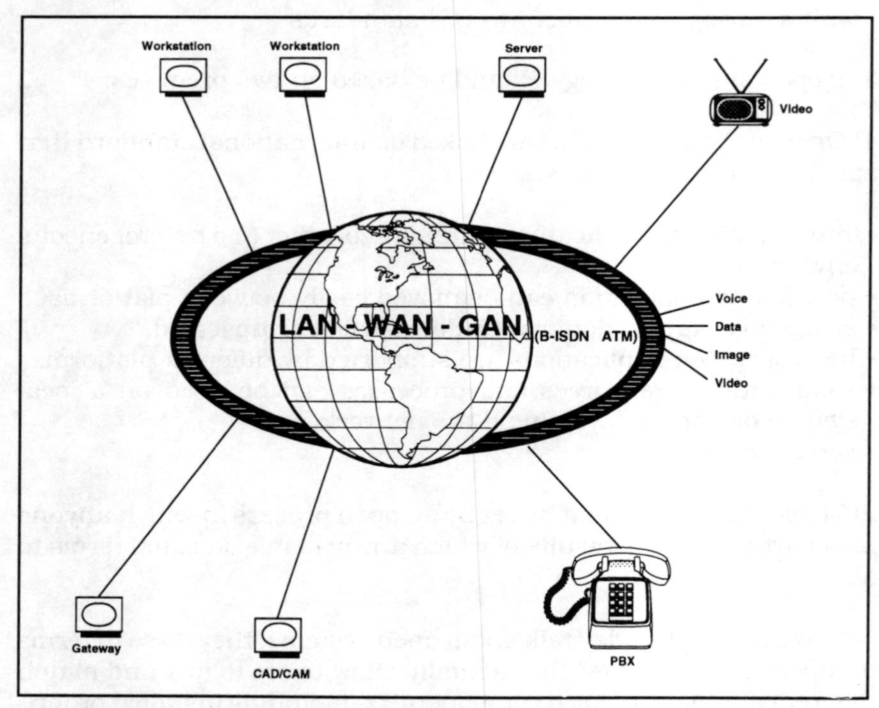


*Figure 5-7: Integrated Enterprise Network*

For example, General Motors is one of the companies studied that now has many incompatibly systems and databases. By achieving interoperability between the finance department's and manufacturing parts database, GM expects to be able to make better use of its data for cross-departmental planning and operations management, and thus help bring new designs to market faster.

McDonnell Douglas is another company that sees open systems as an enabler to business partnering, and the ability to seamlessly communicate and exchange data with suppliers, subcontractors, distributors, and even customers.

Unilever, a global consortium, expects that in the future corporate-wide compliance with open systems would make global information access a reality by seamlessly connecting the supply chain from sales order entry to logistics, manufacturing, warehousing and delivery.

An open system is a compliant implementation of an evolving set of vendor-neutral specifications for interfaces, services, protocols and formats designed to effectively enable the configuration, operation and substitution of the entire system, its applications and/or components with other equally compliant implementations, preferably available from many different vendors [2]. These open environments set up interface standards that establish how computers talk to one another, allow applications software to interact with other applications and operating systems, and allow users to interact with different applications.

In the 1980s, one common approach to solving the problem of application portability and integration was standardization based upon a single operating system. In the 1990s, the challenge is to provide the same services but based on standards implemented across a diverse set of operating systems and hardware. To address the needs created by the current heterogeneous computing environments, it is necessary to develop applications that are open and supported by properly engineered application software to promote:

- Portability across a wide range of systems with minimal changes
- Interoperability with other applications on local and remote systems
- Interaction with users in a style which facilitates user portability.

An example of a design methodology complying with these requirements is Digital's *Network Application Support* (NAS). Through

NAS, Digital implements open, standard-based interfaces between applications and different operating systems. NAS implements an open environment. Specifically, NAS provides the following:

- A model for application interaction, which is called an application dialogue.
- A set of interface standards for the application dialogues
- Products to support the interface standards.

The NAS architecture defines an application integration model, which is a visual representation of an application and of its interactions with its environments. An application must interact with the following elements:

- Users;an application must display information to and accept requests from the user.
- Data; an application must read and write data and access information resources (databases) enterprise-wide.
- Other applications; an application must communicate with other ones.
- Underlying system (computer set); an application must obtain system resources.

The application has, in essence, a dialogue with each of these elements. Traditional programming handles each application dialogue in a platform-specific way. In contrast, the NAS architecture handles each dialogue in a generic way. NAS defines how an application requests a service (data, system resources, etc.) and how the request is to be answered. Handling requests in a platform neutral manner increases programmer efficiency dealing with application distributed across platforms.

Unlike an operating system, NAS standardize only the application programming interface (API) This allows each NAS service to be implemented in a platform-specific way, without requiring the intervention of an application developer, alleviating developers of the burden of unnecessary reprogramming.

The application integration model shown in Figure 5-8 illustrates the way in which application dialogues are handled under NAS.

A set of NAS application services address each of the four application dialogues:

- Applications Access Services (user dialogue), e.g.: windowing, forms, terminal, graphics services,
- Information/Resource Sharing Service (data dialogue), e.g.:compound document, data access, repository, file sharing, print services,
- Communications and Control Services (application dialogue),e.g.:

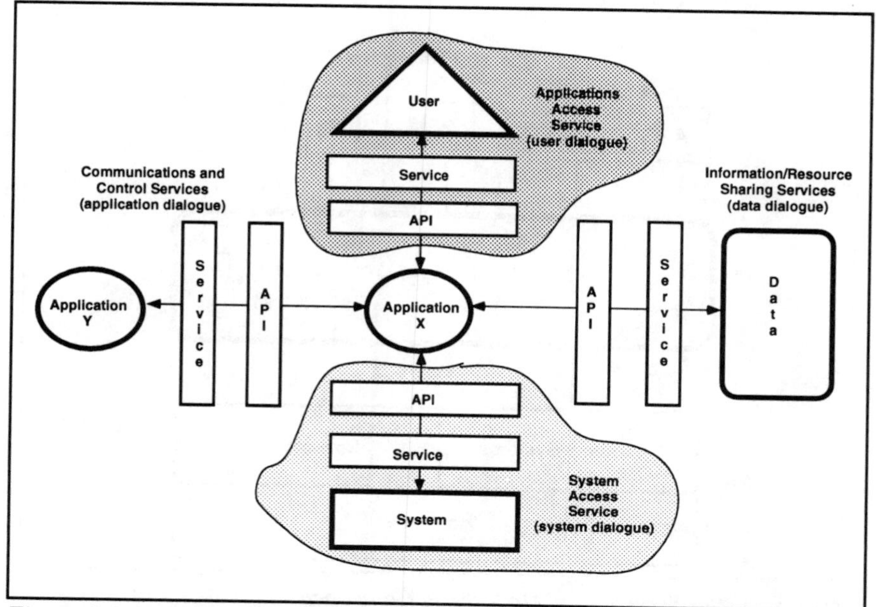


*Figure 5-8: The Application Integration Model in the Network Application Support Environment (API—Application Programming Interface)*

mail, EDI, application control, interprocessqueuing services,
- System Access Services (system dialogue), e.g.: distributed computing.

When using the standard APIs provided by NAS, the four dialogues create an environment that insulates the application from the specific of any hardware or software program. Figure 5-9 shows NAS in relation to the underlying system services.

*Designing a Network Application.* The requirements of today's computing environment do not mean that applications must be designed differently than they have been designed in the past. Modularity is still the key to all good software design. However, in addition to modularity, application designers need to be aware of three new goals that allow applications to take full advantage of the modern computing environment:

- Portability (many platforms)
- Integration (consistency in sharing user interface, data and functions with users)
- Distribution (access of resources from different locations on the network).

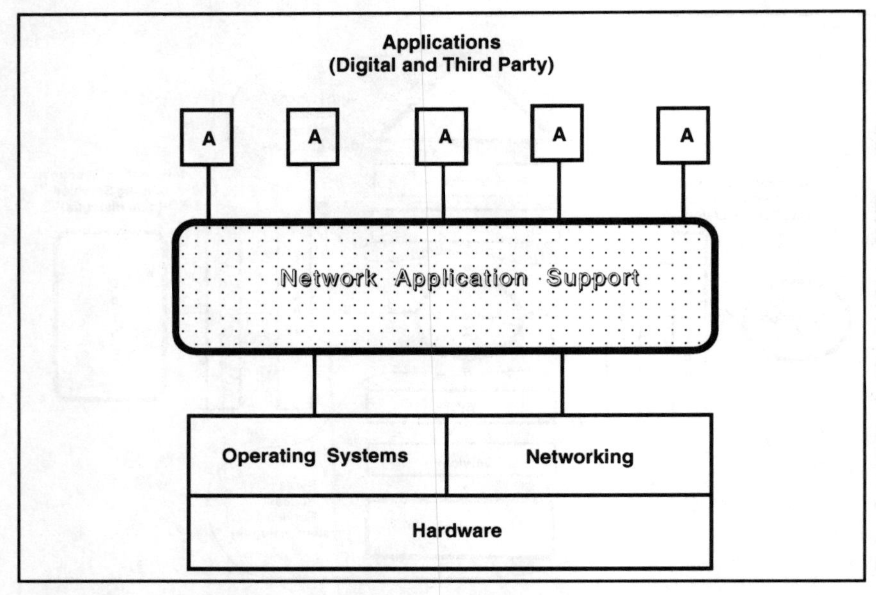


*Figure 5-9: NAS Services and Underlying Components*

*Application integration.* Rather than speaking of an application as being integrated or not being integrated, it is more appropriate to speak of an application's *degree* of integration with other applications.

At the lowest level, there is *no integration*, where two applications cannot be used together. The next level up is *tolerance*, where the applications do not actually share any data or resources, but you can use them together in the same environment. *Integration* begins when the applications start to share characteristics. These characteristics can be as simple as a common style of operation (such as using the same commands for similar functions) or as complex as simultaneous data manipulation.

The three areas of integration having the most effect on the degree of integration an application can achieve are:

- User interface: when two or more applications share consistent look and feel, the applications are easier to learn and use.
- Data: when applications represent their data in a single, consistent fashion, each application can reuse the data from and share data with other cooperating applications.
- Functions: when applications can invoke each other and share functions, allowing groups of applications to perform larger and more complex tasks than any one alone.

The more points at which applications share information and

resources, the more integrated they are, and the more powerful they are in performing complex tasks. Figure 5-10 shows how combining all three aspects of interaction between applications increases the level of integration.

When applications combine function and data sharing with a common user interface, the applications form a seamless environment. The user cannot tell where one application ends and the next begins. The integrated applications are more powerful to the user than any of one application or the sum of the individual applications.

Because integration means "applications working together," an application cannot be integrated by itself. It requires the cooperation of other applications. Integration does not mean that applications themselves must change or that they require an intimate knowledge of each other. For example, you can integrate two existing applications simply by providing a converter that translates the data from one application into a format that is acceptable to the other. This technique is the theory behind the Digital CDA Converter Library. This Library can convert Macintosh MacPaint applications into the DECwrite document. Converters are very useful for integrating existing applications, but they can affect only the data the applications produce.

Traditionally, functional integration has been achieved by having one application control the integration. The controlling application contains hard-coded knowledge of the other applications. In one technique, the controlling application uses a publicly available programming interface to integrate with the other applications. For example, on VMS systems, applications can use the VAX DEC/Code Management System (CMS) callable interface to integrate CMS libraries into the application. Another technique uses messaging to send information between applications.

Integration is communication between applications. Traditional techniques for integrating applications are limiting because the communication is controlling the communication, rather than the communication controlling the application. Consequently, to integrate new applications into the system, the controlling application must be changed to incorporate information about the new application. The solution is to separate the communication mechanism from the message itself; that is, to provide a generic mechanism for communicating between all applications. By using a standard communication mechanism, you can add new applications to the system by adding new messages without having to rewrite the application.

Separating the communication mechanism from the specific message is the basis of the NAS architecture and is reflected in the architecture for the CDA converter. However, to achieve a common communication mechanism, there must be a shift of focus away from

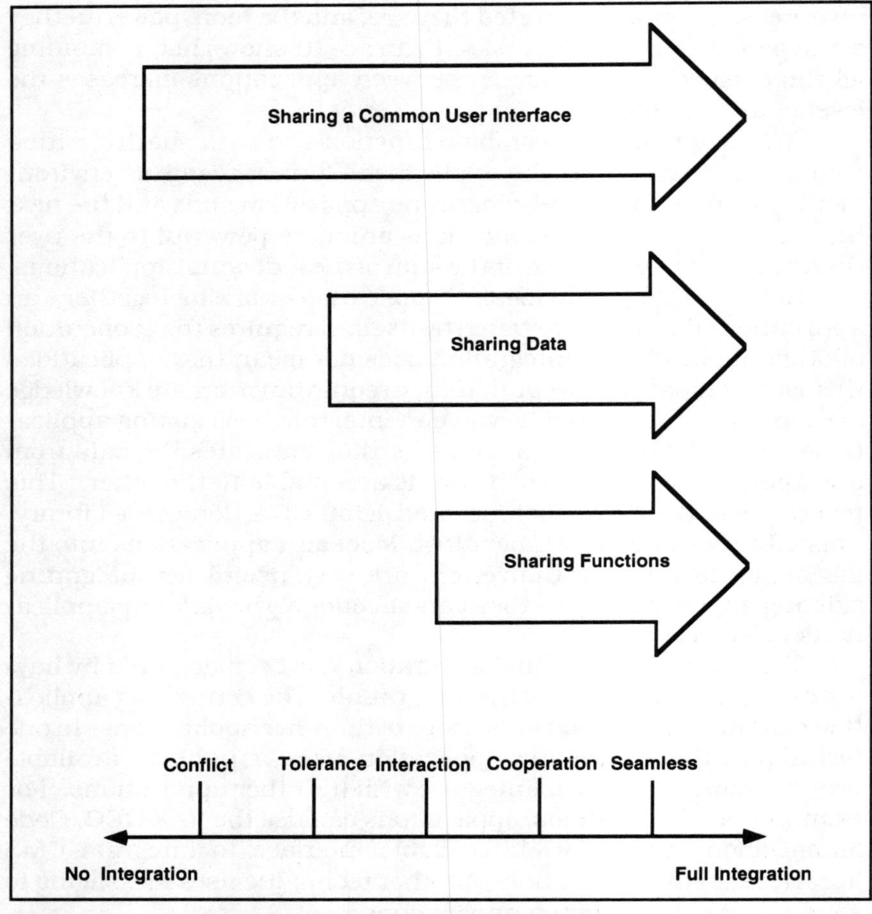


*Figure 5-10: Levels of Applications Integration*

the application as an autonomous, controlling unit and towards viewing them as groups of services controlled by the user [3].

Enterprise Networking Architecture requires a Network Management System, which performs the following functions:

- configuration management
- fault management
- performance management
- security management
- accounting management.

These functions have to support a dynamic, enterprise-wide, multivendor telecomputing environment. The Network Management System has to be adaptive to evolving solutions along with the progressing of business organization, technology, and social awareness.

No matter how extensive an Enterprise Network is, certain issues need to be thoroughly addressed, otherwise the network will not fulfill its promise:

- *Availability*: Perhaps no other management issue is as vital as availability. Access to data is crucial to the daily function of business operations. Network users expect 100% availability of the network.
- *Maintainability*: If an Enterprise Network is to have consistently high availability, it must be maintainable. Maintainability begins with a network comprised of inherently reliable components that meet or exceed clearly defined operational standards. Furthermore, the devices must be fault tolerant, capable of withstanding such insults as power variances and signal distortion. Of particular concern in an Enterprise Network are remote locations. Remote sides, whether "dim" or "dark," present interesting problems which can frustrate any network manager. Because service personnel may not be readily available, equipment at remote sites must be particularly reliable and fault tolerant. Remote site equipment must be designed for remote access and repair, even when it is online and operational.
- *Flexibility*: An Enterprise Network must allow for immediate change. Users and machine processes must be permitted access to databases or applications anywhere in the enterprise network, as necessary, regardless of whether or not that application runs in a local (native) or remote host. The flexibility to reach that application must reside in the Enterprise Network architecture, logically and physically, so that the application may be accessed. If a segment of the backbone is overloaded or out-of-service, you must be able to work around the bottleneck or failed circuit, virtually instantaneously. One key to Enterprise Network flexibility is the network's ability to accept a variety of dissimilar electrical interfaces (i.e., posses the ability to readily interface RS-232, V.35, RS-449, T-1, etc.). Another key deals with network logic. Operating environments and network management systems should be intelligent enough to identify and adjust to continuously varying service demands and disruptions.
- *Network Management Capability:* The complexity of the Enterprise Network mandates a methodology for centralized network control. You must have the ability to manage all aspects of the network from a single point. Current solutions range from corporately developed proprietary systems to commercially available offerings such as IBM's NetView, AT&T's ACCUMASTER Integrator, and System Center's NET/MASTER. Centralized network management sys-

tems can be invaluable for automating the recovery of failed network components or alerting you to impeding service disruptions.

- *Performance*: Performance is the ultimate criterion upon which users judge the network. If they cannot get their tasks done, your network is not effective. As a manager, your measurement is more technical, involving response time for a particular operation or a set of operations to be completed once started. For instance, you might establish as acceptable a minimum service level that calls for 90% of all transactions to be completed in less than three (3) seconds [4].

Effective Enterprise Network Management begins with access and control. You must be able to get into the network to measure performance, isolate trouble, and rearrange devices. Matrix switches allow you great latitude in access and control. Matrix switches are the connection "hub" for the communication resources of the Data Center. Switches should support connectivity to the desktop in order to effectively manage the communication needs of the "Intelligent Building."

Expansion, replacement and/or enhancement to communication equipment or facilities necessitates cable management implementation that minimizes under-floor cabling while providing interface distance extention. These features permit flexibility in the physical location of the communication equipment throughout the Data Center and also provides for physical reconfiguration with minimum cabling effort.

No matter which matrix switch you consider, the issue of access centers on line interface devices is the gateway into the switch domain. The variety of interfaces supported by matrix switching systems differs among vendors. All offer standard interfaces, such as 9.6 Kb/s RS-232 C/D, but if you are following the move toward higher speed processing and communications, your matrix switch plans should include high-speed circuit interfaces (e.g., 256 Kb/s and higher, V35, T-1, and multi-megabyte Token Rings).

Not to be forgotten in the discussion of control is the human interface, the switch console or workstation. Typically, the operation consoles present a convenient menu-driven environment. Some vendors, like Telenex, has progressed beyond simple menu designs to interactive, graphical user interface (GUIs) built around the concept of intuitive operations. Extensive use is made of color and symbols (icons) to represent network devices and switch system functions. Using GUI technology, control is simple as using a pointing device to select an icon that represents a particular matrix function and then initiating its execution with a click of a key. GUIs simplify operation and extend control of the network (which appropriate security restrictions) to operators possesing virtually all levels of expertise, from a novice operator to long-time technicians who are intimately familiar with the network.

The advent of ISDN and other more high-speed network protocols such as Frame Relay will only advance network integration, making Enterprise Network Management ever more demanding [4].

## End Notes

1. The Consultant Forum, no.3, 1988

2. Gardner Group's definition.

3. Digital's methodology: The NAS Handbook, EC-H1190-58, 1991

4. Managing the Enterprise Network, Mount Laurel, NJ.: Telenex Corporation, 1991

## References


*CIM Solutions for Midsized Companies* (1990), White Plains: IBM Corp.

*CIM in IBM* (1989) White Plains: IBM Corp.

Friedman, A. and J. Greenbaum, "*Japanese DP,*" Datamation, February 1, 1985.

Glenn, J.C. (1989) *Future Mind,* Washington, D.C.: Acropolis Books.

Grenier, R. and G. Metes (1992) *Enterprise Networking, Working Together Apart,* Bedford, MA.: Digital Press.

Harmon, R.L. (1992) *Reinventing the Factory* II, New York: The Free Press, p.65.

Harmon, R. and L. Peterson (1990) *Reinventing The Factory,* New York: Free Press.

Harmon, R. and L. Peterson (1990a) p. 233.

"Conference Rooms With View of World, Opening at Sites in U.S., Belgium, U.*K.*" *INTERCOMM,* Volume 28, Number 5, May 1990, Upjohn.

McNeill, R. (1992) "Voice Messaging Goes Hollywood," *Communications News,* March 1992, p. 12.

Pelton, J.M. (1989) "Telepower, The Emerging Global Brain," The *FUTURIST,* September-October, pp.9-14.

Rowe II, S.H (1988) *Business Telecommunications,* Chicago: SRA, p. 157.

Targowski, A. and T. Rienzo (1990) "Managing Information Through Systems Architecture" *The Journal of Information Systems Management, Information Executive,* Vol. 3, No. 3, pp. 40-49

Targowski A. (1990) *The Architecture and Planning of Enterprise-wide Information Management Systems,* Harrisbourg: Idea Group Publishing, p. 119.

Targowski (1990a) p.30

CHAPTER 6

# On-line Government

## Reinventing Government Through Information

There are 82,0051 governmental units in the United States — one federal government, 50 state governments, and 82,000 local government units (counties, cities, townships, special districts, school districts, etc.). About 15 million employees work for government, including 3 million for the federal government and 12 million for local government. In the last decades, Americans here arguing *what* government should do not *how* it should work. As a result of it, we spent $60 billion of new money on the education reform in the 1980s-1993, and tests scores and dropout rates are higher today than in 1980. The same negative result can be cited in environment protection, loan and savings management and other areas of government activities. A question of *what to do* is still important, but it has to be associated with a question *how to do* it also.

"When government does not work, the whole country suffers," said Vice-President Al Gore, head of the Clinton Administration's National Performance Review —an effort aimed at making government work better (interviewed by Garland 1993). He also added: "When government makes mistakes constantly, and wastes money persistently, business and every institution in society — is hurt. Our national government has to make the same kind of transition that many U.S. businesses made as they realized that their old form of organization was no longer adequate."

Complaining is easy. How is it possible to make the federal government, a sprawling institution with a $1.5 trillion budget and nearly 3 million civilian employees, work smarter? Gore's recommended package (September 1993) will save $70 billion to $100 billion over five years. His package is based on the philosophy of the best-seller *Reinventing Government* (Osborne and Gaebler 1993).

Osborne and Gaebler in their book emphasize that a new model of government should be:

- Catalytic (steering rather than rowing)
- Community-owned (empowering rather than serving)
- Competitive (injecting competition into service delivery)
- Mission-driven (not rules-driven)
- Results-oriented (funding outcomes, not inputs)
- Customer-driven (meeting the needs of the customer not the bureaucracy)
- Enterprising (earning rather than spending)
- Decentralized (from hierarchy to participation and teamwork)
- Market-oriented (leveraging change through the market).

The authors claim: "Our map is complete. It is now yours to use." (Osborn and Gaebler, 1993, p.311). It is the good "map," however, it does not include a solution of *how to* put together these wise features of the new model of government. In the best-seller, there is a lack of comprehensive strategy of implementing these features. A key to this strategy is in a concept of working supportive systems. Today, and in the 21st century, these systems are driven by information and information technology systems.

The purpose of this study is to offer a vision and architecture for a system of goals and strategies that will support government goals. Reinventing government will pay off if the renewal is guided by a comprehensive, aims-driven process. The results of reinvented government activities should include the following:

1. Better services at less cost (*What*)
2. Working supportive systems (*How*)
3. Better informed and served citizens with a voice for change (*Who*)
4. Renewed public confidence (*Why*)

In this study we concentrate on the second result: "How to develop working supportive systems." and the third result: "Better informed and served citizens with a voice for change." The remaining goals' strategy definitions lead to the public policy.

A comparison of supportive systems alternatives is provided in Table 6-1.

The development of government supportive systems passes through the four following phases of governmental processes re-engineering:

- Re-engineering I —Automation of data processing routines to launch effective service delivery. The emphasis is on cost analysis through better data batch gathering. It is an introduction of computer information systems into governmental workers practice.
- Re-engineering II — Creation of electronic infrastructure of government affairs. It is a shift from batch processing to on-line, decentralized information generation. The emphasis is on better decision-making in order to anticipate results and provide a spirit of competitiveness to achieve higher service productivity. The electronic infrastructure assures the increase of information scale (enterprise-wide databases) to retrieve added value of information. The end user becomes the master of applications operations and improvements. It is possible to achieve this new process of work via a Graphical User Interface (GUI), which significantly improves user access to information. This is the introduction of new integrated information generation and handling processes into governmental workers *modus operandi.*

| Supportive Systems | Government Model | Government Goal | System Goal | System Strategy |
|---|---|---|---|---|
| Status Quo | Rules-driven | Service Delivery | Data Gathering | Cabinet Files |
| Automated | Mission-driven | Service Effectiveness | Improved Info Generating | Transactions Processing Re-engineering I |
| Electronic | Anticipatory & Result-oriented and Enterprising | Service Productivity | Improved Scale of Information Improved Decision-making | Databases, Systems Integration, GUI Access Re-engineering II |
| On-line | Community Owned | Service Satisfaction | Improved Scope of Information Improved Responsiveness of Government | Internetworked GUI Interaction Re-engineering III |
| Virtual | Customer-driven & Market-oriented | Service Pleasing via Government Choice | Improved Choice of Information | Global Internet-worked GUI Interaction Re-engineering IV |

*Table 6-1 A Comparison of Government Supportive Systems Alternatives*

- Re-engineering III —Empowering citizens in participatory governing of public affairs. A strong inter-networking and user GUI-driven on-line interaction have to be implemented. Government workers and officials have to learn and exercise power sharing in order to democratize equal access to power and seek service satisfaction by customers. Electronic Town House meetings can be one example of on-line government. It is an introduction of customer on-line scope-feedback into government *modus operandi.*
- Re-engineering IV—Empowering a citizen in changing a government, like one which could please customers in services. A global network is required that will lead eventually to a Electronic Global Citizen. This is more political re-engineering than system re-engineering. It is a re-engineering of the customer concept. It is more than re-engineering, it is rather a shift of social paradigm. It leads to spread-out constituency based on pleasing service via global rivalry. The global market place defines a choice of services and, in consequence, a choice of government. It can also require that the traditional scope of governmental services needs to be redefined, eventually expanding into new services or curtailing those that are based on competitive advantage. The customer can do it since he/she has a better choice of information.

Re-engineering is the radical redesign of governmental processes to achieve major gains in cost, service, or time. The key question : "If we could start from scratch, how would we do this?" And the answer is: "Then do it that way, and throw away everything else." Get the goals and strategy first . Do not reinvent things you should be doing in the first place. Re-engineering is about operations; only strategy can tell you what operations matter. Reinventing government means the re-engineering of basic governmental processes and systems to meet new governmental goals and strategy. The re-engineering should be led from the top, create a sense of urgency, and design from the outside in. The point and power of re-engineering is the clean sheet of paper with which it begins. Filling it in begins with citizens: The right question is, "How do they want to deal with us?", not "How do we want to deal with them? (Stewart 1993)."

## Local Governments Aims

Just consider these statistics: In the United States, 12 million employees work in local government, which accounts for 58% of all public sector civilian employment. Local governments are extremely important in the United States—much more so than in many other federally organized systems. They perform an almost endless variety of services that immediately affect our lives. As Michael Tietz (1967) describes:

> The modern person is born in a publicly financed hospital, receives his education in publicly supported schools and universities, spends a good part of time traveling on publicly built transportation facilities, communicates through the post office or quasi-public telephone system, drinks public water, disposes garbage through the public removal system, reads public library books, picnics in public parks, is protected by public police, fire and health systems; eventually the individual dies, again in a hospital, and may even be buried in a public cemetery. Ideological conservatives notwithstanding, everyday life is inextricably bound up with government decisions on these and numerous other local services.

To provide these services, local governments are pervasive in our society. According to the U. S. Bureau of the Census, there were more than 82,000 local governments in the United States in the mid-1980's. There were more than 3,000 counties, 19,000 municipalities, 16,000 towns and townships, 28,000 special districts, and 14,000 school districts. These local governments employ more than three times the number of employees of all the state governments put together [1].

Researchers for *American Demographics* magazine recently summarized some of the ways the United States has changed between 1980 and 1990 (Waldrop and Exter 1990). One significant change is simply the number of Americans—almost 250 million. That is a 10% change in 10 years. Almost 25% of this increase is due to immigration. Between 1970 and 1980, the nation experienced an unprecedented surge of growth in non metropolitan areas. In the 1980's, however, this trend has reversed itself. Some municipalities—large and small—face the challenges of controlling growth and all the problems (and benefits) that a population boom brings. Rapid growth means an increasing demand for services—transportation, water, sewer, and police and fire protection. Other cities lose population. Often this translates into a general cycle of poverty, including unemployment, housing abandonment, and the physical deterioration of the business center. All this change—in fact, human settlement itself—creates the need for government.

In the case of the U.S city, it is a capitalist political economy that pro

Today, a majority of all cities with populations in excess of 25,000 have the council-manager system. With some exceptions, most of these governments include:

- a small city council of five to nine members, usually elected in at-

large nonpartisan elections;
- council responsibility for all of the official policies, including passing ordinances and adopting the budget;
- professionally trained manager, appointed by the council and serving at their pleasure, who is the chief executive officer of the city government;
- an executive budget prepared and administered by the city manager once it is adopted by the council; and
- a mayor who essentially is a figurehead, performing the ceremonial tasks of government.

Almost all positions within the administrative core of government now have technical requirements and often demand professional credentials from applicants. In the late twentieth century, the "boss city" has become the "corporate city," to serve the changing urban economy, sustaining, if not maximizing, the role of economic interests in the private city.

The council-manager system operates best in homogeneous-small cities. One expert recently estimated that of all 3,000 city manager municipalities in the United States, only about 10 consider abandoning the system in a given year, and only 1 or 2 make the change (Mobley 1988).

The global economy has impinged upon local economies. The departure of factories, warehouses, and wharfs precipitated waves of change but did not leave all cities empty. Cities are no longer just nodes for a regional or even a national economy. They have begun to appear in the international arena as critical centers for a global information economy where besides an airport and seaport, new teleport services emerged, supported by the computer-communication way of doing business and living in a highly dispersed environment.

## A Case of the City of Kalamazoo Electronic Government

This chapter presents a case of the City of Kalamazoo (80,000 residents), whose council-manager corporate structure (900 employees) has undertaken a shift from a data processing to total information management paradigm. At that time, the City Manager, Jim Holgersson, initiated a comprehensive overview of actual computer projects in 1990. In 1991, the operation began a conversion from batch processing applications run by out-sourced services to an in-house, network oriented, architecturally planned and designed systems federations complex. The Network Computing Corporation from Charlotte (NC) was contracted with to deliver some turn-key application software. In 1992, a core of applications was operational, and some Electronic Global Village applications will be implemented by the year 2000 [2].

## Corporate City Information Needs and Aims

An organizational structure of the city administration is depicted in Figure 6-1. There are four branches of the Kalamazoo local government:

a. Constitutional services;
b. Community services;
c. Operational services;
d. Supportive services.

Each category of services satisfy the city needs from the perspective of citizens as it is shown on Figure 6- 2. It is an application of Maslow's hierarchy of human needs to the city's needs. The operating services should satisfy citizens basic needs for streets, water, sewer, transportation, and emergency monitoring. The Citizen Action Center should fulfill, citizens' needs for security in the scope of public safety and public administration. The City Hall should satisfy citizens' social needs for parks, golf courses, tennis courts, elections, churches, community service organizations, and cemeteries. The community services should satisfy citizens needs for personal worth in the area of public library, Civic Center, museums, community development, cultural opportunities, citizen needs for achievement, business climate, educational institutions, and citizen needs for self-actualization in the area of civic involvement, festivals, and so forth.

A department of Management Information Systems was established to develop the information management systems that will support the highlighted city services, with the following missions:

1. To provide state of the art, relevant, exclusive, accurate, accessible, and timely multimedia-based information for problem-solving, decision-making, and communicating to all city departments on a cost-recoverable basis through the information utility composed of host and communication computers, intelligent terminals and local area networks.
2. To automate clerical routines, support knowledge workers, and facilitate the decision-making process of executives, commissioners, and citizens via the development of information culture.
3. To stabilize participative management, job enlargement, job redesign, sociotechnical systems, employee involvement, and

quality of work life. MIS should create working conditions under which the city constituency will be learning, developing a vision, and self-actualizing.

4. To support peer-to-peer networking, work integration when virtual resources available are called upon, work dialogue with a shared understanding of products and services, knowledge and skills, geared to results, and with a visionary grasp of the nature of human time, the windows of opportunities, and task-focused teams.

The goals of a new system have been established as follows:

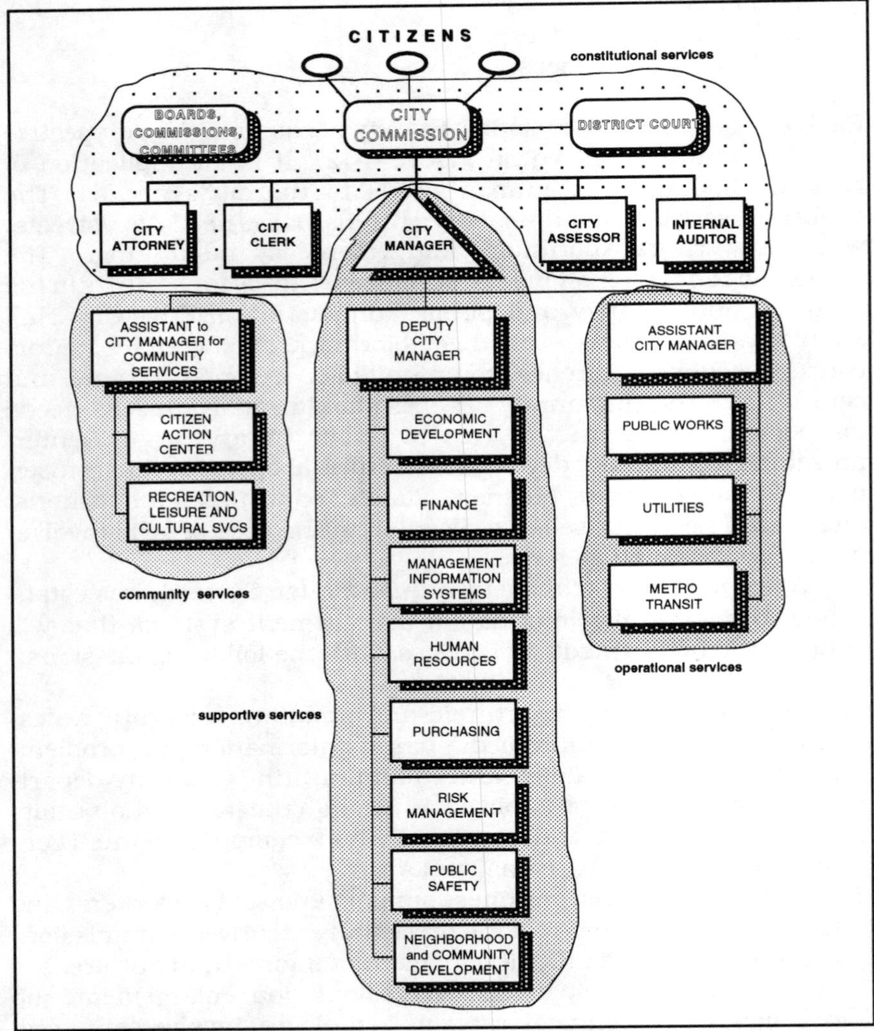


*Figure 6-1: City of Kalamazoo, Michigan Organizational Chart (1992)*

1. To be cost-effective in the use of information resources;
2. To increase the values added and output of each city employee by at least 3% each year through enhanced productivity;
3. To meet users' requirements and expectations for quality and variety of information that is appropriate for quantitative methods in problem-solving and decision making;
4. To be an information leader in all areas of city services, resources, operations, and technology;
5. To rank first among the state's local government leaders;
6. To operate Information Technology Systems in a manner which preserves safety, health, and a sound environment;
7. To support vigorously the reduction of data redundancy, the misuse of information, and the costly duplication of procedures.

The city perceives information resources as a vital optimizing factor among other internal resources such as human resources, money, materials, machines, facilities, management and external forces such as citizens' satisfaction, business community develop-

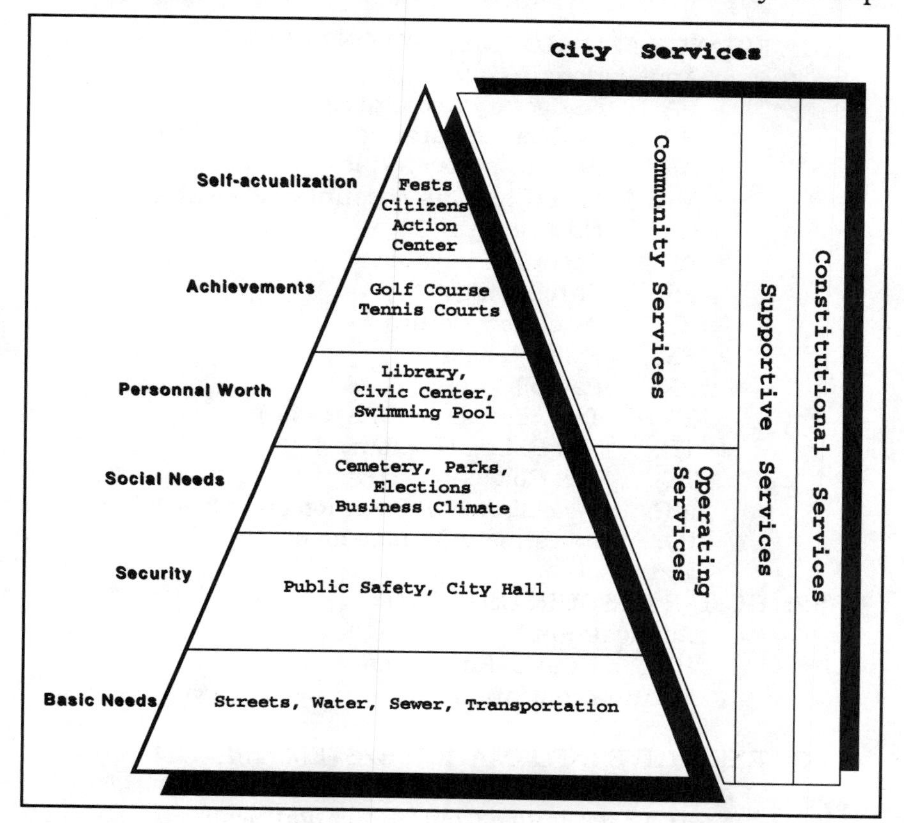


*Figure 6-2: A Hierarchy of Kalamazoo City Needs as Seen by the Citizen versus City Services (The Targowski Model, 1991)*

ment, and relationships with the county and state. Projected goals should attain a higher quality of information, constituting a shift from data gathering to information analysis which will enhance the reservoir of city employee knowledge, leading both to better quality of work life and higher standard for service delivery. The new City Information Culture should be a primary catalyst in moving from the Industrial (Bureau-Technocracy) Era to the Knowledge Era as a prerequisite for the emergence of the Kalamazoo Informative Economy.

### The Static Architecture of City Systems

City computer-based systems called MAGIC (Municipal and Geographic Information Complex) are categorized into federations as it is shown through the static architecture on Figure 6- 3.

The Bill of Systems Processor is as follows (Targowski 1990, p.151):

*MISF—Management Information Systems Federation:*

S1: FINANCE AND ADMINISTRATION (*MIND*)
Applications:
BA: Budgetary Accounting
BP: Budget Preparation
AR: Accounts Receivable
MA: Miscellaneous Accounts Receivable
CA: Cost Accounting
AP: Accounts Payable
PR: Purchasing
IC: Inventory Control
FX: Fixed Assets
PR: Payroll
TA: Tax Assessment (Assessor)
TB: Tax Billing (Treasurer)
TC: Tax Collection (Treasurer)
MC: Miscellaneous Cash Receipts (Treasurer)
IM: Investment Management

S2: HUMAN RESOURCES (*SPIRIT*)
Applications:
HR: Human Resources
Future Applications

S3: EXECUTIVE INFORMATION SYSTEM (*MIRAGE*)
Applications:
ECM: Executive Communication System
PC: Policy control System

ECS: Executive Control System
MCS: Management Control System
EDC: Executive Data Cube
EDB: External Data Bases

*OSF—Operations Systems Federation*

S4: PUBLIC WORKS (*HOBBY*)
Applications
CADD: Computer Aided Design/Drafting
PM: Project Management
GIS/W: Streets and Traffic Signs Database
VM: Vehicle Maintenance
WO: Work Order
FM: Fleet Management
CM: Cemetery Management
Future Applications
S5: UTILITIES [Billing & Collection] (*COMFORT*)
UB: Utility Billing & Collection
UM: Utility Maintenance
US: Utility Services
UE: Utility Engineering
US: Utility Sewer Control
GIS/U: Utility Database
Other Applications
S6: COMMUNITY DEVELOPMENT (*HABITAT*)
Applications:
BP: Building Permits
BI: Building Inspection (Code Enforcement)
HP: Housing Programs
UR: Urban Renewal
GIS/P: Property Records
Future Applications
S7: METRO TRANSIT (*SAFARI*)
Applications:
BM: Buses Maintenance
TS: Transit Scheduling
SS: Supervision Support
GIS.T: Transit Records
Future Applications
S8: PUBLIC SAFETY (*FRONTIER*)
PR: Police Records
TR: Traffic Records
FR: Fire Reporting
JM: Jail Management (County Function)
NC: NCIC Interface

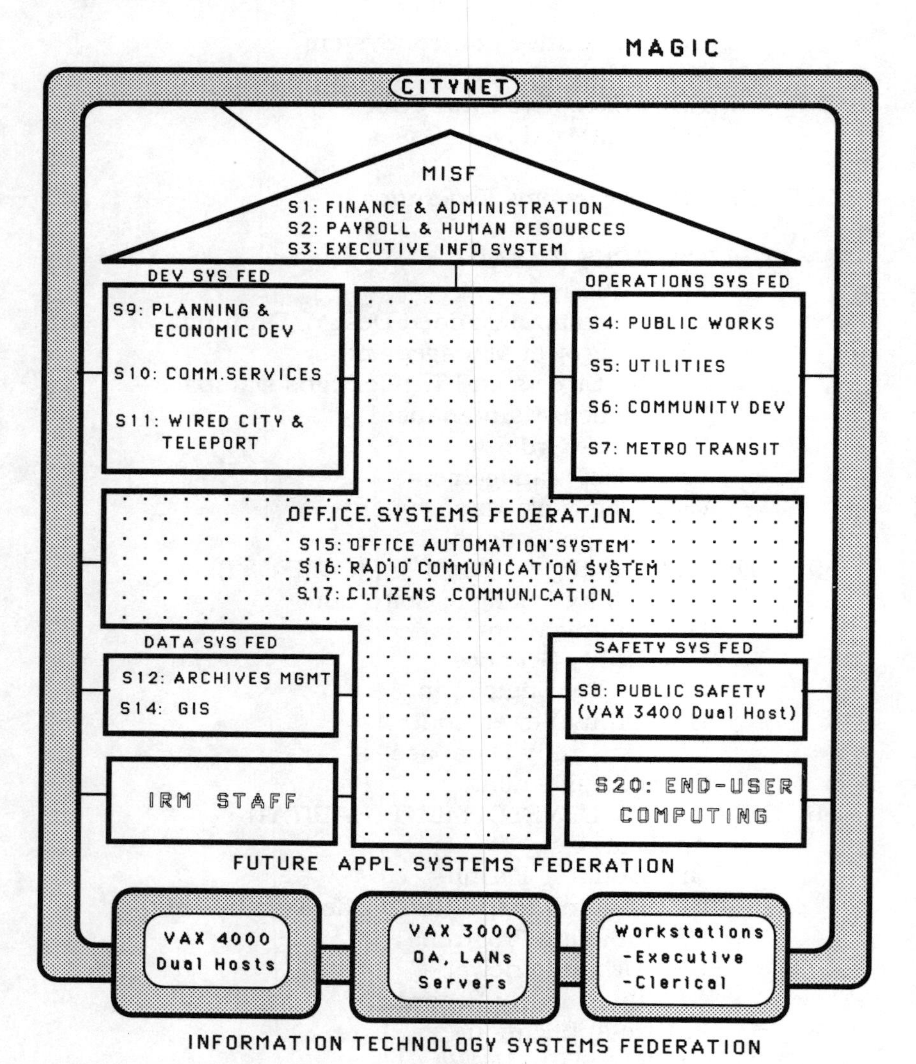


*Figure 6-3: The Architecture of City IRM Environment (NCC Vendor Products: S1, S2, S3, S4) (The Targowski Model, 1991)*

LEIN: Law Enforcement Information Network
CAD: Computer Aided Dispatching
UCR: Uniform Crime Reporting (State)
MDT: Mobile Data Terminals
FMIS: Fire Management Information System
CT: Courts
Future Applications

*DSF—Development Systems Federation*

S9: PLANNING & ECONOMIC DEVELOPMENT (*BOOM*)
Applications:
ED: Economic Development
BS: Business Climate
PS: Planning Services
ZP: Zoning Planning
LU: Land Use
BS: City Building Services
GIS/C: City Owned Property Records
RS: Real Estate & Property services
BC: Business Climate

S10: COMMUNITY SERVICES (*DREAM*)
Applications:
CA: Citizens Action
RP: Recreational Programs
CL: Cultural Programs

S11: WIRED CITY (*VILLAGE*)
Applications:
EH: Electronic Highway
TP: Teleport
EM: Electronic Marketplace
EU: Electronic University
EF: Electronic Forum

*DSF—Data Systems Federation*

S12: ARCHIVES MANAGEMENT (*PAPYRUS*)
Applications:
OA: Optical Archiving (Assessor)
VR: Voter Registration (Clerk)
JS: Jury Selection (County function)
CS: Census—Vital Statistics
CM: Cemetery Management (Public Works)
OL: Occupational Licenses (Treasurer)
GIS/P: Property Records (Assessor)
Future Applications: Knowledge bases

S13: vacant

S14: GEOGRAPHIC INFORMATION SYSTEM (*PARADISE*)
Applications:
CAD Mapping
CAE Mapping
GM: Geobase Management
GI: Geobase Inquiring
GR: Geobase Reporting
GT: Geo Interface for Economic Development, Census and so forth

Future Applications

*CSF—Communication Systems Federation*

S15: CITY AREA NETWORK (*CITYNET*)
S16: OFFICE AUTOMATION (*MISTIC*)
Applications:
AR: Agenda Reviewing (*AGENDA*)
EM: Executive E-Mail (*TOGETHER*)
TM: Calendaring (*RENDEZVOUS*)
CF: Conferencing (*FORUM*)
WP: Document Processing (*SHAKESPEARE*)
FC: File Cabinet (*LEGEND*)
DC: Directories (*FBI*)
COM: Communications (*BRIDGE*)
S17: RADIO COMMUNICATION (*CITY-EXPRESS*)
S18: CITIZENS COMMUNICATION (*DIALOGUE*)
S20: END-USER COMPUTING (*EUC—FUN*)

EUC is already about as large as data processing and may become larger. The City Information Center's role is to act as a bridging function between Organizational Application Systems, Information Technology Systems and freelancing end-users. The IC is fundamentally the support and the information tools used in non-clerical office departments with managerial and professional tasks. The following EUS software is supported by the City IC:

Presentations:
E-1. Word processing software (Word Perfect)
E-2. Presentation software (PageMaker)
E-3. Multimedia software
E-4. Analytical and conceptual graphic software
E-5-10. vacant
Data Management
E-11. File and database software
E-12. Query software (SQL)
E-13. Report generation software (NCC-IQ)
E-14. Commercial databases retrieval navigators
E-15. Uploading and downloading
E-16. User free library
E-17. Library shareware software
E-18-23. vacant
Model Management
E-24. Spreadsheet software (VAX 20/20, Lotus)
E-25. Financial analysis software (DSS>>>)
E-26. Statistical software (SPSS)
E-27. Time management software (Time Line)

E-28. Simulation software
E-29. Programming languages (BASIC, SQL, FOCUS)
E-30. Expert Systems
E-31. Auto-CAD
E-32-35. vacant
System Software
E-36. Operating systems (DOS, Windows, Finder, VMS)
E-37. Security and safety software
E-38. Key utilities software
E-39. Desk accessory software
E-40. File conversion software
E-41. System Transparent software (Mid-Micro link)
E-42-46. vacant.

## The Dynamic Architecture of City Networking Systems

Figure 6- 4 depicts the dynamic architecture of the City Hierarchical Systems for all levels of government and external users, such as citizens. The needs of citizens, particularly for self-actualization, achievement, personal worth, and issues in terms of cost-benefit, are addressed by the commissioners applying the Policy Control System.

The Policy Control System (PCS) plans, controls, and monitors the city aims such as creed, mission, goals, objectives, policies, and strategies. The purpose of PCS is to support the commission judgment on the performance of the city manager. The PCS provides information about the local government structure, nation-wide, state-wide, and region-wide, in addition to competitive positions, and external trends and conditions which my create both opportunities and constraints for the operations and development of the City of Kalamazoo.

The Executive Control System (ECS) supports the activities of the city manager (CEO), who deals with the emerging future and the pressures of change. The ECS measures the city's potentiality, capability, and actuality while controlling latency (the relationship of actuality to capability), and determining performance (the relationship of productivity to latency) in short-term perspective and long-term perspectives with an orientation toward the city's profitable innovations for better quality of life, business climate, and safety.

The Management Control System (MCS) supports the activities of the deputy city manager, and two assistants to the city manager by measuring, controlling, and monitoring the performance of the enterprise units, and departments with an orientation toward maximizing productivity and reducing operating costs.

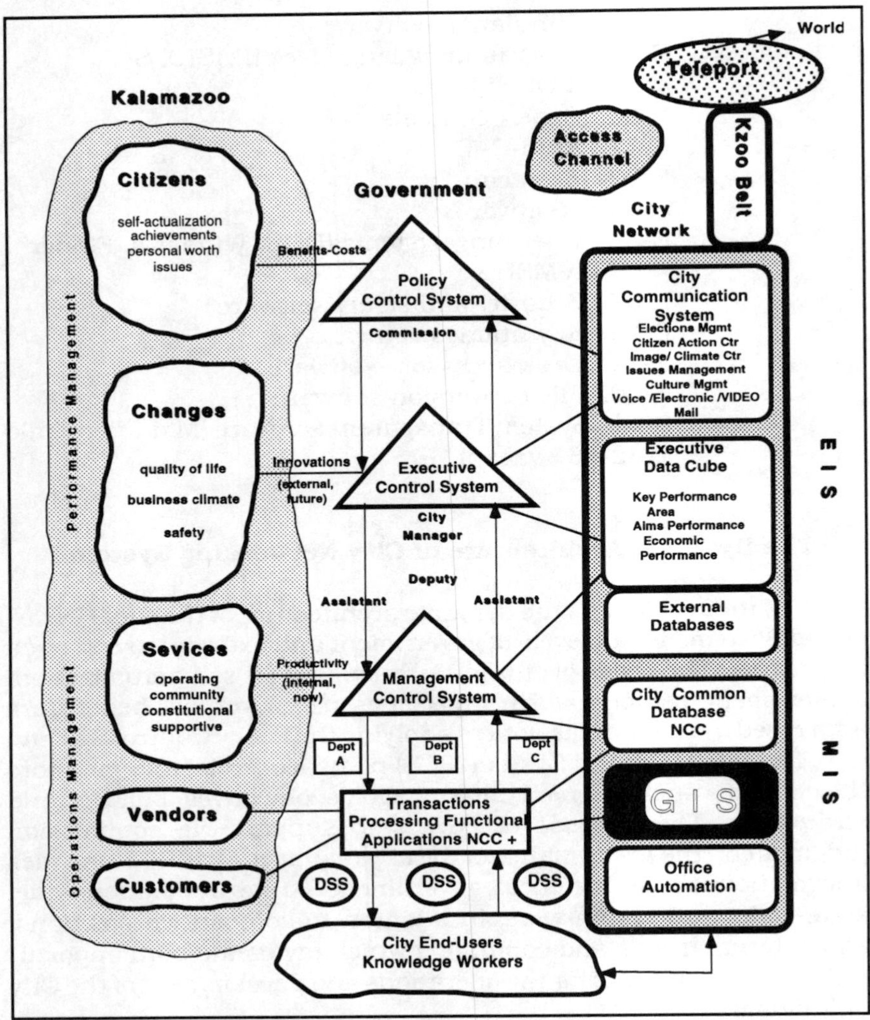


*Figure 6-4: The Architecture of City Networking Systems (The Targowski Model, 1992)*

The Transactions Processing Functional Applications are shown in Figure 6-5. They support the city operations management of services in the scope of performers, vendors, and customers. The applications are run for such functions as budgetary accounting, purchasing, tax assessment, tax collection, payroll, human resources, utilities billing, housing programs, rental and construction permits, voters registration, and so forth.

The Executive Communication Control (ECS) system is a futuristic system which as a management tool, uses professional communication to influence organizational performance and communication climate which affects the degree of openness with which people communicate. The system offers other communicational tools such as voice/electronic communication, future access to Electronic Global Village conferencing, Citizen Action Center, and Image/Issues Management.

The Executive Data Cube (EDC) is a database which updates key performance area data elements, and aims economic performance data elements. It provides access for all executive/management control systems.

The City Common Database for transactional processing of functional-operational applications is updated and used by the applications software (budgetary accounting, purchasing, payroll, and so forth) developed by NCC.

The Geographic Information System (GIS) collects, stores, analyzes and displays spatial (location) data of 200 sets of 20,000 maps. GIS is a set of information layers that include:

- Land Records—parcels, subdivisions, easements, rights-of-way, ownership, valuation and assessment information, maintained at the assessors office;
- Infrastructure Records—sewers, storm drains, street lamps, trees, telecommunication systems, utilities, roads (center lines, intersections, street rights-of-way, and emergency access routes) main-

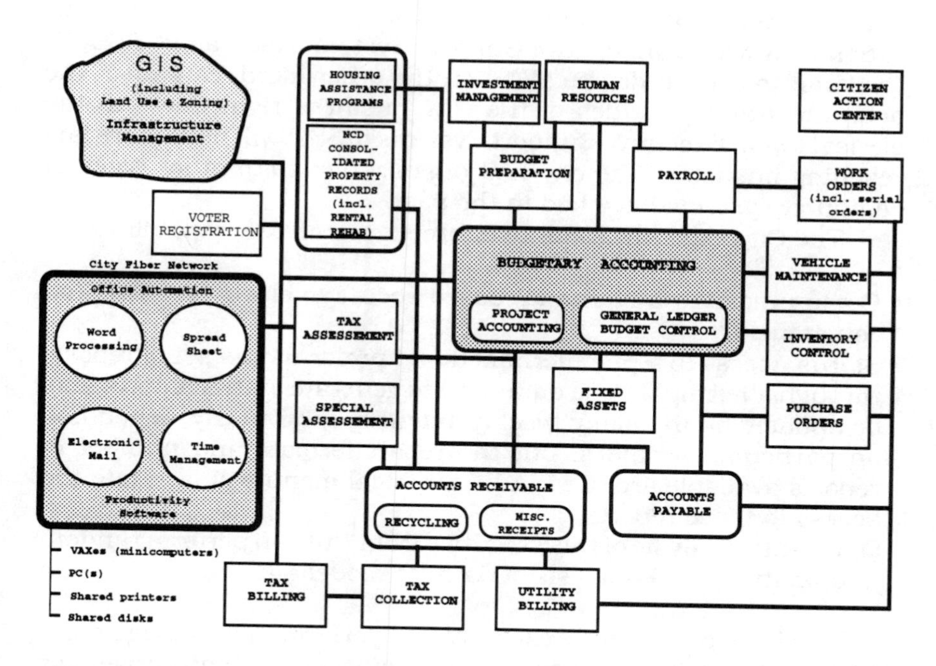


*Figure 6-5: Transactions Processing Functional Applications of the City of Kalamazoo (The Targowski-Taylor Model, 1993)*

tained at the offices of Public Works and Utilities;
- Environmental Records—geology, soil, vegetation, wetlands, hazard areas, and noise controls, applied at different agencies;
- Zoning Records for development proposals and land use plans, applied at the office of Economic Development and Planning;
- Safety Records—maps with locations of accidents transferred to mobile data terminals in police cars and applied at the office of Computer-Aided Dispatching;
- Bus Records—transportational routes, stops, loops for the design and analysis of the Metro Transit System;
- Parks Records—parks and recreational areas for the development and analysis at the office Recreation, Leisure, and Cultural Services;
- Housing Records—for the certification of new buildings and rental accommodations through Neighborhood and Community Development.

The Office Automation System provides such applications as networked document and word processing, time management, spread sheet calculations, and electronic mail communication. The first three applications are also called a Decision Support System (DSS).

Traditional MIS applications (transactions processing) and office automation systems can have significant individual and departmental impact, but rarely do these systems affect the organization as a whole. Executive Support Systems such as PCS, ECS, CCS and to certain degree MCS, on the other hand, can affect the operations and priorities of the City business right down to the clerical level. Executive Support Systems (ESSs) will have more far-reaching impact on the city hall organization than other types of computer systems have had in the past.

The City executives can use three types of ESS capabilities:

- Communication by terminal-based access to electronic mail and conferencing;
- Status Access to a predetermined and preformatted set of reports through a terminal. The data used to generate status reports may be updated hourly, daily, weekly, monthly or quarterly, depending on particular variables, but the report formats and number of reports available are fixed. A hierarchical menu will facilitate the access to these reports.
- Query and Analysis of the Executive Data Cube to perform random and unstructured analysis of data, or modeling.

The ESS tool supports executives in performance management, which begins with the concept of a city as an "enterprise," or "corporation," shapes its image, creates its climate, and sets its aims, from the city level through enterprise units and departments and down to the individual. It is a vertical integration of achievement, linking organizational levels through the use of organizational systems and behavioral tools, such as aims, climate, culture, incentives, and performance appraisal.

### Major Issues and Conclusions of the Kalamazoo City Case

The development of such integrated systems generates some challenging organizational and individual issues which will have to be gradually resolved. Among these issues are the following ones (Targowski 1990, p.89):

a. Setting up motivation, leadership, and a spirit of performance among end-users, systems implementators, and executives;
b. The development of computer skills and expertise among end-users via systematic training sessions and better user documentation;
c. The development of a system strategy supportive culture among end users and systems implementators;
d. The challenge of defining what a state of the art MIS means for the City of Kalamazoo in a very practical, soundly analyzed futuristic, but elegant sense;
e. Balancing the individual departmental needs with the collective community and organizational needs of the City.
f. To organize all participants with diverse skills, attitudes, and visions into one focused project team dedicated to universal corporate success.

The computer and communications capabilities created by the City of Kalamazoo redefines the City's competitive advantage and sets an example for the whole community on how to manage proactively. During the first year of applying Electronic Mail, 24,799 messages were created and sent, which means that an average user sent or received 124 messages during the first year of office automation operations. This tangible sign reflects a dynamic government operation honing its technological organizational, and management competence to more quickly and efficiently respond to community needs.

## Geographic Information System in Government

GIS is a decision-support system using spatially referenced data, which can be entered, updated, queried, analyzed, modeled, and displayed by a computer. GIS combines geographically referenced information (maps) with data (text). This capability allows for the modeling of a new road, as well as analysis of a city's income from real state taxes. GIS is used by :

- Federal Government Agencies
- State Government Agencies
- Local Government
- Private sector Corporations
- Universities and Research Institutions

At the Federal level, GIS is used to track endangered species, predict areas prone to landslides, simulate forest fires in national parks, and allocate emergency equipment in such areas as San Francisco.

At the State level, GIS is used to plan economic development (Georgia), control housing growth (tracing the infrastructure development along with the housing projects in Florida), and so forth.

At the local level, GIS is applied to manage infrastructure maintenance and development, land records, plan economic development, and so forth.

The private sector firms use GIS in mining, forestry (natural resources management), optimal fleet delivery scheduling, and so forth.

Universities and research institutions apply GIS in campus planning as well as in demographic admission analysis (Dickason 1992).

## A Case of the Kalamazoo City

The city GIS was planned to be a strategic tool of progressive infrastructure and tax management [3]. It was also planned to facilitate internal and public access to information on these topics. Infrastructures impacts our life, from switching on lights, to turning the water facet, to driving to work. The public-private relationship has two dimensions:

- The effect on productivity
  Local highways with inadequate capacity produce gridlock and waste time.

- The effect on community growth
  Commercial buildings, residential subdivisions, industrial plants and telecommunications are highly dependent upon an efficient and effective infrastructure.

The City of Kalamazoo maintains about 200 sets of maps or about 20,000 maps total. A list of common problems is as follows:

- Once the manual information is stored it is forgotten
- The labor cost of maintenance and updating maps engineering drawings
- Use of outdated maps and manual records by other agencies is inefficient and leads to error
- The retrieval and verification of map information is slow
- Storage of maps and drawings takes up a lot of office space.

Inaccurate and inconsistent duplicate copies of information can cause confusion among various city organizations responsible for administering government services. This situation could increase the city's liability and exposure. The flow of infrastructure mapping and tax information must be streamlined or the city will face ever-increasing administrative cost.

The possible functions for automated infrastructure management include:

- Automated engineering design
- Rapid retrieval of mapping information and engineering data
- Reduction of map sheet storage and maintenance costs
- Ability to create special thematic maps and graphic presentations
- End user modeling on complex infrastructure variables
- Building plans, permits, and inspection tracking reports
- Construction and utility property easement analysis
- Urban runoff and storm drainage retention analysis
- Coordination of proposed construction with future development
- Street paving and maintenance rating analysis
- Demographic analysis of census data
- Composite street closure and utility repair location analysis
- Automation of maps features, geographic text files, administrative data, and keys to link map and text data.

The support of decisions in policy making and operations should take place in such user departments as:

- City Assessor: tax assessment, parcels, easements, ownership
- Economic Development and Planning: zoning, development proposal and land use plans
- Neighborhood and Community Development: certification of new buildings and rental houses
- Community Services: parks and recreational areas development and maintenance
- Public Works: street, bridges, centerline, right-of-way, emergency

access roads, traffic signs, street lamps design and maintenance, storm drains development and maintenance

- Public Utilities: water, sewer
- Metro Transit: transportation route design and analysis
- Public Safety: maps with location of an accident transferred to a mobile data terminal in a police car
- Citizen Hot Line via the Community Access Channel

The city GIS strategy was based on incremental progress and the following assumptions:

- The application of a comprehensive software package from an established vendor such as McDonnell Douglas-Electronic data Systems (EDS). This package would be VAX Rdb/SQL driven and expanded through the following phases:
  - *Phase I:1991-1992*
    Assessor GIS and Base Map
    Economic Development GIS small projects
    Public Safety small projects
    Public Works small projects
  - *Phase II: 1993-94*
    Metro Transit GIS
    Public Works GIS
    Public Utilities small projects
    Public Safety GIS
    Housing GIS with Image Retrieval
    Economic Development GIS
  - *Phase III: 1995*
    Public Utilities GIS
    GIS Information Kiosks
    Other GIS applications
    Integration of City and County GIS(s).
- The base map would be developed in a scale of 1" = 100'.
- Data attributes for each user layer should be loaded to the geographic Information System relational database by the end-user staff, which would be responsible for data quality. The end user is the master of GIS and the MIS staff which provides support
- The implementation of the GIS applications would be done at the pace of the end user's availability and learning curve.

Future development of the city GIS is planned in the following scope:

- The Kalamazoo GIS applications should be integrated with the Kalamazoo Telecity for electronic access by all interested and eligible users
- The applicational emphasis will change from infrastructure management to environment protection, land redevelopment, and quality of life.

A concept of a GIS application is illustrated in Figure 6-6. It is an architecture of overlapping layers used in Public Works projects.

As a consequence of the City of Kalamazoo's involvement in the GIS project, a Kalamazoo County-wide Committee for GIS (KAGIS) was initiated [4]. A strategy of KAGIS is to generate country-wide cooperation for the common GIS project through the business partnership with EDS, the city MIS, and private developers to pursue township-oriented pilot projects on a cost effective basis. KAGIS transforms into a business entity which is capable of managing and implementing this project.

The following scope of KAGIS interest was defined:

- The Geographic Perspective: Staged Development of Multi-Purpose

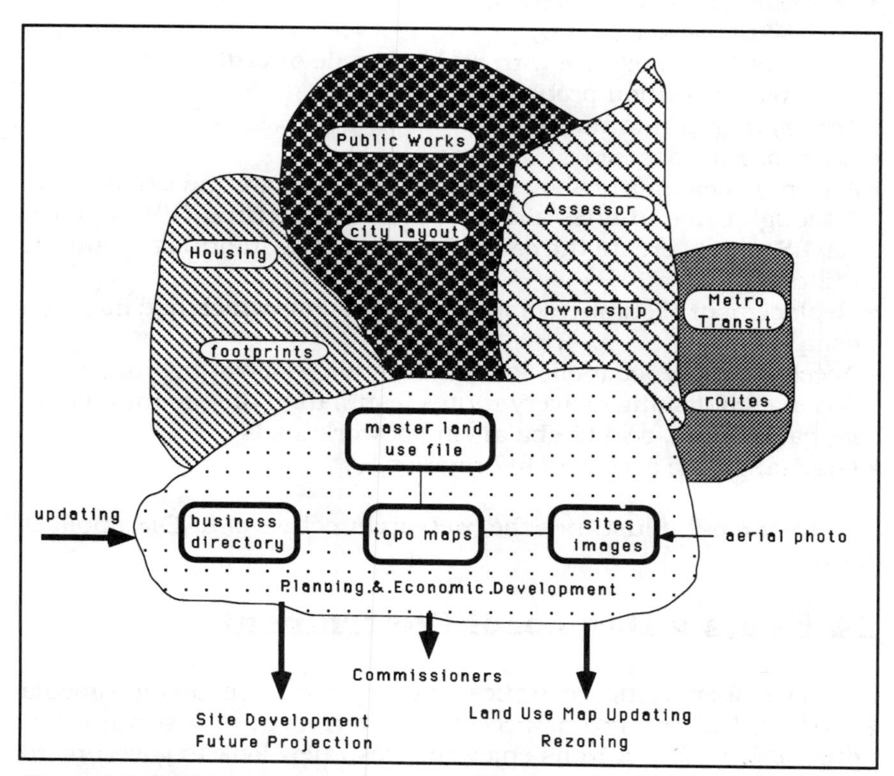


*Figure 6-6: GIS Layer of Planning and Economic Development*

Cadastral Mapping (Land Information System) [5]:
- Invoke a Remuneration Program county-wide
- Adopt a uniform parcel-base recording and reporting system
- Use best existing base map data with starting coordinates
- Overlay parcel maps on this base
- Rectify the map county-wide through photo overflights and local surveys
- Calibrate the coordinate system with the Global Positioning System
- The Business Perspective: the following services should be provided [6]:

For all organizations:
- Appropriate controls over entries, deletions, and changes of data

For the real estate industry:
- Access to lot locations and dimensions
- Access to building layouts (used when fighting fires by the Department of Public Safety)
- Assessed valuation of each property
- Last sales price and sales data
- Warning of those properties:
  - With tax liens
  - Not in compliance with building code or ordinance
  - With pollution problems
- Zoning map printouts
- Locations of easements
- Accurate locations of above ground and underground utility lines. Although some utilities already have computerized GIS systems, KAGIS would create the capability of sharing information among utilities
- Replacement of cumbersome manual systems which had not been computerized
- Access to these locations for construction/excavating purposes
- Aid in establishing delivery routes, estimating time requirements for each route, and the balancing of work loads
- General public access to information.

Figure 6-7 illustrates the web architecture of KAGIS applications.

## 24 Hours a Day Local Government

A number of cities, counties, and states have set up multimedia kiosks that allow citizens access to government and services 24-hours a day. The citizens enjoy the fresh new look to government information.

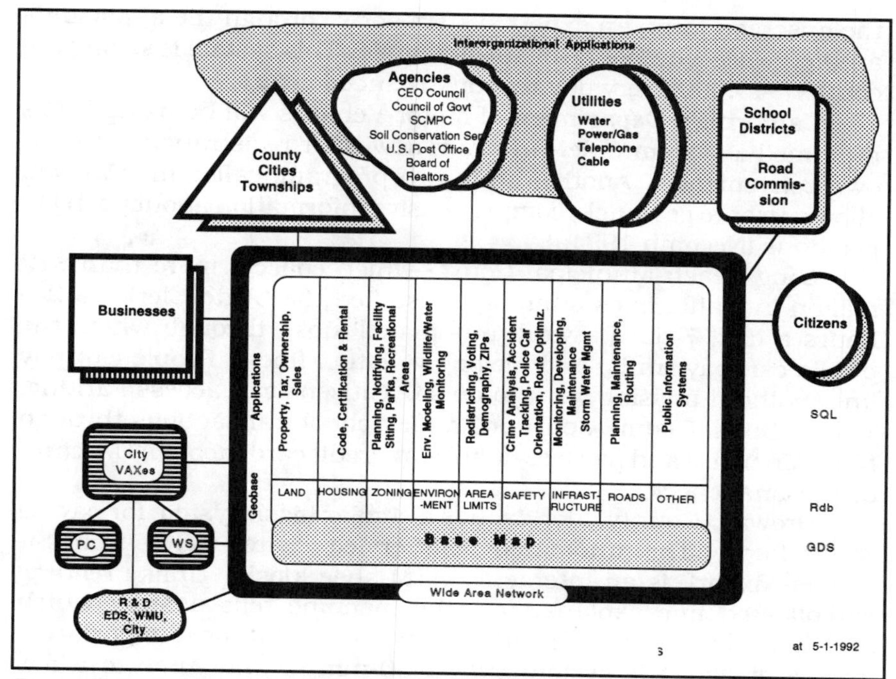


*Figure 6-7: The Web Architecture of KAGIS Applications*

At a senior citizen center in Orlando, Florida, people can get information as to when garbage is picked up on their street or when the next city council meeting takes place. The information is not in brochures or on phone recordings, but in the form of videos, images, sound and words in a computer with a touch sensitive screen housed in a kiosk. From 300 to 400 times every day, the screen is touched by people seeking information on their local government.

The kiosk houses the IBM trademarked 24-Hour City Hall system, a project developed jointly by IBM and Public Technology Inc., a nonprofit association of local governments. By merging a powerful PC with a laserdisk player and sophisticated software, local and state governments have begun using multimedia as a tool for delivering information and services to a diverse range of people.

In Mercer Island, Washington, the city's kiosk is in a local grocery store that is open 24-hours a day. Called the "Island Epicenter" the kiosk contains over 100 files of city and community information, as well as a grocery directory. Like the Orlando system, video clips, images and sound are combined with text to deliver city information (Necombe 1991).

In Tulare County, California, the public is applying for food stamps and filling out Medicaid and MediCal eligibility forms on touchscreen kiosks. While the applicant watches videos and listens to a narrator explaining in either Spanish or English how to fill out the onscreen form, an expert system sorts through the applicant's answer to see whether he or she is eligible. This system is saving the county $40,000 every day through reduced errors.

California's Department of Motor Vehicles will be using IBM's multimedia system to let drivers renew their registrations at their own convenience. Another state-run program, called infoCal, will allow a user to get quick, comprehensive information about open job positions (Necomb 1991a).

Long Beach Municipal Court—which collects more than $12 million in traffic fines annually—installed the "Auto-Clerk," a 24-hours a day, 7-days a week automated kiosk through which the public can pay traffic and moving violation fines. People can pay fines without missing work to do so. Citizens can access Parking, Traffic, Small Claims Court or Traffic School transactions through the main menu and pay fees or fines by credit card, debit (ATM) card, or personal check.

Broward County, Florida, has a similar kiosk system for paying traffic fines. The automated Self-Service Center (SSC), from Siemens-Nixdorf, is an interactive ATM-style kiosk. Unlike reading complicated fine explanations, the machine tells a story which simply guides the user through the process. But most importantly for the citizen, the system cuts a 30-minute wait time to a few minutes. The SSC has definitely been a hit in this county. More and more fines are paid via the SSC out of 500,000 handled annually. This means more revenue and improvements for the county.

The success of governmental kiosks is in human behavior. The citizens do not always like to call city hall for information. As a result, they do not know about services provided by their governments. Using a touch-screen, people do not feel intimidated. They are inclined to start playing with the system and get some real information out of it.

## High-Tech Local Justice

The futurist Gene Stephens warns that emerging technology could lead to constant surveillance and forced behavior change, but at what cost to individual rights? Supersensitive, audiovisual devices, computer networks, genetic identification, electronic monitoring, and other soon-to-be-available products and techniques are a boon to criminal justice agencies.

Electronic monitoring of individuals—used for defendants on bail, probationers, and parolees—is already testing the scope of both Fourth and Fifth Amendment protections. Is monitoring a form of punishment? Stephens asks whether an inmate could be freezed, followed by suspended animation (removing the blood and putting the individual in a "storage"). In a face of massive prison overcrowding, these and other forms of human hibernation offer immediate answers to correctional problems. Many inmates could be "stored" in small spaces using these techniques, but are they constitutional?

Implants offer a whole new array of weapons in the battle to control crime. Convinced offenders could be sentenced to have electrode monitors implanted to keep them within a territory, but beyond this, a subliminal message player might be implanted to give the probationer a 24-hour-a-day anticrime message. Behavior-control capsules implanted in public offenders could change their behavior, like five-year birth control implants in women. Genetic engineering also promises new techniques of removing or altering "deviant" traits. Will these techniques become part of the sentencing of the court, or will they be deemed cruel and unusual (Stephens, 1990)?

High technology could make future police pursuits much more efficient — and a lot less hazardous. The following is an excerpt from an article in *Police Executive* magazine, in the first decade of the twenty-first century (Moore III, 1990):

> The electronics disabler was first used to halt a fleeing automobile on April 12, 2005. Officer John Smith of the Denver Police Department was on patrol in his specially equipped 2005 Ford police cruiser when he observed a 2005 Chevrolet Corvette speeding southbound on Broadway from Interstate 25.
>
> Smith caught up to the Corvette when it stopped for a red light at Broadway and Evans Avenue. Prior to activating the overhead lights of the police car, Smith's partner entered the license plate number of the Corvette into the cruiser's computer terminal.
>
> The response was immediate: The Corvette had been reported stolen about two hours before. An additional message was displayed, in the form of a question: "Disabling requested, Y or N?"
>
> Prior to responding to the question, Smith activated the police car's overhead lights and attempted to stop the vehicle in the usual fashion. The Corvette accelerated in an attempt to elude him. Smith's partner touched the "Y" on the computer keyboard; immediately, a coded radio frequency, identified by the computer according to the year, make, and model of the pursued vehicle, was transmitted from the patrol car. The radio frequency was received by the factory-installed circuitry in the Corvette, activating a relay that disrupted the 12-volt current between the alternator and the coil in the car's engine effectively shutting off the car.

This scenario could happen in the near future. The technology necessary to implement this solution exists today. Moore proposes also that automobiles should have computer-controlled starting systems that require a coded driver's license and additional entry codes. Then, a person with a suspended or revoked license could not start his/hers car, nor could car thieves start any one's else's car, without both a driver's license and each car's unique computer entry code.

We are already accustomed to "smart" cars. On-board computer systems control the engine, prevent the wheels from locking when braked, and adjust the suspension when the car is carrying a heavy load or the road is bad. Many smart cars even guide technicians in making repairs. One of the next major steps in making cars even smarter will be the addition of navigation computers that will act as intelligent assistants to the driver. Future navigation and information systems will automatically keep the driver informed of current location, deduce the best routes to destinations by taking into account traffic and road conditions, speak and/or display turn-by-turn instructions according to the car's position along the route. The systems will also be able to tell the driver the nearest service station that is open, as well as the nearest hospital or retailer of a particular product. Once the vehicle has reached its destination, the navigation system will direct the driver to the closest available parking lot (French, R.L. 1990).

The earliest applications of automated mapping for the public safety application is a map downloaded from the city hall GIS into a police mobile data terminal, as illustrated in Figure 9-8.

A downloaded map provides for the police patrol a route and property ownership data as well as public utilizes' data on water supply, faucets, and so forth. In such a manner the police patrol is well-informed to solve a public safety problem.

Another significant application of High-Tech in justice is in video-teleconferencing services providing a link between court in session, a defendant in jail and his/her attorney at the office. This solution saves on trips, time and resources. It also may lead to a virtual court, solving cases almost instantly.

Any future use of such high technology for justice must consider issues of human rights, individual freedom, and privacy.

## On-line State Government

States affect citizen's lives in a variety of ways. They charter corporations, and oversee banks, insurance companies, railroads, public utilities, and the liquor industry. In addition, they maintain highways, license drivers, and register motor vehicles. Some operate

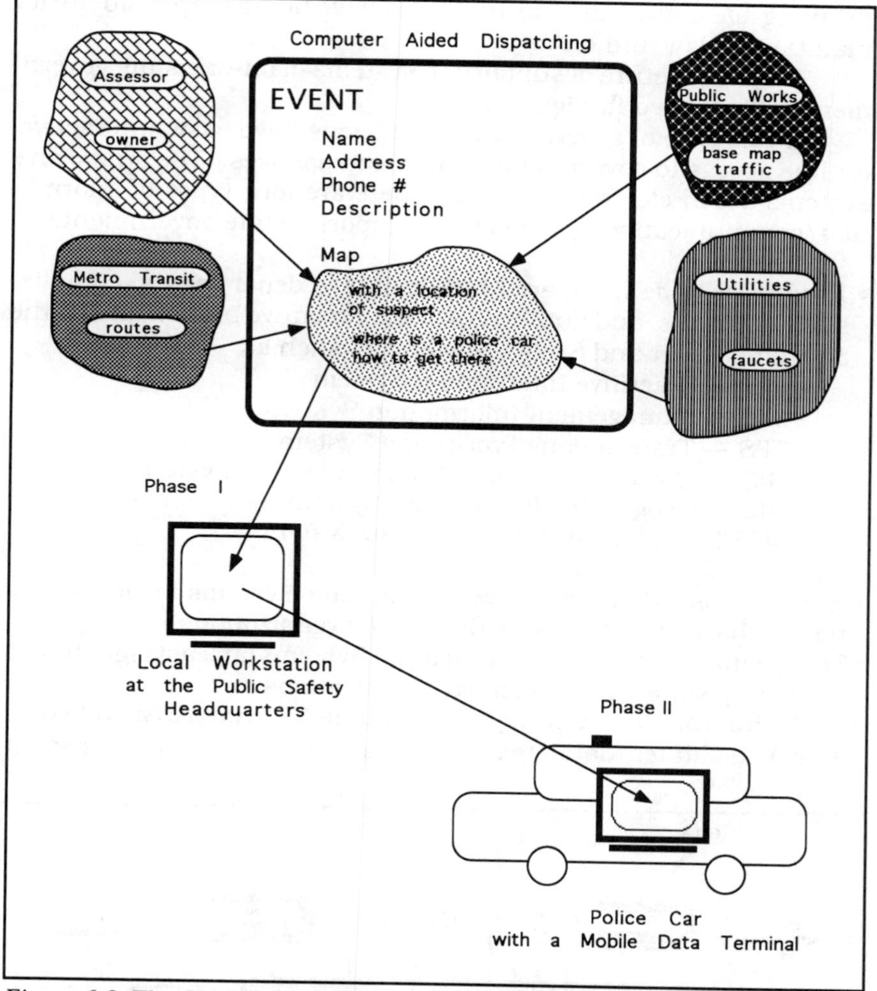


*Figure 6-8 The GIS Layer of Public Safety (The Targowski-Lamberti Model, 1992)*

mass transportation systems, including bus and rail routes. Many states have established offices to protect consumers, the environment and encourage energy production.

All states are active in education. They regulate and help fund local public schools and operate state university systems. States operate hospital systems (especially for the mentally ill), state park systems, and a variety of support services for agriculture and business. They also serve as conduits for federal grants to local governments. In recent years states have created public authorities in increasingly greater numbers (airports, ports, bridges, housing units, parkways, sport arenas, reservoirs, and university dormitories). They regulate marriage licensing, divorces, inheritance, drinking age limit, and so forth. States have broad responsibility for maintaining law and order.

The architecture of supporting systems of on-line state government is illustrated in Figure 6-9.

In this type of government, the citizens have easy, interactive, on-line access to governmental units and services. The supportive systems are in electronic format. There are four types of information/communication systems that support a state government:

- Enterprise-wide IS, shared by the state's departments, commissions, agencies, and authorities of the executive branch and by the Supreme Court and State Legislature, such as:
  - EIS — Executive Information System
  - MIS — Management Information System
  - TPS — Transactions Processing System
  - OIS — Office Information/Communication System
  - GIS — Geographic Information System
  - MDBS — Management Database System

- State Transactions-Intensive Information Systems (at least four types) shared by state-wide dispersed organizations
- State Database-Intensive Information Systems (at least eight types) shared by state-wide dispersed organizations
- State Information Highways, a multimedia Wide Area Network, which facilitates data, text, voice, and video tele-transmission

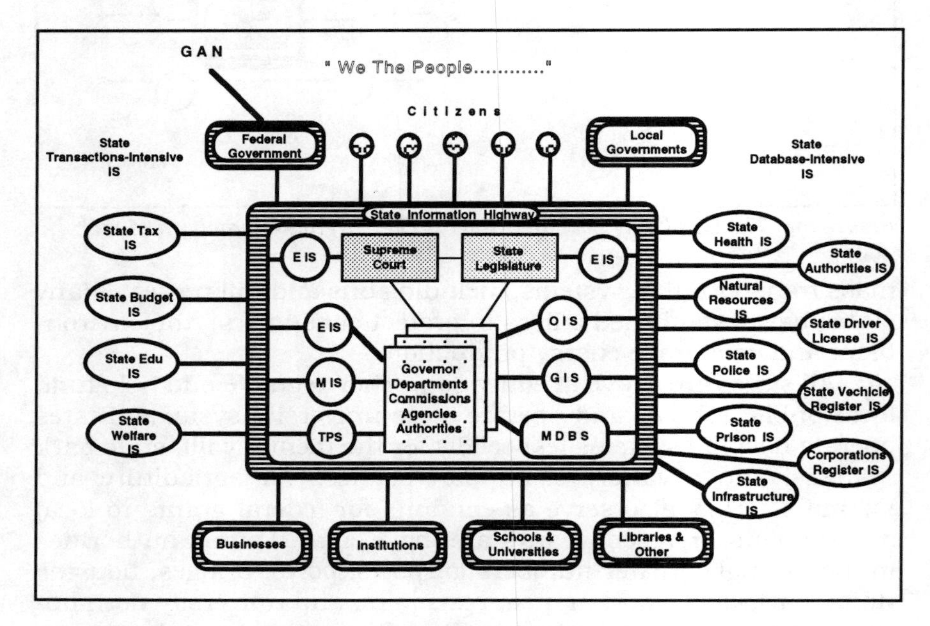


*Figure 6-9 The Architecture of Online State Government*

among state's organizational units, local government, and the federal government.

The strategy of implementing these systems should include tasks of reengineering III. A key to the success of this implementation is the State Information Highway (SIH) which should be reachable at most of the state main locations. The SIH dynamizes the state's systems and is a conduit between local governments and the federal government. A master plan of the On-line State Government Systems should coordinate and standardize efforts to develop all state and local governments systems.

## On-line Federal Government

The Preamble to the Constitution of the United States describes the purposes of the new central government and the federal system: "...to form a more perfect union, establish justice, insure domestic tranquillity, provide for the common defense, promote the general welfare, and secure the blessing of liberty..."

The genius of the federal government is that it has been so adaptable to changing circumstances. A system designed to govern a small coastal, agrarian nation of fewer than 4 million people has, without revising its basic structure, proven adequate for a transcontinental, industrialized, urban country of a quarter of a billion people.... until the 1990s.

This was only possible because the world was driven by rules of *Pax Americana* after World War II. The most industrialized countries were in ruins and complied with American rules. The Cold War, the complexity of the American economy (50% of the world production) and the affluence of the U.S. resulted in the development of the federal government into 13 departments and 3000 agencies, employing 3 million workers with a budget of 30% of the GDP. The federal government is a super corporation, slowed down by internal bureaucracy and strong criticism by business leaders, social leaders, and citizens. Today, some are of the opinion that "this government is designed not to work."

The federal government in the 1990s must be reinvented in order to govern a "perfect union." *How to* develop working supportive systems is depicted in Figure 6-10.

The concept of On-line Federal Government reminds us of the concept of On-line State Government. It is however, a much larger structure with thousands of systems and units of information technology. This government spends about $12 billion yearly for information systems and technology. It is equivalent to the consumption of the whole production of No.2 world computer vendor —

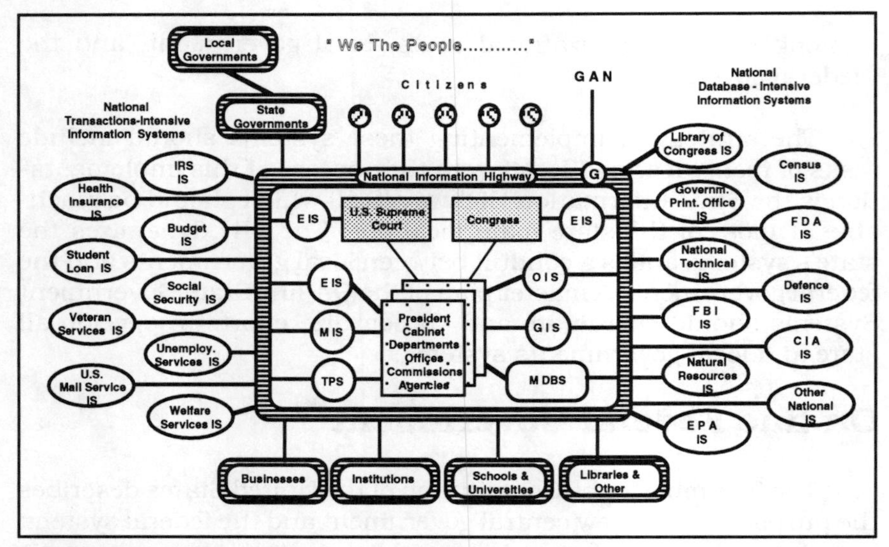


*Figure 6-10: The Architecture of Online U.S. Federal Government*

Digital Equipment Corporation, which used to employ 150,000 workers to achieve this level of sales.

To achieve a level of On-line Government, tasks of Reengineering III have to be undertaken. It will be impossible to continue the bottom-up strategy in systems development and operations. It would only support the existing islands of automation and keep the public sector workers and officials uninformed or even misinformed. Moreover, government customers will be unsatisfied by poor delivery of services.

The key to success of these supportive systems is in the switch from the orientation of information technology "devices" to government-wide applications systems. So far, management of information resources occurs at two levels: government-wide, and at the level of executive branch agencies. At the government-wide level, the following five agencies perform the following tasks (Andriole 1985):

- the Office of Management and Budget (OMB) formulates fiscal and policy controls
- the General Service Administration (GSA) guides procurement, implementation, and management of office information systems
- the Institute for Computer Science and Technology (ICST) promotes standards for hardware and software compatibility and interconnectedness (protocols.)
- the Office of Personnel Management (OPM) develops position classification standards and provides training for office automation
- the General Accounting Office (GAO) is an arm of Congress overseeing Executive Branch agencies in carrying out the law and the effectiveness of their use of appropriated funds.

These types of IRM organization cause the absence of systems leadership and clear delineation of vision, strategy, responsibilities, and coordination among agencies. The lack of a comprehensive strategic system planning at the federal government-wide levels creates problems for the individual executive branch agencies and for state and local government information systems.

The presented architecture of On-line Federal Government should be the systems yardstick in reinventing government. The good news is on the effort of the Clinton Administration (effort led by Vice-president Al Gore) to develop the National Information Infrastructure with a speed of 1 trillion bits per second. It will be based on the National Education and Research Network (NREN) which should provide the most advanced information infrastructure in the nation. It should support long-term economic growth which creates jobs and protects the environment in a competitive global economy. This Information Highway should also support government that is more productive and more responsive to the needs of its citizens (Gore 1993) [7].

# Conclusion

Technology and government in the 1990s should create a new national information infrastructure to promote reforms required to sustain the U.S. leadership in the world.

### End Notes

1. U.S. Bureau of the Census, Statistical Abstract of the United States 1985, Washington, DC.: Government Printing, 1985, pp. 261-292.

2. The case of the City of Kalamazoo was presented by the author during the International Convention of the Information Resource Management Association in Charleston, South Carolina, May 24-27, 1992 (see Proceedings, p. 38-46) with the previous collaboration of J. Holgerson, M. Ott, K. Overly, S. Taylor, and D. Skardarasy from the City Hall of Kalamazoo.

3. A GIS project for the City of Kalamazoo was implemented in 1991-93 by a team of the City executives and consultants: Sammy Taylor (MIS Director), Gisele Green-Czajka (MIS-GIS Support), Valerie Purcell (City Assessor), Teri Schiedel (Planning and Economic Development), Darin Rose (GIS Planner), Andrew Targowski (City Consultant), Jonathan Tarr (EDS) and others.

4. The first chairman of KAGIS was Andrew Targowski (1991-93), the next chairman was David Dickason (1993—).

5. This list of potential geographic applications was developed by David

Dickason, a member of KAGIS.

6. This list of potential business application was developed by Hugo Swan, a member of KAGIS.

7. This undertaking is a part of the High Performance Computing and Communications Initiative Toward a National Information Infrastructure, provided in FCCSET Initiatives in the FY 1994 Budget, National Information Infrastructure Act of 1993, and The High Performance Computing Act of 1991 (P.L. 102-194)

## References

Andriole, S. J., (1985), *The Future of Information Processing Technology*, Princeton, NJ: Petrocelli Books.

Dickason, D. (1992), "Geographic Information Systems: Their Implementation Status in the United States," in ed. A. Targowski, *Vision and Directions of Metropolitan--wide and County-wide Geographic Information System*, Kalamazoo, MI: The City Hall of Kalamazoo Publication.

French, R.L., (1990), "Cars that Know Where They're Going," *The FUTURIST*, May-June, pp.29-36.

Garland, S.B., (1993), "Al Gore: What Business Can Teach the Fed," *Business Week*, September, 13, p. 102.

Gore, A., (1993), *Technology for America's Economic Growth, A New Direction to Build Economic Strength.* Washington, DC: Office of the Vice President of the United States.

Mobley, J. (1988), "Politician or professional? The debate over who should run our cities continues." *Governing*, no.2, pp.42-48.

Moore III, R.E., (1990), "Police Pursuits, High-Tech Ways to Reduce Risks," *THE FUTURIST*, July-August, pp.26-27.

Necomb, T., (1991), "Government at Your Fingertips, 24 Hours a Day," *Government Technology*, vol. 4, No. 10., October, p. 2.

Necomb, T., (1991a), "Multimedia: Will Government Communicate Better?," *Government Technology*, vol. 4, No. 10., October, p. 1 and 44.

Osborne, D. and T. Haebler, (1993) *Reinventing Government*, New York: PLUME.

O'Leary, Ch. and Targowski, A., (1993), *Guidelines For MIS to Support the Planning, Evaluation, and Budgeting of Programs Operated under the Unemployment Law in Poland*, Kalamazoo, MI: The W.E. Upjohn Institute for Employment Research, Project sponsored by the World Bank for the Polish Ministry of Labor and Social Policy.

Perry, D.,C, and L. F. Keller. (1991) "Public Administration of the Local Level: Definition, Theory, and Context," in ed. R. D. Bingham, *Managing Local Government*, Newbury Park: SAGE PUBLICATIONS, p.8.

Stewart, Th.A., (1993), "Re engineering: The Hot New Management Tool," *FORTUNE*, August 23, pp. 40-51.

Targowski, A., S. (1990), *The Architecture and Planning of Enterprise-wide Information Management Systems*, Harrisburg, PA.: Idea Group Publishing, p. 151.

Tiez, M., B. (1967), "Toward a Theory of Urban Public Facility Location," Paper of the Regional Science Association no.11, 1967, p. 36.

Waldrop, J. and Thomas, E. (1990), "What the Census Will Show," *American Demographics*, January, pp. 20-30.

Warner, S B. (1987), *The Private City: Philadelphia in Three Periods of Growth*, Philadelphia: University of Pennsylvania Press.

Weinstein, S. The Corporate Ideal: The Politics of Efficiency. Report.

# CHAPTER 7

# Virtual Schools and Universities

## Society-Economy-Education

### Minimizing the Bifurcation

The bifurcation in the 1990s created two paths of the world development. The most progressive path leads toward the Electronic Global Village with a "password" for 4 billion people, the second path guides toward regional conflicts (based on nationalism, ethnic clashes, religious fanaticism, and so forth) with 4 billion people without a "password". This dichotomy of developmental directions creates an opening for fundamental change. By the decades of the 21st century new visions, directions, ideas, and policies have a realistic chance of being translated into social reality. What are the essential elements of the new social aims?

Neither capitalism nor communism can sustain the world and prevent the Third (Population Bomb) and Fourth Big-Bang (Environmental Bomb) of the next century*. The third option is inevitable. Laszlo (1991) offers the third humanistic and evolutionary strategy with two sets of distinct but interrelated objectives:

* The Second Big-Bang took place when humankind was born.

(1) *Defensive set*: to avert the evolution of the structures of society at the expense of the individual,
(2) *Pro-active* set: to build up, and make effective use of the connections that link people all over the world with each other, with their environment, and with the biosphere as a whole.

The former is to safeguard the development of the individual. This requires that we restrain and control the evolution of hierarchically oriented political and economic systems and processes. The latter is to create a global-level holarchy: a network of cooperative relations in fields and areas where worldwide coordination and cooperation are useful, and indeed imperative.

The promotion of the individual's freedom (set 1) within democracy should minimize the regional conflicts. The restraint of the power of nation-states and politicians should make the majority of regional conflicts a relic of the 19th century. The increased social awareness based on paradigm, shift-sensitive knowledge should minimize old-fashioned, nationalistic ignorance.

The promotion of the global holarchy objectives (set 2) such as cooperation in defense, environment, and free-trade through the global links is just the strategy of the Electronic Global Village.

The Electronic Global Village is the best tool to sustain the individual's freedom, since it allows the individual to create a global support group (Global Human Family) as a substitute for the regional tribalism, and some times, social ignorance which often results in violence.

These noble aims provide a vision for a New World Order. It can be achieved through such change agents as:

- Redefined education systems
- New pro-active leadership in business and public administration
- Global information utility
- Cross-culture communications

The interdependence of these aims-factors is illustrated in Figure 7-1. These aims-factors can be programmed in a sequence of Vision-Culture-Change Agent-Results as are summarized in Table 7-1.

## Crisis in Education

In a *BUSINESS WEEK*/Harris Poll, only 39% of the people polled would rate the quality of American public education overall as "excellent" or "pretty good," and only 60% gave high marks to the schools their children or grandchildren attended. The business community is much heavier in its condemnation [1]. In a 1991

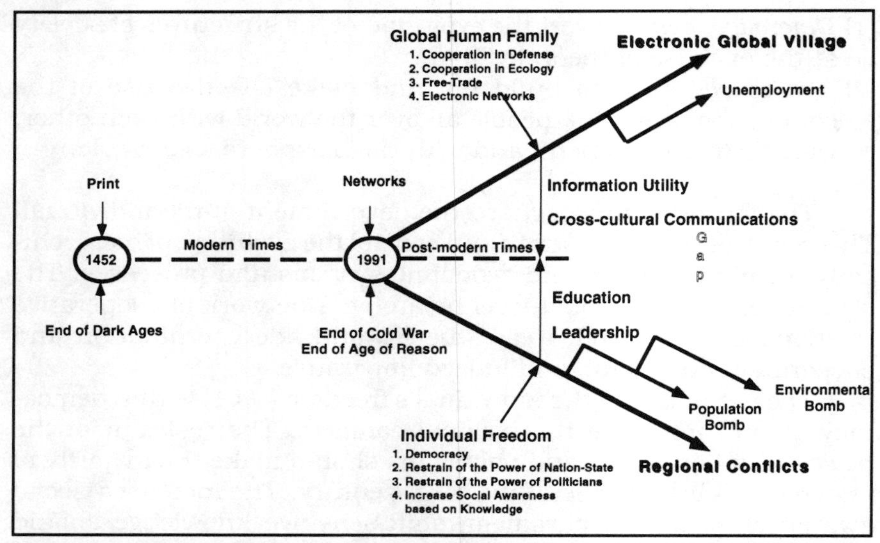


*Figure 7-1: The Architecture of a New World Order and the Minimization Tools of the Bifurcation*

| Vision | Culture | Change Agents | Results |
|---|---|---|---|
| Global Human Family | Cooperation in Defense<br>Cooperation in Ecology<br>Free-Trade<br>Electronic Networks | Information Utility<br>Cross-culture Communications | Electronic Global Village |
| Individual Freedom | Democracy Restrain the power of nation-state<br>Restrain the power of politicians<br>Social awareness based on knowledge | Education<br>Leadership | Fewer regional conflicts |

*Table 7-1 Aims For a New World Order*

survey of employers, educators, and parents sponsored by the Committee for Economic Development, only 12% of employers felt high school graduates write well; only 22% said they had a good mastery of math. Only one high-school junior in five can write a comprehensible note applying for a summer job; among high-school seniors, fewer than one-third know within fifty years when the Civil War took place, one in three do not know that Columbus discovered America before 1750. Among young adults, one government-sponsored study found, well under 40% can understand an average *New*

*York Times* article or figure out their change when paying for lunch, and only 20% can understand a bus schedule.

Nearly 1 million high-school students drop out each year, an average of about 30%, throughout the United States. In some school districts the dropout rate exceeds 50 percent. Perhaps 700,000 more students in each class finish their twelve years hardly able to read their own diplomas. At that rate, by 2000, the literacy rate in America may be as low as 30 percent. In the school year that began in September 1969, American grammar and high schools enrolled 45.6 million students; by 1985, the school population had fallen to 39.5 million, a loss of 7.1 million students. It is still shrinking.

The very assumption of the state's role in providing a standard education for all children—the aim of the "common school" developed 100 years ago—is under siege. When it comes to school reform, there is no organized effort. According to one expert, "We have all these thousand points of light—but no illumination."

Other countries have clear teaching goals backed by tough tests. Some lessons can be learned from abroad:

- Britain:
  National curriculum and testing for England and Wales. Each school can direct its own budget to specific needs of students.

- Germany:
  Corporate dedication to training the next generation of labor through extensive funding of vocational apprenticeships for 70% of all students.

- France:
  Public preschools serve 85% of three-year-olds and 100% of five-years-olds. State funded supplemental classes exist for schools in impoverished districts.

- Japan:
  Well-paid teachers spend 40% of the school day in preparing lessons that encompass a variety of teaching methods to keep student involved.

As in Asia, reverence for teaching lies at the core of the Germanic systems. In Germany, teachers win civil service status guaranteeing job tenure and such benefits as low-interest mortgages. In Switzerland, voters in some counties actually elect instructors to their jobs. This generates strong respect around the teacher as an authority figure.

One clear lesson in studying educational policies around the world is that no single magic key unlocks excellence. The Koreans, who score high on international math and science tests, average among the biggest class sizes in the world (49 in the eighth grade), while the Hungarians, also high scorers, have one of the shortest school years (177 days). The U.S. spends 7.5% of its GNP on education—second highest of the 20 most developed countries—but scored near the bottom in three of four tests.

It is too soon for conclusions, but the formula seems to be obvious. Substantial improvement does not happen until you tackle the big questions: What do you want students to learn, and how can you make sure everyone has an opportunity (Levine et al 1992).

## Economy with Full Unemployment

The main stream of economic activities in the 21st century will be carried out by computers and robots. This will limit the amount of human energy needed for industrial-era jobs. It is inevitable that unemployment will rise in the future because of the impact of computers and robots. Societies will be able to make free time available for the majority of people. Jobs will be the center of life for many people.

The late-twentieth-century confusion was caused, in part, by leisure becoming more important than the jobs that supported it. Few people can or even should continue to work intensively for a lengthy period, otherwise they will lose their outlook on the overall meaning of the life. In the mid-nineteenth century, the average male spent about 40% of his total hours of life on the job, leaving a maximum of 60% for sleep, meals, education, religion, early childhood, and so forth. By the end of the 20th century, the average individual will spend not more than 14% of his life on the job and the percentage is dropping. It means that within the last 100 years, the average man works three times less and this ratio will grow.

Due to the Keynes economic principles, a full employment system, based on balancing ever-increasing consumption desires with ever-rising productive abilities, became the norm after the Second World War. After the Cold War, however, this principle is no longer valid. Full employment, in the contemporary sense, will not be restored, and more and more people are aware of it.

The remedy for this, Theobald (1983) perceives to be in the flexibility of work and learning new patterns. A person will share a job with two other people. More and more people are being given time off to renew themselves and to develop further skills. People will telecommute from homes to workplaces, more people will be involved in longer parenting, and others will do volunteer work for the community. A substitute for jobs as a source of income will be developed either via new entrepreneurship or socio-economic engineering. Perhaps not.

As Alvin Toffler, John Naisbitt, and many others have observed, science and technology are the fastest, most trenchant agents of change that humanity has ever experienced. The rate of knowledge will double every ten years and as a consequence, traditional jobs will shrink. In the 1980s, computers and robotics eliminated half of all manufacturing jobs. In the 1990s another 40 percent will be eliminated. At the same time, the volume of manufactured goods will increase. In the 1980s, three million executives found themselves jobless due to the computerization of middle level management.

According to Cetron and Davies (1989) most of the new jobs that appear, not just in the 1990s, but from now on, will fall into only two categories: the ones you do not want and the ones you cannot get, at least, without extensive preparation.

## Learn or Perish

In classic economics, the sources of wealth are land, labor, and capital. In new economics, knowledge replaces capital as a strategic resource. Knowledge drives competition in business and confrontation in global politics. The most successful war machines and business firms are those which promote the development, improvement, protection, and renewal of products, services, and processes based on knowledge and the learning of new solutions. Learning becomes a new and permanent activity of any worker or manager.

Learning will no longer stop with high school or even college degrees. Specialized knowledge will become obsolete so quickly that adults will be encouraged to take frequent breaks from work, subsidized by their employers, to catch up. "Learning vacations," even for entire families, may become a major part of the travel industry as well as a big moneymaker for colleges whose campuses and faculty would otherwise be idle.

If computers take over so much of the job, what role will be left for the teacher? Rather than just presenting information and issuing instructions, like a coach directing a football team, the teacher will inspire, motivate and serve as referee for the human-to-human discussion that computerized instruction is designed to provoke. A teacher will thus act more like the floor captain of a basketball team, directing the overflow of action, but allowing other team members to take the lead when the situation warrants it.

To be a participant, not a victim of the Electronic Global Village, a student has to learn at these four levels of the following education ("What to learn") (Targowski 1991):

- Level A—Knowledge about humanity and ecology prepares the student to live within accepted values and standards, regulating our relationships among humans and with nature.

- Level B—Knowledge about the past and the future orients the student how to promote cultural tradition and at the same time how to be ready for the challenges of the future. The student learns how to select facts, generate information and evaluate concepts in problem solving and decision-making in harmony with tradition and pressure for change.
- Level C—Knowledge and skills in a given domain prepares the student for a certain profession. This includes: problem-solving, design, assessment, and management of issues, products, services, and processes.
- Level D —Knowledge and skills of cross-cultural communications and telecomputing at the end-user level will make the student a member of the Global Economy and Global Human Family with a "password".

This new scope of education requires new textbooks, labs, and reskilled teachers. The new education should emphasize the comprehensiveness of topics, openness of the world, curiosity of trends and solutions, and motivations for continued education.

The massive growth of technology in the coming decades will render many careers obsolete. Almost everyone will have to face the task of training for an entirely new job at some point in his or her career. Even today, most people change occupations at least three times during their working lives. On the average, the next generation of workers will have to make no fewer than five complete job changes in a lifetime, not counting the multitude of changing tasks associated with each respective job.

In many cases, one job will evolve slowly into the next. One worker might begin his career as a computer programmer at one company and graduate into a systems analyst position for another employer. From there he or she might move on to teach corporate training courses. With that background, he/she might build a part-time career as a technical writer, then strike out as a full-time free-lancer and slowly broaden his/her practice to general-interest journalism (Cetron and Davies 1989).

The goal of lifetime learning is rapidly being institutionalized in the global culture. One force driving this change is that people cannot function effectively with what they learned in school or college a decade ago.

## Careers and Work Trends

### Emerging Skills and Knowledge

A classification of "Major Occupational Groups" in the United States is based on the U.S. Census Bureau approach conceived in 1940 and 1950 (Edwards 1943). All subsequent employment surveys have been based on this same set of categories. However, this classification was appropriate for the Industrial Age of the American Economy, not for the Global Information Economy.

Reich (1992) offers a new classification. According to his proposal, three broad categories of work are emerging. These correspond to the three different competitive positions in which Americans find themselves. The same three categories are taking shape in other nations:

- *Routine production services* - entail the kinds of repetitive tasks performed in old high-volume enterprise by blue-collar workers with low and mid-level managers enforcing standard operating procedures. In this category falls information-processing jobs at data centers, coding sections, building computer circuit boards and so forth. In the 1990s, routine production services accounted for only one quarter of all jobs performed by Americans, and the number is declining.
- *In-person services* - are provided person-to-person and are not sold worldwide. Their immediate object is a customer rather than streams of metals, fabric, or data. Included in this category are retail sales workers, waiters and waitress, hotel workers, janitors, cashiers, hospital attendants and orderlies, nursing-home aids, child-care workers, hairdressers, home health-care aids, taxi drivers, secretaries, flight attendants, physical therapists, and security guards. These services must be punctual and reliable, (as routine production services,) but must also have a pleasant demeanor. They must make others feel happy and at ease. In the 1990s, personal services accounted for about 35 percent of the jobs performed by Americans, and this number is growing. In the United States during the 1980s, over three million new in-person jobs were created. That is more than all the workers who were employed in the automobile, steelmaking, and textile industries [2].
- *Symbolic-analytical services* - include problem-solving, problem-identifying, and strategic-brokering activities. Like routine production services, symbolic-analytic services can be traded worldwide and thus must compete with foreign providers even in the American market. Included in this category are: scientists, engineers, executives, bankers, lawyers, real estate developers, and a few accountants. Furthermore, among them are management, financial, armaments, architectural, and information resource consultants. Organization development specialists, strategic planners, corporate headhunters, systems analysts, art directors, architects, cinematographers, film editors, production designers, publishers, writers and editors, journalists, musicians, television and film producers, and university professors are also included. They all identify, solve and broker problems by manipulating symbols.

They simplify reality into abstract images that can be rearranged, juggled, experimented with, communicated to other specialists, and eventually, transformed back into reality. This is done with such tools as: mathematical models, legal arguments, financial gimmicks, scientific principles, psychological insights, methods of induction and deduction, information systems and services, and so forth. Like routine producers, symbolic analysts rarely come into direct contact with the ultimate beneficiaries of their work. They have partners or associates rather than bosses. Their income depends on the quality, originality and speed of delivered solutions. They work in small teams, which may be connected to larger organizations, including world wide webs. The products are reports, plans, designs, drafts, memoranda, layouts, renderings, scripts or projections and meetings to clarify solutions. Most of the time and cost is spent on conceptualizing the problem, designing a solution, and planning its execution. In the 1980s, about 40 percent of all American jobs were in this category and this number is growing. This category of workers earns 50 percent of all national income. The symbolic analysts' education should include four basic skills: abstraction, system thinking, experimentation, and collaboration for discovery or innovation.

In reality, the goal of transforming a majority of the American labor force into symbolic analysts would be heroic. Other policy is needed. Table 7-2 illustrates some possible directions in modernizing the U.S. labor force. A common denominator for all job level enrichment is access to computerized information. The purpose is to improve productivity, effectiveness, competitiveness and self-management.

In the late 1980s, the U.S. Department of Labor and the

| **Job Category** | **Job Enrichment** | **Result** |
|---|---|---|
| Routine worker | Access to computer-aided routine to gain a new productivity, inovation and empowerment | Routine-symbolic worker |
| In-person worker | Access to computer-aided service to gain a new competitiveness and empowerment | In-person symbolic worker |
| Symbolic analyst | Access to computer-aided problem solving to gain a new knowledge and ocmmunications | Symbolic-knowledge analyst |

*Table 7-2 Directions in Modernization of the U.S. Labor Force*

American Society for Training and Development (ASTD) launched a joint venture research project to identify the "basic" skills employers wanted their workers to possess (Carnevale et al 1988). What employers said they needed most desperately were workers with a solid, basic education plus communications skills and skills in self-management. Reviewing the results, ASTD lumped the most desired skills for workers of the future into seven categories:

- learning to learn
- competence (reading, writing, and computation)
- communications
- personal management
- adaptability
- group effectiveness
- influence

These skills emphasize the power of mind over the brute force of things.

## Demands and Expectations of the New Workplace by Business

In today's Electronic Global Economy, workers must be prepared to change the way they do their jobs in order to capture the benefits from rapidly evolving technology. Education and training go hand-in-hand with productivity, quality, flexibility, and automation in the best performing firms. When measured by international standards, most American workers are not well educated and trained. Larger firms provide more formal training, but most of it is for professionals, technicians, managers, and executives. Our major foreign competitors place much greater emphasis on developing workforce skills at all levels. Experienced production workers at Japanese auto assembly plants, for example, get three times as much training each year as their American counterparts.

American manufacturing and service workers have the skills for yesterday's routine jobs. But, these workers will need new skills to function well in the more demanding work environments that increasingly characterize the competitive industries providing high-wage jobs. Skills and responsibilities are broadening. Competitive manufacturing and service firms will increasingly rely on employees with good higher order skills —reasoning and problem-solving.

Many services now depend on redesigned production systems built around dispersed computing power and on employees with social skills to interact with customers. They need workers who are motivated, managed, and trained in new methods.

Some American companies have emulated Japanese productions systems, which depend heavily on motivated and capable employees to prevent or catch product defects and implement innovations. They emphasize employee involvement and job rotation backed up with substantial, on-going training. American companies that have adapted this model have found that their workers can achieve levels of productivity and quality equal to the best in the world.

Nearly half of all business investments for capital equipment now go to computers and related technology. Personal computers and other inexpensive terminals collect data on the factory floor, track inventories, and help schedule production. Statistical process control reduces variance in production by tracking process parameters over time and examining the trends in those parameters to determine the limits where product quality begins to deteriorate. Computer aided design (CAD) systems automate drafting and specifications. Computer-integrated manufacturing (CAM) slowly introduces flexible enterprising. Service firms rely more and more on decentralized computer systems for data processing, for tracking inventory and sales, and for delivering their products [3].

To be used effectively, these technologies will require workers to learn new, very different skills. Many jobs will remain the same, however, their method of operations will require new skills as shown in Table 7-2. A list of the fastest growing jobs in 1985-95 is shown in Table 7-3.

A list of the 24 largest job categories is provided in Table 7-4. Table 7-5 shows the education requirements for tomorrow's jobs

## Education Technology Trends

In 1990 the Federal administration announced the America 2000 program as a set of six goals to reform the American educational system. America 2000 is an education plan to close the skills and knowledge gap in the United States by the turn of the century. The plan has six goals for the end of the decade:

(1) All children will start school ready to learn
(2) High school graduation rates will increase to at least 90 percent
(3) Students will graduate from elementary school with competence in core subjects
(4) American students will be first in the world in science and mathematics
(5) Every adult will be literate
(6) Every school will be drug-free and violence-free.

Technology, particularly interactive technology, plays a significant role in remodeling the educational system. Interactive technologies are at the heart of an educational method to support students in "learning by doing," rather than "absorbing information" from a teacher. Interactive devices enable children to get responses to their

| Job Title | Percent Growth 1985-95 | Number Needed (in Thousands) | 1995 Median Salary (in Thousands) |
|---|---|---|---|
| ROUTINE WORKERS | | | |
| Computer programmer | 72 | 258 | 40 |
| Computer service technician | 56 | 93 | 32 |
| Actuary | 51 | - | 16 |
| Computer console and equipment operator | 46 | 558 | 21 |
| Medical records technician | 31 | 20 | 26 |
| IN-PERSON WORKERS | | | |
| Travel agent | 44 | 52 | 21 |
| Physical therapist | 42 | 34 | 31 |
| Physician's assistant | 40 | 10 | 25 |
| Podiatrist | 39 | 12 | 57 |
| Correctional institution officer | 35 | 103 | 29 |
| Registered nurse | 33 | 1302 | 23 |
| Occupational therapist | 31 | - | 25 |
| SYMBOLIC ANALYSTS | | | |
| Paralegal assistant | 98 | - | 18 |
| Computer systems analyst | 69 | 260 | 51 |
| Electrical/ electronics engineer | 53 | 367 | 38 |
| Health service administrator | 44 | 220 | 57 |
| Financial services sales | 39 | - | - |
| Engineer | 36 | - | - |
| Attorney | 36 | 487 | 68 |
| Accountant or auditor | 35 | 1047 | 29 |
| Mechanical engineer | 34 | 237 | 32 |
| Computerized tool programmer | 32 | 200 | 23 |

Source: Bureau of Labor Statistics and Forecasting International

*Table 7-3 The Fastest Growing Jobs in 1985-95*

| Job Title | Number Needed 1985-95 | 1995 Median Salary |
|---|---|---|
| ROUTINE WORKERS | | |
| Food service worker | 4436 | 23 |
| Truck driver | 2275 | 31 |
| Housing maintenance | 1750 | 25 |
| Assembler | 1670 | 23 |
| Energy conservation technician | 1500 | 30 |
| Hazardous waste disposal technician | 1500 | 32 |
| Farmer or farm manager | 1485 | - |
| Blue-collar worker supervisor | 1300 | 22 |
| Farm worker or supervisor | 1218 | 25 |
| Auto mechanic | 1197 | 16 |
| Carpenter | 1185 | 32 |
| Typist | 1023 | 14 |
| IN-PERSON SERVICE | | |
| Secretary/stenographer | 3490 | 14 |
| Sales clerk | 2435 | 16 |
| Sales worker, retail | 3300 | 16 |
| Waiter or waitress | 1700 | 14 |
| Teacher, elementary school | 1600 | - |
| Registered nurse | 1302 | 23 |
| Teacher, secondary school | 1243 | - |
| Cashier | 1554 | 19 |
| Sales representative wholesale | 1001 | 38 |
| SYMBOL ANALYSTS | | |
| Bookkeeper | 1904 | 19 |
| Computer software manual writer | 1830 | 34 |

Source: Bureau of Labor Statistics and Forecasting International

*Table 7-4 24 largest Job Categories*

own questions and actions and respond back. At the Carnegie Mellon University (Pittsburgh), an interactive program in physics raised students' grades from a low 30 percent to 80 percent.

In October 1991, NASDC solicited designs for "mold-breaking" schools from educators, businesses, community leaders, and par-

| Schooling Required | Today's Jobs | New Jobs |
|---|---|---|
| 8 years or less | 6% | 4% |
| Some high school | 12 | 10 |
| High school diploma | 40 | 35 |
| Some college | 20 | 22 |
| College or advanced degree | 22 | 30 |

Source: the Hudson Institute

*Table 7-5 Education For Tomorrow's Jobs*

ents. The designers were asked to re-engineer the whole education process and ponder restructuring everything from the way children are taught to the way schools are organized and financed. The result was an avalanche of creative proposals. From 686 designing teams only 11 teams were selected and financed to develop in depth solutions. They will work with dozens of schools in some 20 states, interacting with several thousand K-12 schools (Table 7-6).

In the second phase of the program, 1993-1995, design teams will be awarded two-year contracts to test their proposals. In the final phase, the teams will implement the solutions. Among the team partners are such high-tech companies as IBM, AT&T, and Apple Computer.

The Bensenville, Ill. team design is particularly interesting. A whole town of 17,000 people together planned a school of the future, and they beat all kinds of sophisticated competition. In Bensenville's plan, the teacher's desk is transformed into an electronic teaching center with a videocassette player, laser video-player, and a large screen so the instructor can create interactive instruction. The teacher's computer is connected to each student's computer. Parents communicate with the teacher using modems. Other teams use CD-ROM encyclopedias, electronic bulletin boards, community access TV, and school-specific databases.

Some of these new resources provide access to staggering pools of information. On each side, an interactive video disc may contain 54,000 pictures or 30 minutes of moving pictures, plus audio tracks in different languages. CD-ROMs (for compact disc read-only memory) can store 250,000 pages of text, 7,000 photographic images, 72 minutes of video corresponding in quality to a VHS tape, or 19 hours of speech-quality audio. Access to this information can be far more stimulating than a conventional textbook and lecture method. It is one thing to read about solar eclipses in an encyclopedia; it is far

| Design Team Location | Proposed Sites | |
|---|---|---|
| Los Angeles County, CA | Arizona | Maryland |
| Bensenville, Il | Arkansas | Massachusetts |
| Boston, MA | Colorado | Mississippi |
| Cambridge, MA | Connecticut | New York |
| Lexington Park, MD | Georgia | N. Carolina |
| Minneapolis, MN | Illinois | Pennsylvania |
| Gaston County, NC. | Indiana | Rhode Is. |
| New York, NY | Kentucky | Vermont |
| Rochester, NY | Long Is. | Virginia |
| Providence, RI | Maine | Washington |
| Indianapolis, IN | California | Minnesota |

*Table 7-6: Advanced Schools for High-Tech Education*

more effective, teachers across the country have found, to be able to select that subject on a video-disc or CD-ROM, read it, and then watch the Earth, sun, and moon turn in their courses in an animated, color "movie" of an eclipse (Fisher 1992).

The number of schools with computers expanded from 15,000 to 77,000 between 1981 and 1987, or an average 11 percent a year, with a corresponding increase in the use of laser videodisks [3]. The first attempts to use computers in schools dates back to 1959; early experiments with learning via satellite began in 1973. A dramatic infusion of technology in the American schools began in the 1980s, with the emergence of personal computers. For example, the number of public school computers per 20 K-12 students has risen from less than 0.2 in 1983 to 1.0 in 1991, an increase of more than two million computers in the last decade. In 1988, 429 of the country's school districts had CD-ROMs; by 1991, there were in 1,376 school districts [3].

> Many schools are rethinking their approach to education today. While some are taking only small steps at the moment, Plymouth-Canton Community Schools in Wayne County, Michigan, have made a huge commitment to be at the forefront of technology. That district, serving 15,000 students from kindergarten through high school, passed a $12 million bond issue specifically geared to provide new voice, data, and video technology for its 19 school buildings. The district believed that technology could play a big role in the school's objective to change to outcome-based education, which is based on the belief that all students can learn and succeed, but not necessarily in the same way or at the same time. It also allows the teacher to be the facilitator rather than lecturer and to bring outside resources into the classroom. It was planned that by the start of the 1992-93 school year, the first phase of the three-year project would be complete. Many of the classrooms in the schools have

been wired with fiber-optic cabling, helping the district to utilize information sources that cross community, state, national and even international lines. Teachers are able to present multi-sensory presentations in their classrooms. Each classroom has a control panel on the wall with a host of functions that the teacher can use merely by pushing a button connected to the central media center. With access to such a wide array of information from across the globe, students will increase their ability to solve problems, make decisions, organize, select, edit and write. The expectation is that students will read more, write more and become more active participants in their own learning (Norris 1992).

Today's computer-based technologies go far beyond early word processors. In addition to text, computer-based systems now have access to high-resolution pictures, sound, voice, and full-motion video. These systems can be self-contained in classrooms or can include technology that links classrooms, other schools, other communities, and most important, a universe of information sources, including colleges, universities, and research centers such as national laboratories.

# Distance Learning

### Distance Learning Architecture

Distance Learning can best be described as the provision of education and training opportunities from one site to multiple sites simultaneously —or any educational experience where the learner and the educator are interacting from different locations.

Bi-directional video communications between students and instructors provide a higher quality education than one-way communication. Additionally, it offers numerous benefits, including:

- Improved educational quality
- Specialized instruction
- Greater access to students
- Reduced travel and training costs
- Enhanced curriculum
- Ability to maintain competitive workforce
- Eliminates geographic boundaries
- Sharing resources
- Serves low-incidence classes.

Distance Learning offers greater services with less resources. It is ideal for many applications such as:

- K-12 education
- Adult education
- Police and fire training
- Prisoner arraignment
- Teleconferencing
- Classes in the home
- College credit courses
- Vo-tech education
- Remote medical consultation.

Each interactive video system is custom designed to meet the users specific needs. And there are actually only three different physical elements that make up a system:

- IntraLATA fiber optic transport, which provides exceptional reliability, with virtually unlimited bandwidth and longer travel distances without the use of repeaters.
- Network electronics, which provide quality multi-point capabilities that are fully interactive, with voice, data and video integration.
- Classroom equipment, which is customized to meet specific needs. It includes: monitors, cameras, microphones, video tape recorders, FAX machines, and telephones.

The architecture of Distance Learning is shown on Figure 7-2. Distance Learning has the potential to revolutionize education. Students can study with guest lecturers from the around the world, and they can share their educational experience with classmates in the next town or across the ocean. Distance Learning's greatest impact is in isolated, sparsely populated areas, where students do not always have the opportunity to take the courses they want and need.

> One such place is the Eastern Upper Peninsula Intermediate School District in Michigan, which encompasses 11 local school districts spread over 3,800 square miles. District officials believe Distance Learning will help their schools overcome the educational disadvantages often caused by rural isolation. Seven of those schools have student bodies of fewer than 140 students. There are only 72 students on Mackinac Island and they are totally isolated during the winter except for an icy snowmobile trail across Lake Michigan or a short hop in an airplane. That makes it uneconomical to offer more specialized classes, such as some foreign languages or advanced mathematics, art and business education, which may attract only two or three students at each school. In the past, students could

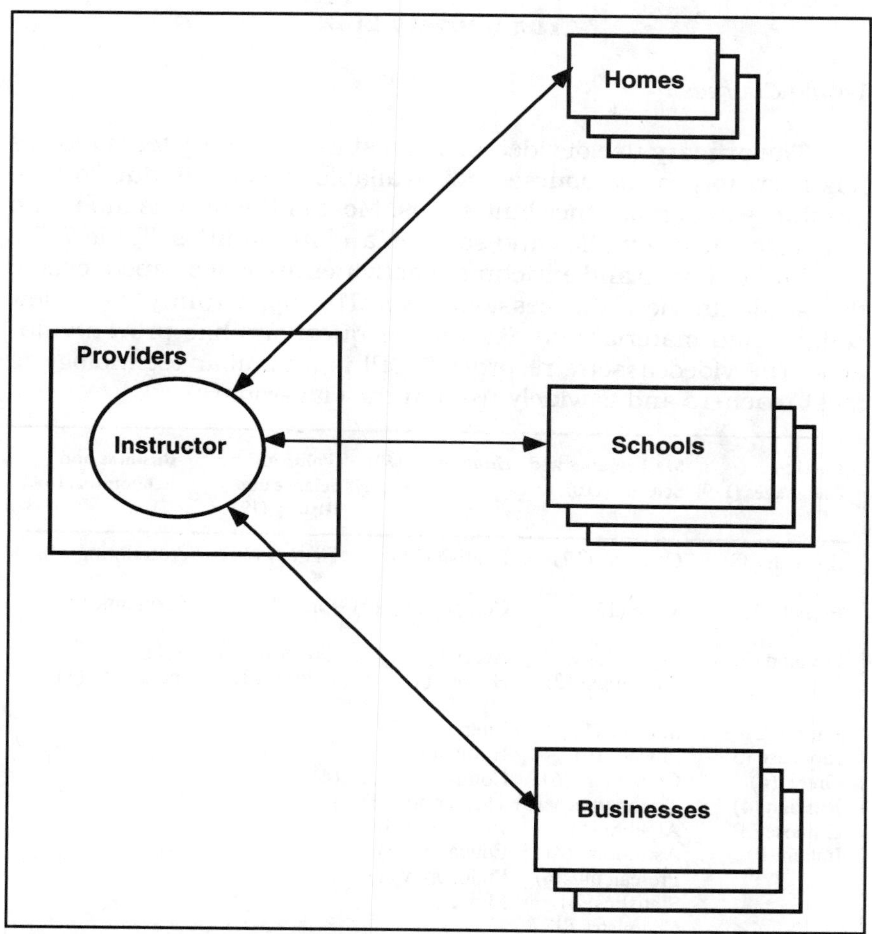


*Figure 7-2: The Architecture of Distance Learning*

be put on a bus to travel 20, 60, even 80 miles to the nearest local district to share teaching resources or teachers could move from school to school. Now, instead of moving kids on buses, the district has signed an agreement with Michigan Bell, GTE, and Chippewa and Hiawatha Telephone to "move minds over miles" with the speed of light. By electronically including students from all those schools in a single class, a class size is such that it justifies an advanced French or Calculus course. It also means that teachers can focus on their specialties, rather than switching from subject to subject to ensure coverage (Butera 1992).

In the face of the many bleak stories about education, Distance Learning is the good story. It is a formidable technology which provides the opportunity for equal levels of instruction.

## The Delivery Link

### *Whole Courses*

The primary use of video-based distance learning technologies has been to provide courses not available to schools due to geographic isolation or other limitations. Most of the need is in foreign languages, mathematics and science, and humanities (Table 7-7).

Many classes and enrichment activities are video-taped, even if the students view the sessions live. The opportunity to review complicated material and ask further questions has proven valuable. The videocassette recorder (VCR) is a familiar technology to most teachers and is widely used in the classroom.

| Foreign Languages (119) | Mathematics and Science (110) | Humanities (69) | Political Science and History (19) | Business and Economics (16) |
|---|---|---|---|---|
| Spanish (38) | Calculus (17) | English (28) | History (11) | Accounting (8) |
| French (28) | Math (12) | Composition (7) | Law (5) | Economics (7) |
| German (26) | Psychology/ Sociology(12) | Art/Art History (7) | Government/ politics (3) | Sales/ marketing (1) |
| Latin (12) | Science (11) | Literature (7) | | |
| Japanese (5) | Physics (9) | Humanities (4) | | |
| Greek (4) | Computers (6) | Communications (4) | | |
| Russian (4) | Trigonometry (6) | Theater arts (3) | | |
| Chinese (3) | Algebra (5) | Journalism(3) | | |
| Italian (1) | Astronomy(4) | Education (3) | | |
| | Pre-calculus(4) | Philosophy (1) | | |
| | Statistics (4) | Music (2) | | |
| | Chemistry (3) | | | |
| | Health (3) | | | |
| | Technology(3) | | | |
| | Geology (2) | | | |
| | Anatomy (1) | | | |
| | Biology (1) | | | |
| | Biomedics (1) | | | |
| | Anthropology(1) | | | |
| | Elementary Analysis (1) | | | |
| | Entomology (1) | | | |
| | Fish and Wildlife (1) | | | |
| | Marine Science (1) | | | |
| | Physical Science (1) | | | |

Source: Office of Technology Assessment, 1989

*Table 7-7: The Most Popular Courses Offered in the United States in 1988-89 Through Distance Learning*

*Partial Course Materials*

Students also receive "modules" or "units" that are integrated into the curriculum. For example, distance learning modules supported under the Technical Education Research Centers (TERC) Star School project encourage students to collect and analyze scientific data and compare it with that gathered by students across the country. Topics include the study of radon, acid rain, and weather. The Jason Project developed a mathematics, science, and social science curriculum for grades 4 to 12 to accompany the live exploration of the Mediterranean sea bottom. Reforms in science and mathematics education partially call for more experiences for students with hands-on activities and cooperative learning; Distance Learning may grow to meet these challenges [3].

*Enrichment Materials*

More and more students receive enrichment activities delivered by distance education technologies. These activities are generally one-time-only presentations designed to inform students (and teachers) on a particular topic. Some are live and interactive, although many schools tape such materials to use at their convenience. In 1989, the Satellite Resources Consortium (SERC) offered six science, technology and society seminars to over 18,000 high school students. Public television stations and independent producers generate a large body of programs that are used as enrichment materials. The Telelearning Project takes its students on "electronic field trips," telephone conference calls to outside authorities, or other classrooms [3].

*Delivering Techniques*

The primary use of the Distance Learning technologies today has been to replicate the experience of face-to-face instruction. The characteristics of traditional instruction retained are instruction in the present (live), and teacher-student and student-teacher exchange (interaction). These qualities distinguish this application of educational technology from previous attempts, particularly educational television, where interactivity was virtually impossible.

Interacting with the teleteacher is the key ingredient in recreating the traditional instruction model. Whether live or delayed, instruction with the instructor is considered by many as a necessary condition for successful Distance Learning. In many of today's systems, interactivity between teacher and students is accomplished via telephone. The video image of the teacher is seen in the classroom, but the teacher cannot see the students in the respective classrooms; this arrangement is known as one-way video and two-way audio. Many of the larger video-based providers are using this model, including Satellite Telecommunications Educational Programming (STEP), Oklahoma State University, Western Michigan University, and SERC.

The two-video, found usually in small, multidistrict systems, is the closest imitation of the traditional classroom that present technology allows. Shar-Ed Video Network, linking four school districts in the Oklahoma Panhandle, is one example. Minnesota has a number that serve four to seven districts each with two-way video and audio. In many of the Minnesota projects, each of the schools offers one class a day to students throughout the system.

Other techniques may include public broadcasting television programs (one-way) and interactivity through computer networks. Computer networks are being used in innovative ways for both student and teacher activities. To encourage a class of writing-resistant seventh graders, a teacher paired her class with a class of fourth graders in another school. The classes exchange letters, and the big brother/sister relationship worked to encourage writing by the seventh graders (Schrum 1988). The Interactive Computer Simulations developed by the University of Michigan School of Education uses telecommunications to study public affairs, such as the Arab-Israeli conflict.

*Providers*

Telecommunications technologies make it possible to aggregate local, state, regional, and even national needs. This aggregation, and the expanding education and technology needs of schools, has brought a widening array of educational services providers to the education market. Among them are: local school districts, regional education service agencies (e.g., The STEP network in Washington State, the Telelearning Project in New York State), states (e.g., North Carolina produces short courses and full courses for their teachers), higher education (e.g., the Rochester Institute of Technology in New York State, Western Montana College,...), public television, other educational institutions (e.g., Sci-STAR satellite program reaches students and teachers in 30 states), commercial providers, public telephone network (Ameritech), hybrid organizations and consortia/joint ventures [3].

Schools pay for these services through different solutions; bonds, legislation appropriation, grants, even tax on rental videos (Missouri).

*Effectiveness*

Much of the research on Distance Learning evaluates only its effectiveness in higher education and business. The effectiveness studies have been quite consistent. When used in business, military training, and adult learning, there is no significant difference in effectiveness between distance learning and traditional instruction methods, and student attitudes are generally positive about the experience. This conclusion, however, does not necessarily extend to the K-12 settings. Little research exists that specifically addresses K-12 distance learning, and what does exist, is limited in scope and often anecdotal [3].

## The Technology Link

Choices for technical solutions in Distance Learning Systems should be defined by the specific demands of each distance learning situation. In the most fundamental sense, what Distance Learning Systems try to do is connect the teacher with the student when physical face-to-face interaction is not possible. Just as highways move vehicles or pipes carry water, telecommunications systems carry instructions, moving information instead of people.

The technology at distant locations, including computers, videocassette recorders (VCRs), fax machines, television monitors, cameras, and even the telephone is critically important. Together, these technologies affect how interaction takes place, what information resources are used, and how effective a distance learning system is likely to be. Some systems allow simultaneous, two-way audio and visual interaction plus an exchange of print materials. Other systems limit interaction to the one-way communication of voice, images, or data. Still others permit only delayed (asynchronous) interaction.

Many technologies are being used to provide education over a distance. Transmission systems include: satellite, fiber optics, Instructional Television Fixed Service (ITFS), microwave, the public telephone system, and coaxial cable. Any of these technologies can be interconnected to form "hybrid" systems. No one technology is best for all situations and applications. Different technologies have varying capabilities and limitations, and effective implementation will depend on matching technological capabilities to educational needs.

The information and communication services of the Future School are shown in Figure 7-3.

The costs of Distance Learning Systems are difficult to analyze because technological options are so varied and changing so rapidly. There are no simple formulas to help estimate the cost of a technological system. The cost of technology is declining steadily. At the same time this technology becomes more powerful than in the past, when it was more expensive. These two trends produce systems that are increasingly less expensive and more capable for teachers and students. Local schools and districts are finding that they can now

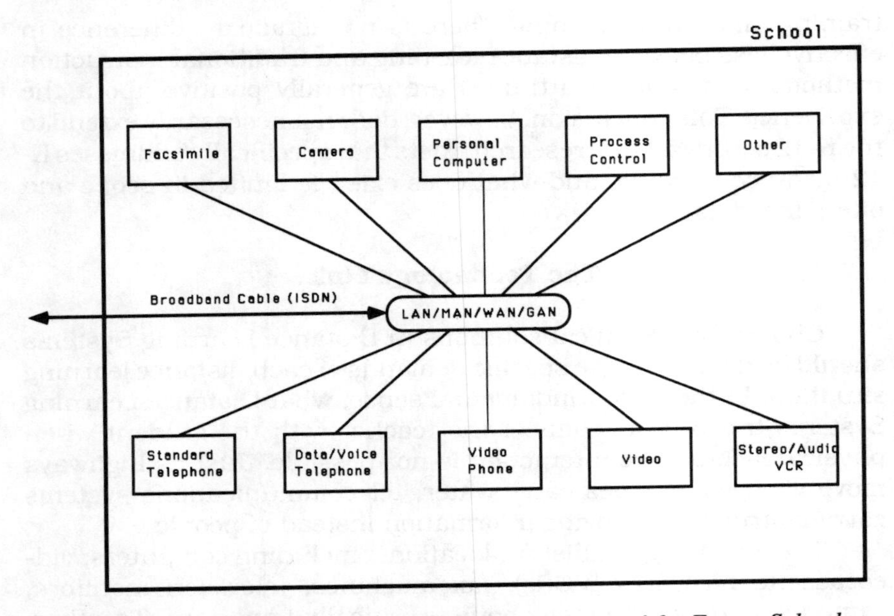


*Figure 7-3: Information and Communications Services of the Future School*

afford (often with some help) the technology tools they need. The following examples are illustrative [3]:

- The Panhandle Shar-Ed Video Network was installed at a total cost of $340,000. This included laying fiber between four schools, outfitting schools with all necessary hardware, maintenance and operation for 5 years. Annualized costs come to $17,000 per site. After 5 years the local telephone cooperative will charge the schools a minimal fee for continued maintenance and use.
- In the Missouri Education Satellite Network, participating schools pay a one-time fee of $8,000 for equipment and an annual programming fee of $1,000. Tuition costs for student courses and staff development are extra. Participating schools that choose to lease hardware and services pay an annual fee of $3,500 plus tuition.
- The Kentucky Educational Television initiative will construct a satellite uplink, install satellite downlinks at each of the State's 1,300 elementary and secondary schools, and build a new Telecommunication Center, at a cost of $11.4 million.
- Houston's InterAct ITFS transmitting equipment cost $330,000. This system reaches participating schools within a 50-mile radius around the city. Participating schools have invested an average of $12,000 for hardware.

Schools implementing Distance Learning Systems have two

types of costs: 1) initial cost that includes equipment and development; and 2) ongoing costs that include programming, transmission, operation and maintenance, and system expansion. The most visible costs are the startup costs, which can be quite high, especially if resources are scarce and the system has to be built "from the ground up". High startup costs can make some technology options too expensive for school districts to afford on their own [3].

## The Teacher Link

While the intellectual and social demands on teachers have escalated at an astonishing rate since this century began, the nature and organization of teachers' work have changed little since the middle of the 19th century. We now live in an age when many elementary school students have their own microcomputers. These situations can put the most amazing achievements of modern science and technology to work in support of their learning. Yet, there are teachers still working in the same job descriptions that teachers had in the mid-1800s when McGuffey's Readers and spelling slates were the leading educational technology [4].

Distance Learning technologies can bring the teaching profession out of the age of McGuffey's Readers and into the 21st century. The very technologies that can bring better resources into the classroom to help students can also improve the quality of the teaching work force. To improve teaching via technology, enthusiasm and excitement for technology must be matched with careful attention to three critical factors:

- Involving teachers' in the planning and implementation process. Teachers' concerns must be factored into any planning for Distance Learning. This may instill initial skepticism and apprehension among parents, students, and teachers. Like the introduction of computers, some of this apprehension subsides once technology becomes "demystified". Teleteachers consistently report that it takes them more time to prepare for distance teaching lessons, and more time to follow up with students (Barker 1989). Part-time or reduced teaching loads should be considered.
- Accounting for the change in the nature of teaching and the teacher's role. Teaching in a Distance Learning setting challenges teachers to rethink their interaction with students and redesign course structure.
- Educating teachers to take advantage of these tools. The key to any Distance Learning is the teacher. Not every one makes an effective teleteacher. It is imperative that teachers get adequate training not only in the technical aspects of the system, but also in the educational applications of the technology. Some universities, such as Iowa State University and Harvard University have imple-

mented computer networks for their own graduates, who majored in education, to support information sharing, give professional advice, and provide help for new teachers in the difficult first year of teaching.

## Distance Learning Activities

The number of Distance Learning projects operating or being planned in the United States has grown from only a handful to many.

### *Federal Activities*

Federal Government funds have accelerated the growth of distance education in the United States, through direct purchasing power as well as the leveraging power of the federal dollar. The Star School Program (Department of Education) and the Public Telecommunications Facility Program (Department of Commerce) are the primary federal programs directly affecting distance education in elementary and secondary schools.

Other federal agencies have interests in distance learning through their responsibilities for technology development, training and education. Yet, no agency-wide strategy or interagency coordination is now in place [3].

Experiments in interactive Distance Learning in American public schools dates back to 1971, when the National Aeronautic and Space Agency (NASA) offered the Office of Education free time on its satellites. Three demonstration projects were founded, one in Appalachia, a second in the Rocky Mountain region, and another in Alaska. Transmission began in 1974. The Appalachian Regional Commission (ARC) evolved into The Learning Channel, a cable television educational provider. The Rocky Mountain project formed the basis of the Public Service Satellite Consortium. The format for class instruction used by many of today's providers was developed during the ARC project. (Arundel 1989) (Grayson 1974).

The Federal Government is not the basic provider of K-12 education in the United States; this role has been traditionally exercised by states and localities. The federal role in education has been to address particular issues, most prominently equity, access, and national priorities, through targeted funding and research. In 1988, Congress created the Star Schools Program (The Omnibus Trade Bill and Competitiveness Act passed by the 100th Congress), a comprehensive federal effort to develop multistate, multi-institutional K-12 distance education. It is not surprising, then, that the Federal Government has not articulated any kind of comprehensive Distance Learning policy. Many Department of Education programs, however, allow use of these funds to support Distance Learning.; because the resistance to, or ignorance of this technology on the part of some state and local officials, very little of this money has been used.[3]

Today, virtually every state is actively planning for distance education, already administering a statewide plan or system, or has a local distance learning projects in place. States are also beginning to look beyond their borders to share resources and respond to national programs.

# Electronic School

### Distance Learning School

As far as Distance Learning is concerned, the question is no longer *whether* to pursue this market, but *how* to pursue it. A Distance Learning School is the provision of education and training opportunities from one site to multiple sites simultaneously. Any educational experience where the learner and the educator are interacting across space.

Technology links can be chosen among the following solutions:

- fiber optics
- satellite
- IFTS (Instructional Television Fixed Service)
- microwave
- coaxial cable.

Some hybrid solutions can be recommended, since no one technology is best for all situations and applications. A mature solution will contain the following features:

- interactive 2-way video and audio
- full motion - near broadcast quality
- remote customer control of scheduling
- easy to use
- ability of every location to broadcast

System components are:

- Class room equipment
  - screen monitors

- cameras
- microphones
- microcomputers (optional)
• Transport electronics
• Local telephone loop (for wire-based solutions)
• Telephone Central Office (ISDN)
• Interoffice fiber cable
• Other ends (schools) of the linkage system.

The classroom equipment can be a modern teleconferencing facility with other ends at locations of :

• cooperating schools
• specialized centers (Math & Science)
• cooperating businesses
• international schools abroad
• others

Ameritech, the Midwest corporation of the Ma Bell Companies is building intraLATA Distance Learning Networks throughout its five-state Great Lakes region, using communication links to transmit data, voice and video over the same network. The basic technological application is not new. Many corporations use similar video-conferencing networks to control travel-related costs, improve managerial productivity and be more competitive by enabling people to work together in spite of geographic distance. In the 1990s, Ameritech is bringing this advanced technology to schools. The effort is focused on two-way interactive video-conferencing because it comes closest to reproducing the traditional classroom experience. Ameritech offers a range of solutions, from a compressed-video alternative to near broadcast-quality transmission.

> One of the most ambitious networks was developed by Ameritech Information Systems and Illinois Community Unit School District 300. The network was designed to help broaden educational opportunities in a district that serves more than 11,000 students in 13 urban, rural and suburban communities comprising a wide range of socioeconomic, ethnic and racial backgrounds. With distance learning, 200 students at Hampshire High will have the broad range of courses and the cultural diversity commonly associated with large high schools, without losing the personalized attention found in smaller schools. Students make friends and learn about people through electronic communication just as they do through face-to-face contact. After a while, students in different locations forget about the equipment and truly interact as if they are in the same

room. A by product of this system are professional development programs for teachers. Teacher are very busy people and they are not eager to spend time traveling to and from staff meetings. Now teachers use the distance learning network to hold meetings. All of the district math teachers, for example, can get together for a discussion after school.

In May 1992, almost 70 guidance counselors located at five Indianapolis public schools participated in a seminar on drug dependency with experts from Milwaukee and from Indiana University in Bloomington. The seminar was a two-way test of a new distance learning network that Indiana Bell is building for the Indianapolis Public Schools. With 90 locations and 600 miles of fiber optics, this is one of the world's largest fiber-optic interactive distance learning network.

In the face of the many bleak stories about education, Distance Learning is good news. Distance Learning proves especially useful for addressing problems of scale (not enough learners in one location) and convenience (elimination of travel and scheduling constraints), allowing groups to communicate without having to "be there together." Today's "distance learning" applications are likely to evolve over the next two decades into diverse, "virtual school" communities with large national and global student bodies of all ages and with supporting industries. Short courses and institutes conferences, whole courses and full degree programs are already available to learners participating from home and work. Programs of this kind are likely to grow rapidly as information technology advances improve the effectiveness of learning from home (Olson *et al* 1992).

## Telematic Super School

Increasing awareness of how communications technology can benefit education is the primary reason why Ameritech funded and spearheaded the effort to create SuperSchool. When educators think of technology, they tend to think about computers instead of telecommunications. As it was demonstrated in SuperSchool (on June 15-18 1992 during SUPERCOMM/ICC '92 in McCormick—Chicago), Ameritech can provide schools with total telecommunication solutions. The applications showcased have resulted from the company's effort to integrate products from more than 70 vendors with its communication technologies.

Visitors observed an array of applications in five SuperSchool rooms. Those rooms include:

- Administrative Office

- Distance Learning Classroom
- Home Learning Center
- Library/Learning Resource Center
- Advanced Studies Center.

Virtually everything demonstrated in SuperSchool is already being used by schools throughout Ameritech's five-state region. None of these services are out in the future. What visitors have seen can be implemented right now.

The pipelines of communications technology are expanding the walls of the traditional classroom, taking children across the country and around the world on an unparalleled journey of discovery. In a Telematic Super School, students use this pipeline to:

- Take a computer voyage through the chambers of a dog's heart,
- Create a thunderstorm in a computer weather lab,
- Share a face-to-face language lesson with students in Montreal, and much, much more.

The architecture of the Telematic Super School is shown in Figure 7-4.
A list of applications of the Telematic Super School is as follow:

(1) Administrative System
Goal: To increase administrative effectiveness
- Homework Hotline
- Absence Verification/Notification
- Emergency Announcements
- Teacher Substitution Voice Mailbox
- Information Listing on Voice Bulletin Boards
- Absence Reporting System
- File Transfer of Student Information (e.g., grades) and Electronic Forms
- Electronic Bulletin Board Access
- Cellular Dispatch (e.g., school bus routes)

(2) Office Automation System
Goal: To increase productivity and communications
- Word Processing
- Desktop Publishing
- Presentations Design
- Electronic Mail
- Bulletin Boards
- Computer Conferencing

(3) Distance Learning Classroom

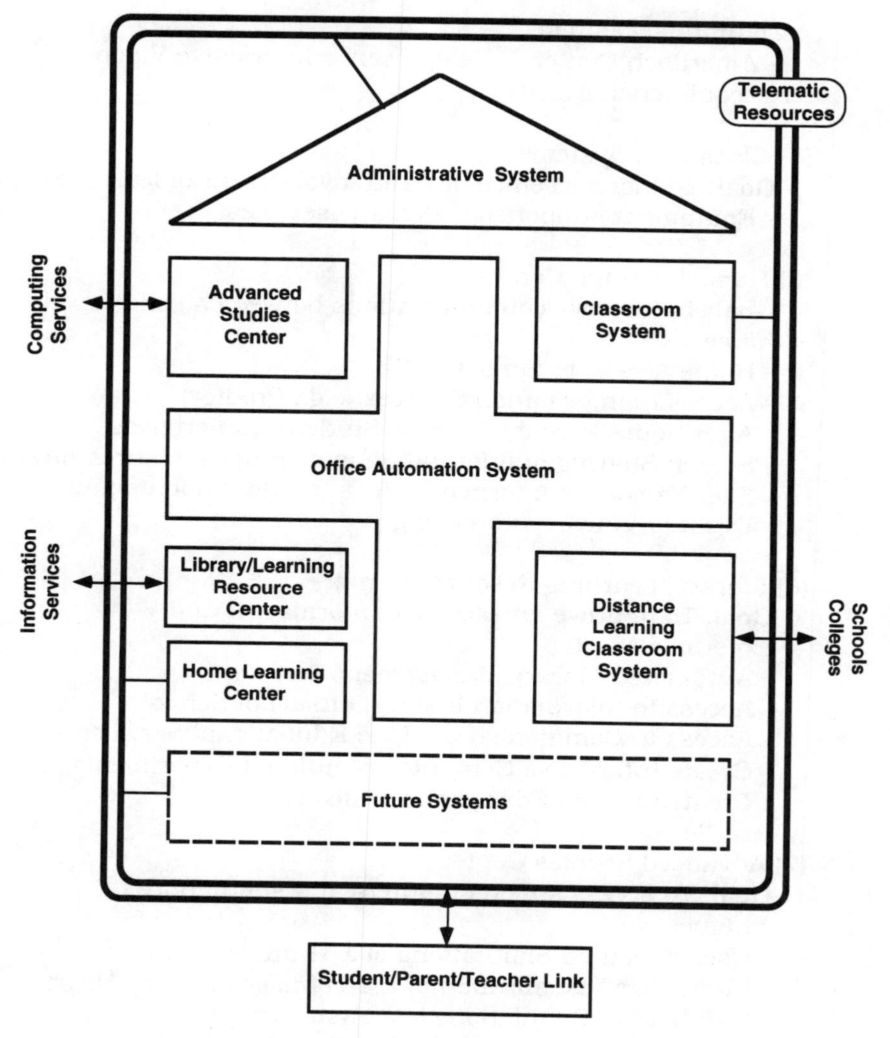


*Figure 7-4: The Architecture of Telematic Super School (The Targowski Model)*

Goal: To increase the opportunity of learning new things
- Ohio University/Appalachian Schools Distance Learning Project
- Teacher Guided Lecture
- Cultural Exchange Seminar from World Trade in Montreal
- International Relations Class from Jacobs High School — Carpenterville, Illinois
- French 3 on the Georgia Distance Learning Network —

Columbus, Georgia
- Ameritech Center — Telepresence Interactive Video Teleconferencing

(4) Classroom System
Goal: To increase efficiency and involvement of learning
- Equipment supporting telematic services

(5) Home Learning Center
Goal: To improve communications between a school and home
- Home Access to School LANs
- Access to Information Sources (e.g., Prodigy)
- After hours Parent/Teacher/Student Conferences
- Screen Sharing of Information (e.g., grades, test results)
- File Transfer of Information (e.g., student documents)
- Desktop Video-conferencing

(6) Library/Learning Resource Center
Goal: To improve an access to information and experiments
- Automated Library Management Services
- Access to Information Inside/Outside of School
- Access to Commercial On-Line Educational Services
- Create Interactive Computer-Simulated Experiments
- Create Custom Educational Video tapes

(7) Advanced Studies Center
Goal: To access supercomputers and simulation techniques
- User-Executed Simulations of a Thunderstorm
- Interactive Visualization of a 4-D Image of a Dog Heart
- Database of Simulations (e.g., astronomy)
- Collaborating Learning in Mathematics, Physics, ...

(8) Telematic Resources Center
Goal: To increase information, communication and problem solving capabilities
- Computers
- Telecommunication Equipment
- Classroom Equipment.

Traditional teaching emphasizes the transmission of information through lectures. At the Telematic Super School, with new technology as an ally, teachers can use their skills fully to help students analyze and process information. Information and communication technologies (telematics) entice students into exploring databases, videotapes and CD-ROMs. The teacher becomes a coach, moving from student to student, making suggestions and piloting students' journeys through an abundance of information.

A goal of the Telematic Super School is to use technology to make education more student-directed. It is not a goal to replace teachers, but to free them to do what they do best —teach.

> At Jefferson Elementary School in Sterling Heights, Michigan, a computer called "OZ" drives a $13 million multimedia teaching tool installed by Warren Consolidated Schools in 1992. The goal is to hype the learning experience for Nintendo-generation kids raised on playing video games, while providing instant access to an amazing amount of information. There are 16 elementary schools in the Warren Consolidated district, which covers parts of Warren, Troy, and Sterling Heights, and each has an OZ of its own. A product of San Diego-based Jostens Learning Corp., OZ is an integrated-learning system designed to snag kids' interest and keep them from yawning in the classroom. It is packed with real-life and animated programming, full- motion video and professional-quality sound and can route all of those between classrooms. OZ's nerve center is about the size of three four-drawer file cabinets. If the wiring installed in Jefferson Elementary to make the system work were placed end to end, the resulting wire would stretch about 100 miles.

> The Jostens system has turned Jefferson into the school of 21st century. It allows a student to quickly become acclimated to high-tech learning technology, while providing teachers a standardized way of tracking students' progress. The Jostens network currently links Jefferson Elementary with 15 other elementary schools plus the Macomb Math/Science Technology Center at Butcher Community Education Center. Warren Consolidated officials boast that their's is the first school district in the U.S. to link so many buildings in such a sophisticated network. Michigan Bell in 1992, put the homes of 115 specially selected fourth-graders at Jefferson and Thrope Elementary Schools into the exclusive learning loop. In the 1990s, it was the only home-learning network of this kind in the U.S. These youngsters have on-line access to some of the best educational video programming in the world without even leaving their homes. By the 1996, thanks to a $44 million bond issue, systems like OZ will juggle learning programs in a district-wide network, linking all elementary, high schools and six middle-level schools.

The dominant feature in a typical classroom is a 27-inch television-type screen. This is a window to the world. The classroom teacher is in command. By tapping a touch-tone dial at a control center by the teacher's desk, he/she selects programs that fit the class work. One main computer houses all programs. It is at their fingertips. The teacher can call up a wide range of videotape or laser disk programs, live television, in-house camcorder productions or closed-circuit TV. Name it and OZ has it. For example, a teacher in social studies might dial up a National Geographic videocassette or a map or article from an encyclopedia. In history, it could be an audio tape of a speech of President Kennedy. The computers tutor the students on a one-to-one basis in math, language, arts and so forth. For children born in the Television Age, the ability of the TV monitor to bring the world into the classroom seems no big deal, but going hands-on with computers and telecommunications is a significant achievement for the majority of students. "I like computers because of the mouse and the headphones," a fourth-grader said. "I love the printer because you can print and not write." "Also, you can talk to the computer and it talks back."

### Teacher-Parent Link

Repairing the severed connections between parents and teachers is a high educational priority, and a difficult challenge given the time pressures confronting families, especially single parent and dual income families. Parents are not a replacement for skilled teachers. Their most effective role in their child's learning is often that of a role model, motivator, occasional tutor and disciplinarian —a complementary role of the K-12 teacher.

To help parents perform this kind of role more effectively, a number of schools around the country have recently set up homework hotlines that allow parents to get clear instructions from teachers about class homework assignments. Expanding this service requires wider school use of the voice messaging services already available. Personalized parental alerts such as notification of attendance and tardiness problems, low grades and missing assignments are soon likely to become a part of these hotline services.

By 1995 or shortly thereafter, with the spread of "smart phones," which are already becoming available in the corporate environment, parents will be able to dial up assignments, class schedules, school menus, notices of meetings and other information displayed on-screen. More widespread use of voice messaging services will also support voice mail consultations, which can allow parents and teachers to communicate without the usual difficulties of arranging "live" conversation during working hours.

Expert systems that respond to some of the more routine questions parents ask about their children's progress and development will also be available by the late 1990s. These systems not only can save parents and school officials and teachers the time and burden of trying to meet face to face, but they can also identify parents whose questions require more than routine answers, and flag a school official to send information to that parent, telephone them or arrange an appointment.

Problems of arranging meeting dates, places barriers of social or class differences and other obstacles to face-to-face parent contacts. These problems are substantially minimized in the electronic communications world although basic written communications skills are needed to use these communications systems. By 2010, voice to text and voice recognition systems will eliminate even this barrier.

Schools will also be able to make available to parents, on-demand, detailed students' performance assessments, identifying specific skills that need to be improved. Teachers would regularly add information to detailed running assessments of all student's performance, much of this generated automatically by intelligent tutors and class information systems.

Video parent tutorials with suggestions on how to work with their children at different grade levels in various subjects can be available on-demand. Parent-teacher video conversations, including video messaging, could create a more vivid sense of social presence than voice mail and forge a stronger bond between parents and teachers.

Parents will be able to participate electronically in interactive electronic meetings among parents or between parents and teachers, principals, school board members or even legislators. Meetings with school boards or legislators might involve concerns about school restructuring or about specific policies or practices, curriculum or the like. Meetings between parents and teachers could deal with personal concerns, such as how a child is behaving, or what learning materials in the home could help a child catch up in areas where he/she has fallen behind.

Some parents may also desire to contact other parents in more spontaneous and informal settings than PTA meetings about their children or about specific school activities or general school policies. Such parent-to-parent dialogues are an important reinforcement tool for those concerned with their children's education but fearful of their own ability to raise these issues with school authorities.

Today, cable and low cost camcorder technology can enable parents to watch and listen to school board meetings. Asynchronous electronic contacts without video can also take place given access to a telephone line and a terminal. As optical fiber reaches into homes, it will be possible to provide interactive participation by parents from their homes with full-motion video (Olson *et al* 1992).

> In New York City, when parents threaten to "pick up that phone and call your teacher" there is no idle hope that maybe the teacher will not be around. Voice messaging makes sure that teachers, parents and students communicate better. Parents call P.S. (Public School) 75, located at 96th Street and West End Avenue in Manhattan, and listen to prerecorded messages from teachers discussing daily activities, homework assignments and upcoming school events. The voice messaging systems is a part of Project TELL (Telecommunications for Learning), a $3.5 million education program conceived and funded by New York Telephone— and designed and implemented by the City University of New York (CUNY) (the Graduate School's Stanton/ Heiskell Center for Public Policy in Telcomm and Information Systems). The three-year program also includes initiatives for personal computers in the home and classroom. P.S. 75's Voice Messaging service is accessed by dialing (212) 866-7046. A caller is then instructed to press a certain digit on the touch-tone pad to hear school announcements, a personal message from a teacher, or other types of information such as the weekly cafeteria menu. Callers can leave messages for teachers and other school personnel as well.

## Electronic Global University

### Education for All

According to Rossman (1992) we see signs (practical simplementations) of an electronic global university even before all of its institutional forms exist:

1. Students in one country are taking courses in another via computer conference and/or television or using combinations of other technologies.
2. Catalogs of courses from many universities and countries are available electronically to prospective students around the world.
3. An international faculty at first perhaps simply consists of all teachers who offer courses electronically with colleagues in other countries.
4. University administration or governance involves networks of those in colleges, universities, government agencies, professional associations, and business corporations who are assigned to plan and administer such electronic education programs.

5. Electronic classrooms and other facilities exist where students and faculty meet in "cyberspace."
6. Student activities, coffee houses, clubs, and action projects begin to involve students electronically from and to more than one country.
7. Provision for guidance and counseling. (A student in Singapore meets with his advisor in British Columbia on a computer network).
8. An emerging electronic global university library.
9. "Co-laboratory" facilities, through which students undertake lab work together, and connect electronically across national boundaries.
10. Special event lectures and student action/convocations (such as global "live aid" concerts) are shared from country to country.
11. On-line electronic bookstores are available for students in another country to use in ordering print books or downloading electronic ones.
12. Faculty meetings and faculty training can be shared from country to country electronically.
13. Many scholars, without leaving home, participate electronically in international conferences of associations of scholars who meet by discipline and profession. A few international scholarly and scientific journals begin to appear on-line, computer cross-indexed for instant search and retrieval.
14. Continuing education conferences and workshops exist (as in the National University Teleconferencing Network) in which participants from more than one country participate electronically.
15. Electronic bulletin boards exist on the cyberspace campus.
16. Electronic global university press.

The above mentioned initiatives are not yet well seen by the academia and students, since they are taking place on "invisible wire." Sooner or later these activities will be more popular as higher education institutions will perceive the electronic global university as an answer for their lack of resources and diminishing quality of education.

The birth of the electronic global university has roots in the following historic events that guided the development of the higher education institutions (Rossman 1992):

- Correspondence courses provided from one country to another.
- "Open universities" provide courses to a million students via radio and television. The first one began in Iowa in 1933.
- Advanced Instructional TV (AITV) was initiated by Stanford University in 1969 and California State University in Chico developed ITV for the northern part of the state by the end of 1975.

- Hewlett-Packard Corporation launched satellite instruction for 200 sites in North America in 1981.
- In 1982, the National University Teleconferencing Network (NUTN) was established at Oklahoma State University to offer "a wide variety of programs by satellite."
- In 1984, National Technological University (NTU), a consortium of university departments of engineering, began provide graduate courses. In the 1990s, it offers courses for East Asia students.
- By 1990, fifty-four electronic networks were involved in educational programming via satellite throughout North America. Instruction was offered in medicine, law, banking, insurance, law enforcement, auto repairs, and much more.
- Students in Scandinavia, for example, when taking courses in Ph.D programs, do not have to visit campuses in other countries.

The list of historic events indicates that the education system is spreading out, reaching far places of the globe and that the technology is the guiding force of this process.

The purpose of the electronic global university is to:

(1) Provide affordable education in those countries and national locations that are far away from the academic centers (Becker 1989).
(2) Provide telematic access to advanced, competitive education in those countries and national locations that are looking for such solutions.
(3) Offer to many people the chance of updating their education with refresher courses that could be taken without having to drop out of the work force.

The electronic global university becomes a cyberspace of knowledge centers and workers who will gain and exchange knowledge to pursue better awareness of social and environmental issues determining the survival of the humankind on the Earth.

## Global Lecture Hall

In 1972, Takeshi Utsumi, Chairman of the Global Systems Analysis and Simulation Association in the USA (GLOSAS/USA) initiated the GLOSAS's project on energy, resources and environment for global peace gaming.

In effect, GLOSAS developed a concept of the Global Lecture Hall and implemented its several global teleconferencing sessions. The demonstrations encompassed more than two dozen universities linked together, from the East Coast of North America to Japan, the Republic of Korea, Saipan and Guam, from Fairbanks, Alaska to

Caracas, Venezuela, to Brisbane, Australia, to Western and Eastern Europe, and the Mediterranean countries.

These demonstrations have helped GLOSAS discover and compensate for the technical, regulatory, economic and marketing impediments to the creation of a Global University. Considerable interest in these Global Lecture Halls has been expressed from various organizations around the Pacific Rim, Latin America, and Europe. The associates of GLOSAS are working on the establishment of Global Pacific University (GPU), Global Latin American University (GLAU) and Global European University (GEU).

The GU is conceived as a worldwide educational network and a permanent organization of international education exchange via various telecommunications media. The GU is to be a broad collaborative partnership of universities and businesses; of governmental and non-governmental, and community organizations; of students, workers, and individual citizens; working towards an educational and non-profit telecommunications network which will span the globe. It seeks to provide on a global scale all kinds of educational, cultural, information, knowledge, vocational and community activities, rather than being confined only to traditional educational offerings.

The GU is a network that will eventually provide a technical and administrative infrastructure for exchange of education through telecommunications across national boundaries as easy as it now is within many developed countries.

Recognizing that mankind is faced with a wide range of critical problems, the Global University is directing itself to four essential goals (Utsumi and Magalhaes 1992):

(1) The globalization of educational opportunities to make possible the highest quality of education for all the world's learners.
(2) Support of research and development, including such projects as:
   a. globally networked "think tanks" for creating philosophical assumptions, creating new models of educational exchange, and collaborating on problems of global concern;
   b. research on new technologies that will improve the quality of educational endeavors;
   c. global coordination of research results and the accomplishments of educators around the world.
(3) Use of global scale tools such as peace gaming and global village meetings so as to explore new alternatives for world-order capable of addressing the problems and opportunities of an interdependent globe.
(4) Globalization of employment opportunities to enhance the job flexibility of the world's workers.

(5) Empower under-served people of the Third World countries by giving them access to the educational excellence of many countries via various telecommunications media.

The Global University is being developing through series of teleconferencing of the Global Lecture Hall. It is a very ambitious idea and project, which has great chances of implementation. Particularly, the Third World, Eastern and Central European countries should be the most interested in GU, since they seek solutions from the most developed countries.

# Electronic University

### Campus of the Future

Higher education in the U.S. is big business—a $100 billion business, representing 2.7% of gross national product. No other nation can boast of so many and such different institutions: 156 universities, 1,953 four-year colleges, 1,378 two-year colleges and technical schools. Collectively, they employ 793,000 faculty members —not to mention an army of deans and other administrative personnel — and accommodate 14 million students. One sign of the astonishing increase is in part-time students; only about 20% of these part time students annually receive one or more certificates of graduation, from A.A (Associate of Arts) to Ph.D.

Anticipating a surge in "distance learning" cable entrepreneur Glenn Jones in 1987 founded the Mind Extension University. Based in Englewood, Colo., it beams college-credit courses to 36,000 students across the country, under the aegis of such established institutions as the University of Minnesota and Penn State. In 1992, a branch of the University of Maryland began offering the nation's first four-year Bachelor of Arts program via Mind Extension; 60 students are enrolled. Today students are often working. They need to be able to compete, and they want a flexible learning format. Because of time constraints—children, jobs, commutes— they cannot go to the typical campus.

It is not only the students who have changing needs; so do the various communities that colleges and universities are trying to serve. Inside what was once the ivory tower, there is a growing interest in new kinds of alliances with business. For example, IBM and Western Michigan University created a CIM lab not only to teach Western Students in Computer Integrated Manufacturing but also to provide teaching and demonstration for IBM customers. On campus at the University of California, Irvine, Hitachi has built a high-tech research lab, which it shares with U.C.'s top-flight biochemistry department.

During the great expansion that took place after World War II, American colleges and universities sought to be all things to all people. In the new age of austerity, schools are being forced to rethink their missions, decide what they can do best and—in a form of academic triage —abandon certain fields of learning to others. At some schools, there are too small classes and too many courses. For example, at Boston University there are 150 courses that study the human mind. However, what we know about the human mind could be taught in 30 classes. It is fairly common for neighboring colleges to share talents and facilities, particularly in arcane specialities. But institutions need to collaborate with international schools, for example from Europe or Japan.

At the campus of the future there will be TV consoles that could beam-up taped lectures by any professor on campus or even let students monitor courses from other schools. Built-in computer terminals will tap into the card catalogs of half the college libraries in the country, call up encyclopedia articles or scan the daily papers.

There is in fact, no need for a crystal ball to envision the university of the 21st century. Dozens of U.S. campuses from Kansas' Sterling College to Ohio' Youngstown State, from the huge State University of New York system (total enrollment: more than 369,000 on 23 campuses) to tiny Alaska Pacific University in Anchorage (639 students), officials are deciding not only how to do the same with less money but also how to do less with less, very often with help from technology.

Too many higher education institutions have been run like the government, and that means that they have been run badly. The result is bureaucratic bloat with a self-perpetuating *nomenclatura* of assistant deans, development officers and other office-bound personnel. Some innovative schools—such as Rice—have chosen to dismantle their bureaucracies to devote more resources to labs, libraries and classrooms (Blackman et al 1992). If universities and colleges perpetuate only their own past glory, they will be judged irrelevant.

## Telematic University (NTU case)

An example of the University of the Future is the National Technological University (NTU) with the headquarters in Fort Collins, Colorado. It is a private, non-profit institution founded in 1984 to serve the advanced educational needs of today's busy, highly mobile engineers, scientists and technical managers. On a nationwide basis, NTU offers a wide range of instructional television courses taught by the top faculty of 43 of the nation's leading engineering universities and other organizations and institutions selected because of their special expertise.

NTU's functions are to:

- award master's degrees in selected disciplines;
- provide research seminars in each discipline;
- operate an instructional television network (ITV) via satellite for convenient, flexible, on-site service nation-wide;
- offer Advanced Technology & Management Programs in the form of non-credit short courses and workshops to introduce new advanced technology concepts to a broad range of technical professionals; and
- established a sophisticated satellite network infrastructure between industry and the university community.

NTU began regular satellite delivery of advanced technical education in August, 1985. During the 1991-92 year of satellite networking, NTU offered 24,765 hours of academic credit instruction and 2,880 hours of state-of-the-art Advanced Technology & Management Programs.

The network operates on G-STAR 1 with one modern Ku-band transponder to provide up to 12 compressed digital video channels throughout the day and evening. The signal is received by subscribers through the day and evening. The signal is received by subscribers through medium-size (generally 3.6 meters or larger) downlinks located near the viewer.

Telephone lines from the receiving sites to the campus classroom provide for faculty-student interaction. This interaction is supplemented by electronic mail, computer teleconferencing and telephone office hours. Many courses are broadcast in real time, so that NTU students can interact with the instructor by telephone or fax instantaneously.

Most students, however, view the programs asynchronously, using videotapes of broadcasts because this mode provides the needed flexibility for adult learners. For these students, interaction with the instructor is by telephone during office hours, electronic mail via the NTU computer and INTERNET, and the regular flow of hard-copy assignments either by mail or increasingly, by fax.

Enrollees report that the courses are challenging and applicable to their work environment and that the instruction is provided by the above average teachers. The students are overwhelmingly interested in taking additional NTU courses.

A coordinated, national delivery system for advanced education of engineers and scientists is clearly in the nation's best economic and defense interests. Top faculty are in very short supply. Modern telecommunications provides a delivery system to launch the cooperative effort by NTU universities. Each participating university that delivers courses has or will have an earth station or uplink.

NTU's academic programs (Master of Science) are:

| | |
|---|---|
| COMPUTER ENGINEERING | MANAGEMENT OF TECHNOLOGY |
| COMPUTER SCIENCE | MANUFACTURING SYSTEMS ENGINEERING |
| ELECTRICAL ENGINEERING | MATERIALS SCIENCE & ENGINEERING |
| ENGINEERING MANAGEMENT | SOFTWARE ENGINEERING |
| HAZARDOUS WASTE MGMT | SPECIAL MAJORS PROGRAM |
| HEALTH PHYSICS | |

The 1992-93 NTU Bulletin lists 892 courses from the participating universities in the above curriculums. Undergraduate bridging courses for non-majors wishing to enter the M.S. Programs in Computer Engineering, Computer Science and Electrical Engineering are also available. These courses are taken by approximately 100,000 technical professionals onsite in Fortune 500 companies and government agencies.

In 1991-92, 24 new sites were added outside of the United Sttes. These were primarily locations of U.S.-based member corporations, although several independent organizations were also included. Except for sites in Canada and Mexico, none of these non-U.S. locations are currently eligible to participate in the NTU graduate programs. They do, however, have complete access to the non-credit short course offerings on videotape. The new international sites are located in the following countries: Australia, Austria, Britain, China, Hong Kong, India, Japan, Korea, Malaysia, Mexico, New Zealand, Philippines, Singapore, Spain, Taiwan, and Thailand.

The future of the NTU is in the integration of television and computer technology into true multimedia. Instruction on-demand, at an individual's desktop workstation (NTU LAN-to-customer-LAN via satellite DCV), will then supplement the group methods there are now employed. Interactive, international linkages should grow rapidly because of the new economies of transmission by satellite and fiber optics. The principal challenges will be human and organizational.

As the co-founder and the first President of NTU, Lonel V. Baldwin states: "most engineering graduates enter the work force with a baccalaureate degree today, as they did in 1990, despite the enormous complexity of modern technology. Furthermore, accelerating technological change makes career-long learning a necessity" (Baldwin 1991). NTU is the answer of the visionary leaders from the corporate sector, leading engineering schools and the federal government to improve engineering education via modern tools of the Electronic Global Village.

# Electronic Classroom

## Virtual University

In today's world of fast-moving global knowledge and fierce competition, the windows of opportunity are often frustratingly brief. Few universities boast in-house expertise to quickly launch diverse and complex courses in different curricula. Ever hear of the virtual university? It can be a model of the next millennium education.

The virtual university is a temporary network of independent education centers—courses, providers, students, instructors, even former rivals—linked by information and telecommunication technology to share knowledge, skills, laboratories, costs, and access to other programs. It will have neither central office nor organization chart. It will have no hierarchy, no vertical integration.

Instead, this new evolving university model will be fluid and flexible—a group of collaborators that quickly unite to exploit a specific educational/research opportunity. Once the opportunity is met, the venture will, more often than not, disband. It is not only a good idea, it is inevitable, if excellence in education and research is still a goal.

In the concept's purest form, each educational/research institution that links up with others to create a virtual university will be stripped to its essence. It will contribute only what it regards as its "core competencies," the buzz phrase for the key capabilities of an institution. It will mix and match what it does best with the best of other institutions and service providers.

The education centers will design courses and global advisers will guide students where the courses should telematically come from. Most universities put undue emphasis on owning, managing, and controlling every educational/research activity. If something was worth doing, the university did it itself. But there is just not enough time in the day to manage everything anymore.

Partnering—the key attribute of the virtual university will assume even greater importance. More universities are waking up to the fact that alliances are critical to the future. Knowledge is changing so fast that nobody can do it alone anymore.

The virtual university may now exist mainly in the imaginations of a few education thinkers and theorists, but similar structures have long characterized several industries. In businesses as diverse as movie making and construction, companies have come together for years for specific projects, only to dissolve once the task is done. The leveraged-buyout firms form virtual-style combinations when assembling lawyers, accountants, and investment bankers to do a specific deal.

What is different now is that large and small universities have begun using elements of the virtual concepts to gain access to new students or courses. Telematic technology will play a central role in the development of the virtual university. It is easy to envision a world which could make the creation of a virtual university as strait forward as connecting courses from different education centers to assemble a final degree of a given curriculum or its customized version. The Electronic Global Village's communications superhighway would permit far-flung students and advisors to quickly locate course providers through the education/research electronic global clearing house. Once connected, they would sign "electronic admission contracts" to speed linkups without legal headaches.

Teams of instructors and administrators in different centers would routinely work together, concurrently, rather than sequentially, via computer networks in real time. Artificial intelligence systems (like expert systems) will advise how to use the system.

If excellence and flexibility are the obvious benefits of the virtual university, the model has some risks too. For starters, education centers joining such a network lose control of the functions ceded to their invisible partners—who may drop the ball. Proprietary knowledge and information may escape. And the structure will pose stiff new challenges for administrators and faculty members, who must learn to build trust with outsiders and manage beyond their own walls.

It is also possible that the virtual university can become a hollowed university, which will be looking for a balanced budget but will not be developing knowledge on location. It can outsource, for example, science courses to low-wage countries at the risk of losing local faculty. On the other hand, the idea of a rapidly formed virtual university composed of the best of everything will have the competitive advantage for all the students.

The architecture of a virtual university cyberspace is shown in Figure 7-5. Its components are as follow:

(1) University Systems & Services Environment
(2) Distance Learning
(3) INTERNET
(4) Electronic Global University
(5) National, state, and local Telematic Networks
(6) Student body (local and global)
(7) Faculty members (local and global)
(8) Administrator (local and global)

As far as University Systems & Services (1) are concerned, these are the following components:

- Classroom Systems & Services equipped with the telematic technology to inform rather than automate learning via distance

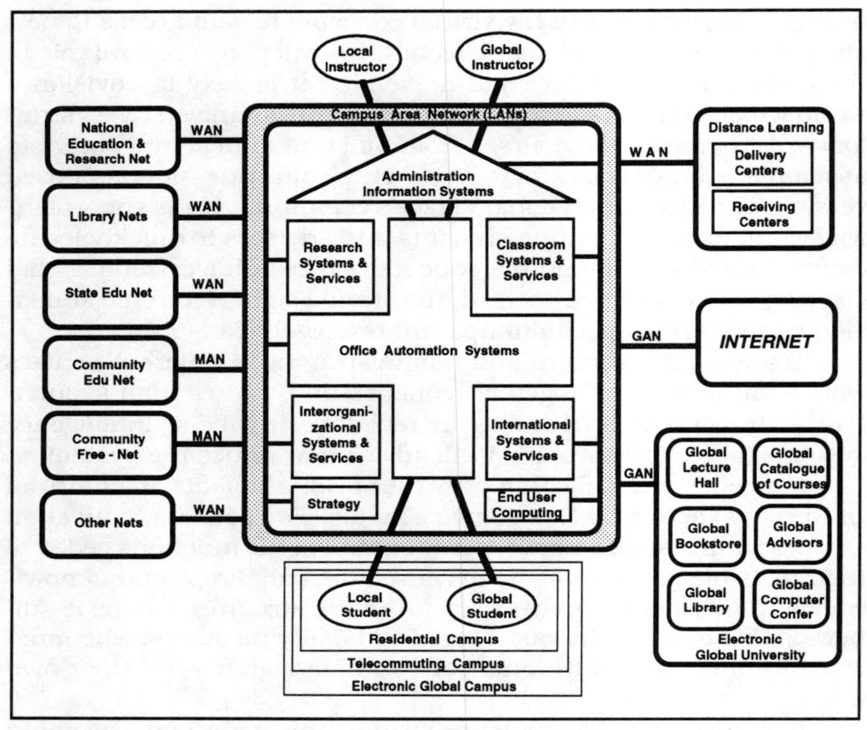


*Figure 7-5: The Architecture of a Virtual University Cyberspace (The Targowski Model)*

learning techniques and networking class material with remote sources.

- Research Systems & Services interconnected withlocal/national laboratories, super computers, and libraries to demonstrate scientific experiments and access appropriate sources of information. Among providers of such solutions are National Education and Research Network, State Edu Net (e.g., Michnet), Libraru Nets, and so forth.
- Interorganizational Systems & Services interconnected with local/national cooperative institutions and businesses, such as laboratory co-users, advisory boards, Community Free-Net, and so forth.
- International Systems & Services delivered via the Electronic Global University and INTERNET.
- Administrative Information Systems, such as admission, registration, registrar, schedules, catalogs, payroll, inventory, fleet, facilities management.
- Office Automation Systems integrate all other components of the virtual university via electronic mail, computer conferencing, bulletin boards, word processing, time management and calendaring, individual Decision SupportSystems, and so forth.

- Campus Area Network is a set of the backbone network and local area networks which supports multimedia telematic communications among all members of the university community.

A distance learning element (2) consists of delivery centers (exporting) and receiving centers (importing) with the help of teleconferencing (including shared computers), telephone, and fax equipment.

The Internet (3) is an international data highway. It is a global network of computer networks. A connection to the INTERNET allows you to interact with computing sites around the world.

The Electronic Global University (3) consists of the following components:

- Global Lecture Hall, an emerging distance learning virtual classroom,
- Global Catalogue of Courses, such as which is being developed at the International Center for Distance Learning of the British Open University about courses offered electronically by Commonwealth institutions. Another database, ECCTIS 1000, informs about 60,000 courses worldwide.
- Global Bookstore of electronic textbooks, customized to given course requirements.
- Global Advisors who will support global students in designing and finding an appropriate curriculum and courses, that can be transferrable to a given college.
- Global Library which will allow the global students to retrieve and download electronic books and information services.
- Global Computer Conference for the global students who study at the Electronic Global Campus, as a mean of communications.

National, state, and local Telematic Networks (5) is an evolving family of telematic resources being developed at all levels of the globe. From the community level (Free—Net, EduNet) via the state level (EduNet) to the national level of Education and Research Network (NERN).

Instructors as well as students can be "local" and "global." In general a student can study at residential, telecommuting or the electronic global campus.

• • • •

When the telegraph was invented, the first reaction of the Pony Express...was to try to buy faster horses. Then they tried to hire better riders. They did not realize that the world had changed (Mabus 1991). The emerging electronic global university changes the *modus operandi* of many institutions. Some of them will blossom; others will disappear. The electronic global education perhaps will influence noncollege towns and poor neighborhoods which will have a new access to vast educational resources. The rich neighborhoods and university towns will improve their own education into a more sophisticated and global system. All this can shift the global society from a trade-oriented into a learning society. Society will be better educated, and ....wiser, for less money.

## Endnotes

1. Source: BUSINESS WEEK, September 14, 1992, p. 85.

2. U.S Department of Commerce, Bureau of Labor Statistics, various reports.

3. U.S. Congress, Office of Technology Assessment, Worker training in the New International Economy, OTA-ITE-457, Washington, DC.:U.S. Government Printing Office, September, 1990.

4. "Tomorrow's Teachers: East Lansing: A Report of the Holmes Group, April 1986, p.6.

## References

Arundel, K. (1989) "Personal Communication" Washington, DC., U.S. Department of Education, September.

Baldwin, L., V., (1991), Higher-Education Partnerships in Engineering and Sciences, *The Annals of The American Academy of Political and Social Science,* March, volume 514, Newbury Park: SAGE Publications, pp.76-91.

Barker, B. (1989) "Distance Learning Case Studies," Office of Technology Assessment, May.

Becker, J. (1989) The Concept of a University of the World. *The Information Society* (6) no. 3, pp. 83-92

Blackman, A., J., Reid, J., Wilwerth (1992) "Campus of the Future," *Time,* April 13, pp. 54-58.

Butera, P. (1992) "Distance Learning: Teaching Beyond the Classroom Wall," *AMERITECH Magazine,* vol. 4, issue 2

Carnevale, A. P., L.J. Gainer, A.S. Meltzer (1988) *Workplace Basics: The Skills Employers Want,* Alexandria, Va.: American Society for Training and Development

Cetron, M. and O. Davies (1989) *American Renaissance,* New York: St. Martin's Press, p.48

Dewitt, P. (1992) "The Century Ahead — Dream Machines," *Time,* Fall

Special Issue, pp. 39-41.

Fisher, A. (1992) "Edutech," *Popular Science*, October, pp. 68-71

Gregory, S.S., (1992) "Tomorrow's Lesson: Learn or Perish," *Time*, Fall Special Issue, pp. 59-60.

Grayson, L.P., (1974) "Educational Satellites: The ATS-6 Experiments," *Journal of Educational Technology Systems*, vol. 3, No. 2, Fall, pp. 89-124.

Henry III, W.A. (1992) "Ready or Not, Here It Comes," *Time*, Fall Special Issue, p 34.

Kamlani, R. and K. Mihok (1992) "How the World Will Look in 50 Years," *Time*, Fall Special Issue, p36-38.

Levine J.B., R. MacKinnon, I. Reichlin (1992) "Why Akio and Yves Beat Out Johnny," *BUSINESS WEEK*, September 14, pp. 79-80.

Mabus, R., (1991) "A New Light in Education," *T.H.E. Journal* 19, no. 1 (Spring), pp. 48-51

Martin, Th., L., (1988) *Focus on the Future: A National Action Plan for Career Long Education for Engineers*, Washington, DC., National Academy of Engineering

Norris, E.(1992) "Ameritech Connects Education To The Future," AMERITECH Magazine, vol. 4, issue 2.

Olson, R., M.G. Jones, C. Bezold (1992) *21 St Century Learning and Health Care in the Home: Creating A National Telecommunications Network*, Washington, DC.: Consumer Interest Research Institute.

Rossman, P. (1992) *The Emerging Worldwide Electronic University*, Westport CT.: Greenwood Press.

Segal, T., Ch. Del Valle, D. Greising, R. Miller, J. Flynn, and J. Prendergast (1992) "Saving Our Schools," *BUSINESS WEEK*, September 14, pp. 70-78.

Schrum, L. et al (1988) "Today's Tools," *The Computing Teacher*, vol. 15, No. 8, May, pp. 31-35.

Targowski, A. (1991) *Chwilowy Koniec Historii*, Warsaw: Nowe Wydawnictwo Polskie, p.167.

Theobald, R. (1983) "Toward Full Unemployment" ed. by H.F. Didsbury, Jr. *The World of Work*, Bethesda, MD.: The World Future Society.

Waterman R. (1987). *The Renewal Factor*, New York: Bantam Books.

Utsumi, T., and M. R. A., Magalhaes (1992) " Global (Electroni) University for Global Cooperation, *IAMCR conference in San Paulo*, August.

# Part III

# Local Information Infrastructure

# CHAPTER 8
# TeleCity

## "The Rise of Urbanization and the Decline of Citizenship"

Man has lived in cities for ten thousand years, though at the start they were very small and primitive. However, even small and primitive cities have always held people closely together. In cities, people communicated with each other, interacted, developed their ideas, and in general advanced more quickly than people living in isolation or in smaller groups. The city-dwellers themselves always felt "advanced," with a high opinion of themselves. In the cities, culture moved foreword most rapidly and most of the new and supposedly better originated there. The city dwellers even controlled the development of language, using it to praise themselves.

The Greek word for city is *polis.* The behavior of a city-dweller is "polite." He/she has "polished" manners. These became complimentary words that could not be applied to countrymen. Countrymen worked in the fields of country villas, so they were "villains." The Teutonic word for a farmer is *Bauer* or *Boer,* so countrymen are "boors." These are not complimentary words. The Latin word for "city" is *urbs,* so city-dwellers are "urbane." A later Latin word for "city" was *civitas,* so city-dwellers are "civil" and "civilized." The word "civilization" comes from the same roots.

Throughout most of history, city-dwellers were a small minority of humanity, and considered themselves an elite. About two hundred years ago, however, cities began to grow rapidly as more and more people flocked into them. Cities are still growing, more rapidly than ever, and have changed greatly in the process. The opportunities in cities, the excitement - and the problems - have increased enormously. By the year 2000, half the people in the world will be living in cities (a staggering 3000 to 3500 million), some of them containing as many as 30 million people (for example, Mexico City).

Cities have changed more in the last 200 years than in all the previous 5000. The first cities in Mesopotamia were tiny, with populations of 2000 or less. What caused this sudden spurt of growth? When farming was developed in Mesopotamia 10,000 years ago, enough food could be grown for the small population. People then began to build more permanent houses and to make and sell utensils and decorative objects. As more goods were produced, trade spread between areas, and cities became centers of trade. People came to these growing cities looking for work, and to be part of the life in settlements that now boasted squares, monuments, fine houses and sometimes palaces and temples.

A wish for power and wealth affected the history and growth of early cities. Later, people came to cities because important new ideas were formed there. In Athens, thinkers and playwrights created works which laid foundations for much of today's science and art. In Rome, ideas of government and law were formed. In London, a modern parliament was created and practiced. In Paris, art was mastered. In Washington, liberal democracy was formed.

Cities with ports were well placed for trade, and some like Venice, Antwerp, and London grew in wealth and power. Cities changed little until the Industrial Revolution of the 18th, 19th, and 20th centuries. Industries grew up in cities and around the coal fields. Some whole new cities grew up around new industries (Detroit, for instance, grew up around the automobile industry). The inventions of the Industrial Revolution began to transform the cities in another way.

Crowded, dirty and unhealthy conditions at the center made people move out. New, faster trains and buses allowed them to create suburbs where the air was fresh and there were trees and gardens. Old country villages were swallowed up by spreading suburbs. The latest phase of the Intellectual Revolution is in computers and telecommunications. This will produce an enormous effect on our lives in cities and satellite villages (subdivisions).

In examining the many problems of cities, it is easy to forget that cities are very often exciting and enjoyable places. While there are slums and traffic jams, there are also theaters, concerts, amazing buildings, beautiful parks and galleries, libraries, zoos, cinemas and sporting events. Even the streets of such poor cities as Cairo,

Calcutta and Bombay, though chaotic, are fascinating to stroll on because so much happens. The cities of the future should not be made up of empty stretches of concrete, but of bustling streets (Royston 1985).

Bookchin (1987), in his book, *The Rise of Urbanization and the Decline of Citizenship*, challenges contemporary prejudices by revealing that the city can function as "an ecological and ethical arena for a vibrant political culture and a highly committed citizenry." This is not the city that has gone out of control. Unchecked urbanization - the industrial, commercial, and urban sprawl that we call urban belts — destroys the social fabric, brutalizes the landscape, and sets the society and nature at odds. Drawing on a wealth of little-known historical sources, as well as modern experiences, Murray Bookchin uncovers the "hidden" history of the city, which was once conceived as a nurturing environment that educated the individual while creating a rich legacy of popular, democratic, and participatory institutions. Historic experiments like the Swiss communities and America's New England colonies—particularly the direct democracy embodied in the town meeting—have shown that economic and political coordination within and between communities can transcend the nation-state.

If citizens allow themselves to be disenfranchised by the growing power of the state and a centralized corporate economy, Bookchin warns, urban dwellers will be reduced to "passive taxpayers," rural people will function as "a laboring class for industrial agriculture," and the landscape will continue to be afflicted "with over-built, densely populated, and over-extended structures." Murray Bookchin indicates that we must recover "a new participatory politics structured around free, self-empowered, and active citizens." He aims for nothing less than the restoration of a healthy balance between city and countryside, the reconciliation of humanity and nature, and the preservation of the age-old appeal of urban life.

## The Forces of Change

Unprecedented changes taking place in American cities have made it necessary to replace traditional planning and management practices with new strategic-planning techniques. Old techniques based on continuous growth and historically known solutions are replaced by new ones based on new urban patterns and rapid technological changes. America's cities are not alone in confronting economic dislocation, overcrowding, exhausting of building land, industrial and highway development, racial and ethnic change, central-city deterioration, growing affluence, and the attraction of suburbs made possible by the car and rapid train service. Frankfurt, Hamburg, and Hannover in Germany, Liverpool, London, and

Manchester in Great Britain are among those large cities also experiencing substantial population losses.

New urban patterns will be determined by the availability of transportation energy. Van Till (1980) argues that today's spread-out residential patterns, characterized by urban sprawl and heavy commuter traffic from remote areas into urban centers, were built on the assumption that cheap and abundant gasoline would always be available for transportation. His most pessimistic scenario envisions a loss in transportation energy of 52% by the year 2000. What impact will it have on the organization of our cities and suburbs in the years ahead? In the 1960s, by combining two dimensions in her analysis, the degree of concentration, and the degree of region-wide specialization or sub regional integration, Catherine Bauer Wurster derived four models of the future of American urbanization:

- present trends projected (Northeastern metropolis),
- general dispersion (Sun Belt cities),
- concentrated super-city (visionary architects, like Soleri and Small who propose higher density-oriented community than in Manhattan),
- constellation of relatively diversified and integrated cities (people live near their places of employment).

Only the last model is appropriate for the fuel-efficient scenario. Workers will be strongly motivated to reside within walking or bicycling distances of their jobs, supplemented by bus transportation and telecommuting.

Therefore, in the energy-short future, the quoted author and the author suggest the following pattern for American settlements:

- Residential and economic patterns emerge around a number of regional nodes, within which most transportation, except what is required for production, is confined.
- The most desirable housing within these sub-regional centers is located within two miles of the node, with other housing concentrated within a mile radius of the node.
- Location of these nodes emerge on the basis of the current land use, and they occupy, in large part, existing shells. Node development occurs where there is a high density of building space (residential, commercial, or industrial) and where there is proximity to transportation.
- Homes will be designed so that people can choose to learn and work, as well as live at home.
- Office and commercial spaces will be mixed with living spaces.
- Urban developments will be a mosaic of villages separated by forests, lakes, and open fields, and linked with each other. The villages will offer a life-style that draws on the best of past urban

design while embracing the opportunities of the future.
- These new villages (subdivisions) will be accessible and friendly to people young and old, able or disabled, resident or visitor. Traffic management is one aspect of this principle—everyone should be able to walk and ride in safety within the village.
- Cars have dominated urban design in more recently created suburbs. The Global, Healthy TeleCity will be redeveloped to maximize the use of in-village transport, with cars parked in secure, serviced interchanges on the periphery of each village or subdivision.
- The design and operation of the villages will demonstrate the use of alternative sources and much more efficient use of energy, recycle storm water and waste water, and improve the management of waste in general. Each of these functions presents a challenge for improved environmental management, and has important design implications. In this way the villages (subdivisions) will be instructive models to influence city management world-wide.
- Land outside the 6 mile radius of sub-regional nodes is available for agricultural uses.

These premises of future cities will generate energy for the redevelopment and application of new technology in social engineering. Cities in the Northeast and Midwest are in open competition with both adjacent communities and those more distant for people, employment, and financial resources. A mid-size city like Kalamazoo (80,000) can either invest heavily in public improvements and encourage private investments, or avoid needed outlays. While a move to economize may have immediate political appeal, this invites further deterioration. Only by creating a new confidence in the future, through their own examples, can cities remove the image of impending doom and encourage a sustainable private commitment (Culvert 1979).

The phenomenal success of Boston's Faneuil Hall Market, Chicago's Water Tower Place, and Philadelphia's Gallery at Market East has stimulated the planning and implementation of similar in-city shopping centers in a number of other northern cities, including Baltimore, Detroit, Milwaukee, and St. Louis.

Another new creation that is changing the character of the city core is the mixed used development. The MXD, as it is referred to by developers, integrates residential, hotel, shopping, and entertainment functions into a single complex of structures. The downtown boom generates jobs mostly for suburban dwellers. Due to the mismatch of skills, the inner-city remains in poverty.

In the future, a city like Kalamazoo will witness greater "in-fill development" in already urbanized areas. Older land uses, such as outdated industrial plants and commercial centers, will be upgraded or retrofitted with new amenities to make them more marketable.

Rapid technological changes will influence new ways of social communications and transportation (Kemp 1990):

- More public meetings will be aired on public-access cable television stations. These stations will also be used to educate citizens on available services and key issues facing their community.
- Advanced telecommunication systems, such as systems with conference calling and facsimile transmission capabilities, teleconferencing, computer conferencing, and business TV, will reduce the number of business meetings and related personnel and travel costs. They will allow the city government officials and businesspersons to better communicate with their peers across the community, state, country, and globe.
- In densely populated high-traffic areas, the public will press for more efficient mass-transit.

New strategic planning for city development must consider the shared vision, committed aims, and motivations to mobilize all available resources. This strategic planning should foster a city spirit. It is an asset that is intangible but potent: will power. We must break away from the syndrome of tragedy. The much-predicted Malthusian population bomb has not yet exploded. Poverty is widespread but not overwhelming. Tendency is not destiny. Citizens should realize that solutions are possible. The dream of a better city is always in the heads of its residents.

## Principles of Healthy Cities

Studies of revolutionary cultural and political changes through history suggest certain typical symptoms tend to appear one to three decades ahead of the central change. These advance indicators include (Harman 1977):

- Decreased sense of community
- Increased sense of alienation and purposelessness
- Increased frequency of personal disorders and mental illness
- Increased frequency and severity of social disruptions
- Increased rate of violent crime
- Increased use of police to control behavior
- Increased public acceptance of hedonistic behavior (particularly sexual), of symbols of degradation, and of lax public morality
- Signs of specific and conscious anxiety about the future
- In some cases economic inflation or structural stagnation

The news stories of the past decades suggest that many of these advance indicators are observed in today's society. However, a new force is transforming our physical environments and communities into more livable place, and that force is the Healthy City Movement. The first meeting of the Healthy City planning group was in January 1986. It was organized by the World Health Organization with the help of ideas formulated by Leonard Duhl from the University of California, Berkeley and Trevor Hancock from Toronto. The Healthy City model has now been applied in over 1000 cities and communities world-wide. It provides a post-modernistic solution to the very serious social, economic and environmental problems that cities face today.

The Healthy City movement focuses on an integrated process for community improvement embracing three fundamental principles:

- collaborative partnership among community residents of diverse disciplines and sectors
- empowerment of community residents to direct their overall "health," or quality of life
- involvement of local government officials as participating key players.

The Healthy City offered by Duhl and Hanckok (1986) is characterized by the following parameters:

1. A clean, safe, high quality physical environment (including housing quality)
2. An ecosystem which is stable now and sustainable in the long-term
3. A strong, mutually supportive and non-exploitative community
4. A high degree of public participation in and control over the decisions affecting one's life, health and well-being
5. The meeting of basic needs (food, water, shelter, income, safety, work) for all the city's people
6. Access to wide variety of experiences and resources with the possibility of multiple contacts, interaction and communication
7. A diverse, vital and innovative city economy
8. Encouragement of connectedness with the past, with the cultural and biological heritage and with other groups and individuals
9. A city form that is compatible with and enhances the above parameters and behaviors
10. An optimum level of appropriate public health and sick care services accessible to all
11. High health status (both high positive health status and low disease status).

The TeleCity solutions can contribute to parameters 1 (safety and infrastructure management), 6 (socializing and resource sharing), 7 (innovative economy), 8 (tradition through information access), and 9 (city system). Doxiadis (1974) suggests that there are five human needs that the city must satisfy:

1. Freedom to move (so as to maximize potential contacts)
2. Safety
3. A quality of life that satisfies the citizens' aspirations
4. Human contacts
5. Creativity and human development

All these needs are strongly based on communications, information handling and processing. Dioxiadis (p.87) sees that the city should "bring people together to benefit from their contacts, but at the same time to form a proper structure that can keep them sufficiently apart, so that the exposure to and the danger from each other is minimized." One of the advanced solutions that can provide such a structure is a telecommunications and computer network. It is unlikely that one can recreate North American cities like those in ancient Greece and Italy, or in modern Africa or in Latin America with a common market, a *Forum Romanum*. It will be possible when the *tum-tum* culture replaces the car culture. The Information Society provides an avenue for socialization via electronic infrastructure.

## Early TeleCities

The concept of a wired city (Telecity) was developed in the United States in the context of President Lyndon Johnson's so called Great Society of the 1960s. The presidential task force charged the telecommunication industry with a clear mission to "encourage the growth of communications of all kinds within localities" as an approach to improving urban life.

In the early 1970s, ideas of the wired city became instrumental to both governmental and corporate investments on interactive cable projects in the United States and Japan. In 1972, the Japanese initiated experiments in Tama New Town and later in Higashi-Ikoma. In 1974, the United States National Science Foundation supported a series of experiments with interactive cable. In 1977, Warner Communications introduced QUBE, a 3-channel interactive cable television system, on a commercial basis in Columbus, Ohio.

These ventures suggested that a variety of public and commercial services could be effectively provided over two-way cable systems. But the same experiments also cast doubt over hopes of widespread consumer interest in new information services. We must remember that this was before the 1980s when the quiet and creeping revolution of personal computers shifted the civilizational paradigm "from one to many." The PC base in the 1990s is about 60 million computers, more and more of which are connected into networks.

At least four interconnected developments took place:

1. The visionary aspect;

*early visions:*

- electronic democracy
- education
- huge list of interactive services
- shopping
- working and even producing television programs from the home

*later visions:*

- electronic communication
- transactional services (travel, shopping)
- integration of telephone, television, and computer services
- significance of telecommunications to the economic development of localities and nations
- furtherance of national interests and cultural values

2. The technological aspect:
   - early projects were based on coaxial cable networks
   - later projects are based on fiber optics, satellite, and micro-electronic technology
3. The market forces aspect:
   - awareness that market forces represent a constraint on the stipulation of communication services
   - attention was shifted toward long-range planning development of broadband media
   - in the near future utilization of narrowband media, such as telephone for video services (with compression)
4. The developers aspect:
   - the first wired cities experiments were developed by cable companies, mostly to provide clearer reception of broadcast signals to the home and eliminate the need for rooftop television antennas
   - the later projects are undertaken by telephone companies which look for delivery of information services to residents and institutions of local communities.

Projects that approach the image of the "advanced wired city" are shown in Table 8-1.

QUEBE, in Columbus, Ohio, is an illustrative television experiment. QUEBE came into the home with a "black box" home terminal that resembled a small calculator. The terminal had 30 buttons for channel selection, plus five "response" buttons that viewers used to communicate with the cable operator. The viewer could select yes or no, choose among five multiple-choice answers to a question or order

| Location | Initiated | Terminated | Households | Developers |
|---|---|---|---|---|
| Higashi-Ikoma (Japan) | 1972 | 1985 | 156 | MITI, 8 agencies New Media |
| Columbus (USA) | 1977 | 1984 | 50,000 | Warner City Cable |
| Alameda (USA) | 1977 | continuing | n.a. | n.a. |
| Biarritz (France) | 1979 | continuing | 1,200 | 300 agencies French PTT |
| West Berlin (Ger) | 1981 | continuing | 20,000 | Deusche Bundespost |
| Milton Keyns (UK) | 1981 | continuing | 20,000 | British Telecom Cable |

*Table 8-1 Advanced Early Wired Cities*

a pay-per-view program. The central communication computer had "pooling" capability, scanning each home terminal every slice of time, similar to the way that time-sharing computers operate. The subscriber paid $10 more per month over basic cable. During "live and interactive" events, the viewer at home pressing buttons in response to questions could give a congressman an opinion about national security or make a reservation at a restaurant. Higher-capacity multiple-strand cable systems were later installed in Cincinnati, Pittsburgh, Houston, and the suburbs of St. Louis and Chicago. In 1983, these cities were connected to Columbus via satellite for about one year. During that year, the company maintained the national QUEBE Network, with national interactive pools, advertising strategies, and interactive game shows. QUEBE innovated the entire cable industry; it could sell movies and participation but could not cover the cost of two-way programs. Warner Communications failed to use TV interactively to produce a product that can compete with other available programming and generate sufficient revenues (Dutton et al. 1987, Davidge 1987).

In May, 1992, IBM discussed a $500 million investment in Time Warner's movie, TV, and cable division as part of a plan by the companies to merge IBM's computer technology with Time Warner's video technology. IBM's investment would equal the 12.5% stake in Time Warner Entertainment being sold to two Japanese High-Tech companies, Toshiba and Itoch.

After color TV, the wired city is a quest for a new strategic product for the mass market. The following 20 services were tested by the first wired cities in Japan (Murata 1987):

1. TV retransmission channel
2. Radio retransmission channel
3. TV local origination channel
4. FM local origination channel
5. Video request service
6. FM program request service

7. Guide for the system
8. CAI—Computer Assisted Instruction
9. Data request in all still picture form
10. One-way facsimile (local government PR, reservation confirmation, news bulletin)
11. Data request by facsimile
12. Shopping guide
13. Cashless transactions, with bill paid by automatic bank account deductions (charge cards)
14. Reservation service (for travel tickets, hospitals, and beauty parlors)
15. Health and welfare services
16. Telemetering
17. Emergency service (gas and fire prevention)
18. Emergency broadcasting (TV sets are turned by the System Center)
19. Televotes
20. System showrooms (all services are available to residents outside of the experiment site)

In the 1990s, the Japanese wired city concept is the subject of controversy and competition between MITI and MPT. The concept involves a combination of CATV network and digital communications network. MITI is promoting the New Media Community Plan (multi-info services) while MPT is supporting the Tskuba ACCS and Teletopia Plan for 10 models cities, selected from 100 applicants. Among information services, CAPTAIN-videotext is the most popular one.

The French PTT appears solidly entrenched as a telecommunication monopoly. This policy reflects the belief that information technology and services are fundamentally important to the French Society, and thus remain firmly under public control. The following services contribute to broad social goals (Weingarten 1987):

- The smart card, a card with a microprocessor that can be used as a sophisticated ID card, as a mechanism for administering government benefit programs, as a vehicle for electronic fund transfer and for other purposes.
- The Mintel, a device that attaches to the telephone lines and provides access to online databases such as an automated yellow pages.
- The Antiope system, one of the first operational videotext systems, which features high-quality graphics.

The 1991-92 Kalamazoo TeleCity plan took shape when the telecommunications and computers state of the art were quite advanced in comparison to early wired city developments. The real question is not whether we should build a TeleCity, but rather how to build it in a rational, less utopian manner. We base the approach presented here on comprehensive planning and careful implementation, involving all potential partners in the city, including the citizens (Community Access Channel, Citizens Committee) and businesses of Kalamazoo [1].

## The New Urban Landscape Through Telework

Despite a slow start, there is a little doubt that telecommuting is in our future. Los Angeles County is implementing an aggressive traffic and emissions reduction regulation that requires employers to reduce single-occupancy travel by their employees. Other urban centers are considering similar regulations, and their constituents will discover that telecommuting is essential to any meaningful solution.

If telecommuting catches on, it will not be the first time telecommunications has reshaped the urban landscape. The telephone helped to separate office from factory, allowing knowledge workers to be concentrated in urban skyscrapers, and creating our modern pattern of moving suburban workers to the urban center.

We are now attempting the reverse: moving jobs to workers. Corporate headquarters are being relocated to suburban office parks. This merely brings urban gridlock closer to home, forcing even the urban escapees to consider telecommuting options like suburban satellite offices linked to their headquarters. Telecommuting from home will also increase. However, satellite offices or telework centers will increase more quickly, because they offer a better fit with existing corporate work styles. In the long run, telecommuting is certain to uncouple where we work from where we live, making the electronic cottage vision a practical option in a few decades (Saffo, 1991).

### What is Teleworking?

What would you do with an extra 2.5 hours a day? Spend it with your family? Visit with friends? Resume an old hobby, or develop a new one? Simply enjoy life? Thanks to the Telework Center, a group of lucky people have an extra 2.5 hours daily for activities of their choice. That adds up to 12.5 hours per week!

Teleworking is a transportation demand-reducing measure to curtail work commuting. It allows employees to work in a community telework center (TWC) equipped with computers, modems, fax, copying machines, and other state-of-the-art equipment. The concept is to move the work and not the people. By teleworking, people will no longer commute to downtown every day. They will send their work to their companies main offices along "electronic highways." Teleworking closes the gap between where people live and work.

### Telework Center

The Hawaii State Department of Transportation dedicated the nation's first telework center on July 14, 1989. The Hawaii Telework Center Demonstration Project is located in the Mililani Technology Park on Oahu. The projects is a public/private sector joint venture to pilot-test the use of electronic highways to move work instead of people. The project is a transportation innovation, capitalizes on modern technology, and integrates the private sector as an important project partner.

Nine private sector employees are currently working as telework participants along with seven state employees. All participants live in adjoining communities. Teleworking can produce significant social, economic and environmental benefits by reducing work-related travel, commuting costs, air pollution, fuel consumption, and driving stress associated with peak-period commuting. Teleworking also benefits employers and employees by reducing office space rent and parking expenses, and providing increased employment opportunities for residents of neighboring islands, parents with young children, and handicapped citizens.

Within the Hawaii Telework Center Project, the state will also experiment with "business on-line access." The Title Guaranty Escrow Company will be electronically linked with Hawaii's Bureau of Conveyance, the state's property Tax Assessment Division, and the state's civil courts.

The final evaluation report on year one of the Hawaii Telework Center found the following results ( SMS Research, 1991):

1. The Telework Center project was easily understood and supported by both government and business. Throughout the project, support ($450,000) was always in excess of what was requested.
2. Teleworkers unanimously support the Telework environment. Employees stationed at the Center saved as much as eight hours per week in travel time. They also saved fuel and vehicle maintenance costs, parking fees, and incidental expenses associated with commuting. They developed their own work culture, and quickly mastered the techniques of performing their jobs from remote locations. The situation was particularly rewarding for working mothers.
3. Supervisors developed strong support for Telework Center. The experience was good for their employees and caused no significant

problems in work management.
4. Employers were satisfied with Telework Center. All of them said that they would send more workers to telework centers if the space was available.
5. Telework Centers will contribute meaningfully to travel, cost and infrastructure maintenance reduction. In 2010, telework may be appropriate for up to 9.3% of the total work force. In other words, about 64,000 workers will enter Telework Centers.
6. Telework Centers are consistent with social movements in the 1990s. As populations move away from urban centers, telework centers will provide effective, low cost alternatives for the rapidly increasing cost of office space in urban centers.

The Hawaii Telework Center demonstration project was a success, with clear indications that telework centers are an important component of a comprehensive transportation program in the Interactive Age.

**State-wide Telecommuting Experiments**

While telecommuting offers many potential advantages for employees, employers, and society in general, the energy and environmental benefits of reducing vehicle miles traveled (VMT) motivated the Washington State Energy Office's (WSEO) interest in telecommuting. Here are some of the reasons why (Quaid, 1992):

- Since 1970, the volume of traffic has more than doubled on major freeways in Washington's Puget Sound region. Seattle now has the fourth highest traffic congestion in the nation.
- The transportation sector is the largest consumer of energy in the state. Washington's residents spend more than $2 billion annually on gasoline, more in times of crisis.
- Motor vehicle are the major source of air pollution in the state, and pollution levels are beginning to exceed air quality standards. While significant improvements have been made in automotive fuel economy and vehicle emissions standards, they have been overwhelmed by increases in travel. In order to reduce pollution and energy consumption from personal transportation, the number of vehicle trips must be reduced.
- The pollution of the region is expected to increase significantly in the next 20 years, seriously exacerbating existing transportation-related problems.
- WSEO calculates that if 15% of the Puget Sound workforce telecommuted two days per week, work trips would drop 6%. The reduction would save approximately 14 million gallons of gasoline and eliminate 7,000 tons of carbon dioxide emissions annually.

- In 1990, WSEO launched the Puget Sound telecommuting Demonstration to explore the environmental, and personal side of telecommuting.

From January, 1991 to January, 1992, the Washington State Energy Office operated a telework center for state employees. This component of the demonstration enabled the state agencies to provide a professional work environment closer to their employees residences. For example, several agencies located in Olympia used the Seattle-based center for employees who lived in Seattle. A total of nine public agencies used the telework center.

Each employee using the center had access to a professional workstation and a computer linked to a Local Area Network (LAN). The LAN provided word processing, spreadsheet, and database application software. Each employee could store additional required software on their local computer's hard drive. In addition to telephones, some employees used modems to connect with their main offices, and had access to other support equipment including fax machines, photocopiers, and laser printers. A conference room was available to agencies using the center and, on a reserved basis, to other state agencies. The subsidized cost for each workstation was $1,500 for the one-year period. The TWC also had access to public transportation and adequate parking facilities. It was located near shops, restaurants, and day care facilities. Employees could use the center on evenings and weekends, as well as during usual business hours.

From a preliminary analysis of their experience with telework, there are many indicators of improved organizational functioning. On the other hand, the interactions between telecommuters and other people in the organization always appear to be smooth and should be explored further.

### Who Does Telecommute?

Many types of jobs are suitable for teleworking:

- Occupations requiring little face-to-face contact
  Architects
  Engineers
  Computer analysts/programmers
  Accountants/bookkeepers
- Task-oriented occupations
  Data entry personnel
  Word processors
- Occupations requiring high daily or weekly use of telephone
  Insurance adjusters
  Reservationists

Securities agents
Selected government workers
Travel agents
Public relationship workers

- Occupations accomplished largely by use of computer terminals
  Writers
  Business communicators
  Faculty members
- Tele-immigrants from overseas

Telecommuters choose certain kinds of tasks for working at home. Task preference for teleworkers (Quaid, 1992):

| | |
|---|---|
| Reading | 70% |
| Writing | 68% |
| Text/word processing | 66% |
| Analysis | 58% |
| Problem solving | 42% |
| Computer programming | 40% |
| Speaking on phone | 38% |
| Record keeping | 36% |
| Design | 32% |

The projected growth of telecommuters in the United States:

| | |
|---|---|
| 1990 | 3.4 million |
| 1992 | 4.6 |
| 1995 | 11.0 |
| 2000 | 22.0 |

Source: Link Resources, NY

The same study estimates that 4.5% of the 18 year and older civilian work force is telecommuting in 1992. Telecommuting continues to grow rapidly, particularly among business executives, managers, engineers, and scientists. Public sector commuters are up to 240,000 (1991) from 160,000 (1990). Who is telecommuting ? Consider the following examples from the private sector (Mokhartian, 1992):

- The Traveler's Insurance Company finds that offering telecommuting to several hundred underwriters, claim processors, software developers, researchers, managers, and others gives the company an advantage in recruiting and retaining the best workers in the highly competitive labor market in Hartford, Connecticut.
- JC Penney has 120 catalog order takers working from home in several cities in the United States, resulting in savings in office

space costs and flexibility in staffing to meet peak-period demand On-call telecommuters can be at work in minutes instead of hours.

- Trans World Airlines and the Best Western hotel chain maintain reservations centers in penal institutions in California and Arizona, respectively. The employers again enjoy just-in-time staffing flexibility and save money on office space and lower-cost labor while the prisoners earn wages, learn marketable skills, and in some cases are assured continued employment on release.
- There are many companies experimenting with telecommuting, including Digital Equipment Corporation, Data General Corporation, Arthur D. Little, Inc., Blue Cross/Blue Shields of South Carolina, Chase Manhattan Bank, and Control Data Corporation.
- The state of California is expanding the availability of telecommuting to its workers following the successful completion of a landmark two-year pilot program in 1990. Statewide, an estimated 1,000 employees in at least 25 departments are participating.
- In Fort Collins, Colorado, a city with a population of 85,000 people located 60 miles north of Denver, all municipal employees can telecommute, if they so desire.
- Word processing pools in Haiti are already servicing a range of United States based insurance companies. Workers in Barbados are being paid $3.50 per hour to input data into American Airlines computers that had previously been done by Arizona and Oklahoma based workers at a rate of $8.50 per hour plus worker benefits. Hong Kong-based typesetters are also being used by publishing concerns in Great Britain, Australia, and North America. Offshore banking operations in the Bahamas and the Cayman Islands already employ thousands of teleimmigrants telecommuting to work in the United States and Europe. (Pelton, 1990). Swissair sends transactions for data processing to India.

## Telework Benefits

While traffic reduction was the primary objective for testing teleworking, quality-of-life improvement is emerging as the most significant and often discussed benefit of teleworking. Workers report less stress since leaving the two-and-a-half hour daily commute, are more productive, and are now enjoying quality time with their families. Employees also have more time for themselves, and have increased their involvement in neighborhood, school, and community activities. Working close to home has had an incredible impact on these workers and their families. Hirata, from the Hawaiian experiment says "Our ultimate goal is to establish telework centers in all major communities in the state. We want to provide jobs near homes."

Telework reduces business travel and provides businesses with increased opportunities to decentralize from congested downtown areas. With a corresponding effort to establish satellite offices to better serve the general public in their own communities, traffic congestion from employee, business and citizen travel can be greatly reduced.

These first projects prove that communications can be substituted for transportation. Transportation specialists realize that we cannot build enough highways to meet all of the demand of our growing population, and even faster-growing vehicle population. Building highways today is becoming increasingly difficult, physically, environmentally, and socially. An agenda of "conventional transportation approaches only" is unhealthy and could easily lead to transportation gridlock in many areas of the country. Innovative transportation concepts and approaches are critically needed. The automobile has congested our cities, but computer and telecommunication technology can help to decongest them by reducing work and business travel.

Telecommuting can save much productive time. Table 8-2 shows how daily commuting times that we experience add up.

Telework centers have all of the advantages of the solutions brought by the new technology in general and avoid the major disadvantages that plaque work-at-home solutions. Work-at-home has been opposed by some because of its potential for exploitation of labor and the possibility of increasing the isolation and withdrawal of low-income and disadvantaged groups. The major advantages and disadvantages for employers, employees, and society at large are presented in Table 8-3 (SMS, 1991).

## A New Information Landscape of North America

The application of networking information technology traditionally had impact on urban areas. However, networking technologies may have decisive impact upon midsize and nonurban (rural) areas of the United States. Many states have one or two major metropolitan areas. For example, the State of Michigan has one large city, Detroit, with a population of 1.2 million, or 2.5 million including suburbs. However, Michigan has about 20 midsize cities with developed nonmetropolitan areas that can benefit from networking technologies.

### Rural Area Network

Rural America was transformed by two major events, the Industrial Revolution and the American Civil War. Both events greatly increased the demand for agricultural products. As the commercialization of farms progressed, the size and value of farms

| Round Trip (Minutes) | Hours per Year | Equivalent for 40-Hour Weeks |
|---|---|---|
| 20 | 80 | 2 |
| 40 | 160 | 4 |
| 60 | 240 | 6 |
| 80 | 320 | 8 |
| 100 | 400 | 10 |
| 120 | 480 | 12 |

Source: Paul and Sara Edwards, Working from Home, Los Angeles, CA: Jeremy P. Tarcher Inc. 1987

*Table 8-2 Time Spent Commuting*

| **Employers** *Advantages* | **Employees** *Advantages* | **Society** *Advantages* |
|---|---|---|
| • Increased productivity<br>• Decreased turnover | • Cost savings<br>• Freedom (time, location) | • Decreased traffic<br>• Decreased highway cost |
| • Better hiring<br>• New labor pools | • Work autonomy<br>• Greater access to jobs | • Better local economy<br>• Employment opportunity |
| • Lower overhead<br>• PR value<br>• Hiring incentives to new employees | • Fewer distractions<br>• Better lifestyles<br>• Increased safety & security | • Stronger family & community ties |
| | • Better family life | • Flexibility in land use planning |
| ***Disadvantages*** | ***Disadvantages*** | ***Disadvantages*** |
| • Initial capital investment<br>• More urban sprawl<br>• Data security more difficult<br>• Changes in managerial style<br>• Possible loss of corporate identification for employee<br>• Possible loss of productivity due to ineffective supervision | • Employee isolation<br>• Promotions more difficult,<br>• Less interaction with management<br>• Lack of support services | |

*Table 8-3 Advantages and Disadvantages of Telework Centers*

increased, while the number declined. The disappearance of the family farm undermined the viability and independence of rural communities.

For example, Michigan's rural economy prospered throughout the 1970s. With a booming national economy, the demand for natural resources was quite high. Rural manufacturing was a special beneficiary of this growth. With labor costs and land values increasing, many manufacturing firms, especially those in low-tech industries producing standardized goods, moved to rural areas where their input costs were lower. The relatively low education level of this workforce attracted more low-tech manufacturing facilities than high-tech ones. As a result, manufacturing grew more rapidly in rural than in urban areas, with rural unemployment rates dropping below those in urban areas (Bloomquist, 1988). Farmers also benefited from higher prices and high rates of inflation, which allowed them to make greater investments in productivity-enhancing technologies.

This prosperity came to an end at the close of the seventies. The rise of OPEC and the sudden oil crises, bad loans to the Third World, the United States grain embargo, deflationary policy, and heightened foreign competition had the greatest impact on rural areas. Unlike the more vigorous complex manufacturing industries that completely recovered from the 1980 recession, these routine manufacturing industries had almost 12% fewer employees after the recession (United States Department of Commerce, 1988). This decline in employment is probably permanent; many of these jobs have actually disappeared. Jobs had moved to the manufacturing and service facilities in urban areas.

The restructuring and decentralization of business operations could also benefit rural Michigan. Depending on the particular case, a firm might decide to manufacture a product at its central headquarters, but transfer downstream activities like distribution, sales, marketing, and service elsewhere. Rural areas could benefit from this development, to the extent that they can effectively compete for these newly externalized jobs.

In an information-based economy, many factors determine communication needs. To evaluate a rural community's technological requirements, consider not only the community's own economic activities, but also examine its economic aspirations and, increasingly, look at the activities of its competitors, whether they are in businesses in urban areas or in other countries. Rural areas will be unable to compete if the pace of technology deployment lags greatly behind that in other areas. Indicators suggest that a technology lag is likely. The history of the telephone, for one, points to such an outcome: first came major trunks linking Northeastern cities, followed by lines to smaller towns in their immediate vicinity, then connections to major Midwestern cities, and so forth: a sequence of connecting ever lower-order cities. Although patented in 1876, it took 12 years for the telephone to reach Chicago, and transcontinental service was unavailable until 1915 (Kielbowicz, 1987).

Together, the trend towards unbundling and decentralized intelligence will allow rural communities to have greater choice about and control over the configuration of their communication infrastructure. This is an important advantage since communication technology defines communities. As John Dewey (1915) has pointed out, "Society not only continues to exist by transmission, but it may fairly be said to exist in communication. There is more than a verbal tie between the words common, community and communication. Men live in a community by virtue of things they have in common; and communication is the way in which they come to possess things in common."

Just as many large businesses, universities, and local and state governments chose to build their own telecommunication networks, some rural communities are finding that creating their own network is the easiest option. These communities avoid wading through lengthy regulatory procedures and convincing the telephone company that the community could generate sufficient demand for service to justify the investment in sophisticated telecommunications equipment. The town of Bloomsburg, Pennsylvania is taking such an approach. In conjunction with Bloomsburg University and the Ben Franklin Partnership, the town has proposed the construction of a high-capacity digital "highway" to Harrisburg, where it would link up with the access points of all long-distance telecommunications providers. (Figure 8-1).

Bloomsburg decided to establish its own telecommunications system largely because alternative strategies were unavailable, or unworkable. Because the town cannot demonstrate sufficient existing demand conditions, long-distance carriers are unwilling to invest in technology that would make the link between Bloomsburg and Harrisburg unnecessary. Because this route crosses a LATA boundary, Bell of Pennsylvania cannot carry the traffic. The "electronic

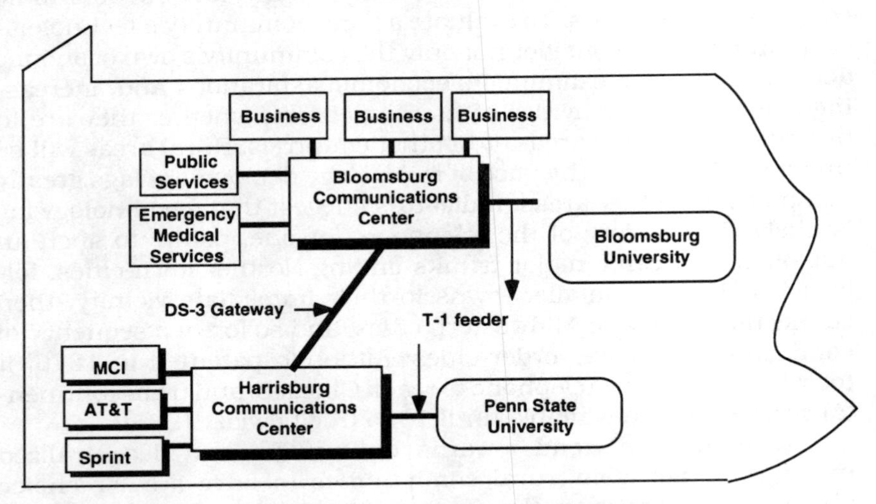


Source: Dovetail Systems Corp., "Telecommunications Opportunities for Bloomsburg," Bethlehem, PA, June 1989.

*Figure 8-1: PA Bloombsburg Telecommunications Network Configuration*

highway" between Bloomsburg and Harrisburg consists of a 45 megabit per second digital microwave link with the capability of providing a broad range of telecommunication transmission services, such as high-speed data, high-resolution graphics, and compressed motion video. The total cost is estimated at $800,000. The town considers this price to be modest when weighed against the potential long-term economic benefits.

Networking can generate whole new economies. As noted above, networking allows like-minded people not only to communicate with one another, but also to share common resources, thereby benefitting from significant economies of scale and scope. This kind of networking could be especially fruitful in rural locations, where people and facilities are few and far between.

Just as businesses are embracing these developments to create their own customized networks, so too might rural communities. Instead of being established along functional lines, like many business networks, Rural Area Networks (RANs) would be configured around geographic boundaries and the needs of entire communities. Designed on the basis of a ring, or campus type, architecture, a RAN would link up as many users within a community as possible, including businesses, educational institutions, health providers, a telework center, and local government offices (Figure 8-2).

As another example of RAN configurations, many small rural Scandinavian towns have established a number of telecottages to enable local residents to prepare for, and access the benefits of, the Information Age. Among the kind of services provided are (Qvortrup, 1989):

- information services
- data-processing services
- information technology consultancy
- distance working facilities
- training and education
- telecommunication facilities
- village hall facilities

Some of these telecottages, despite being set up using outside funds, operate on a self-sustaining basis. In other places and communities, they continue to be subsidized.

Rural network configurations are as diverse as rural America itself. Experimentation is in order. Sudi Nazem, for example, proposed the idea of creating nucleus cities or hubs, throughout the State of Nebraska, each of which could serve communities within a 30 - 40 mile radius. He points out that while "it is inconceivable that all small communities in Nebraska could be connected by the costly network any time soon, it is not...inconceivable to install a high technology network for approximately twenty-two townships" (Nazem). The Maine Research and Productivity Center, in Presque Isle,

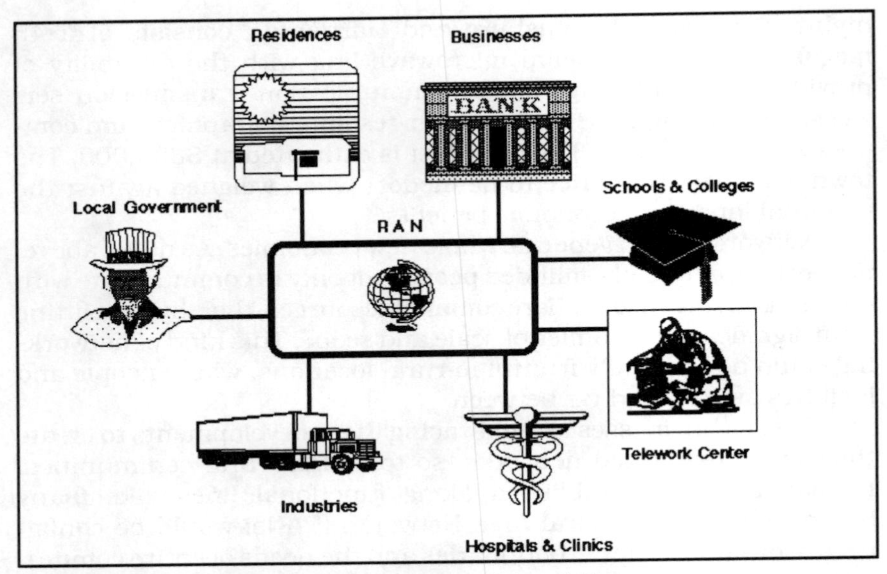


*Figure 8-2: Rural Area Network (RAN() and Rural Players*

already serves as a hub for small businesses in the surrounding area. The Center provides access to comprehensive information services as well as shared CAD/CAM system, among others.

An arrangement like the RAN described above requires that the community, the telephone company, and regulators agree on a tariffing arrangement that considers these various users as a defined group. For example, all the state colleges in Vermont are treated as one users' group, with the same rates. These rates are lower than would be possible if each entity were treated as an individual user.

## Metropolitan Area Network

Metropolitan Area Networks (MAN) provide switched data networking services at very high speed (50 to 600 megabits per second) within a geographic area of at least 50 miles. MANs connect a Local Area Network (LAN) to other LANs, as well LANs to Wide Area Networks (WAN). As designed by Bellcore, MANs will provide switched multimegabit data services (SMDS) providing 45 Mbits in bandwidth. WANs allow users to set up virtual (or logical) private networks, and provide individual services on demand. These networks are intended for shared usage (see Figure 8-3).

A metropolitan area network is essentially a huge LAN that encompasses an entire city, providing data transport at fiber optic speed. A MAN links other lower speed LANs, such as a token ring, Ethernet, fiber distributed data interface (FDDI), workstations, CAD/CAM transmission, slow-scan, full-motion, host-to-host computers interconnection, PBX interconnection, telework centers, and

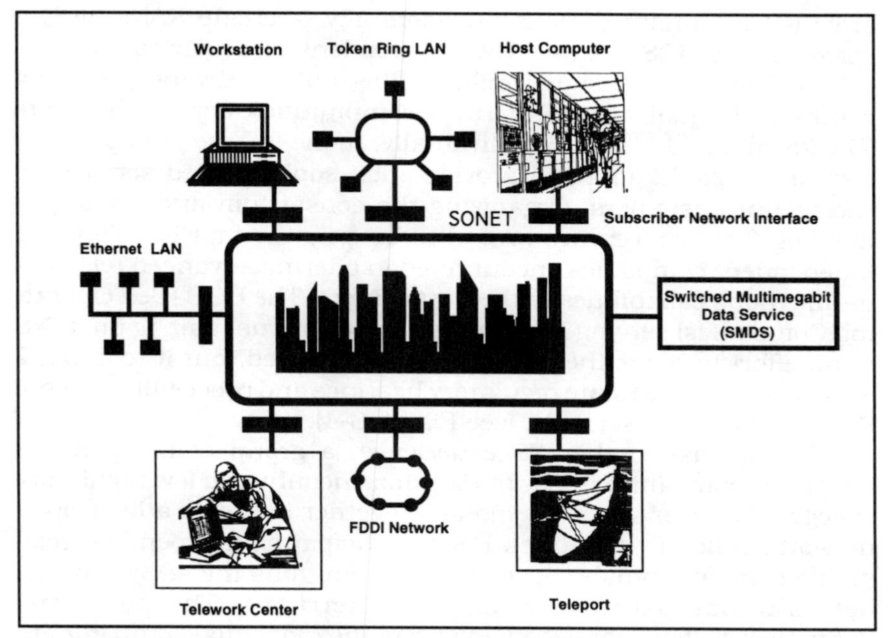


*Figure 8-3: Metropolitan Area Network—MAN (SONET-Synchronous Optical Network, 150-600 Mbps, FDDI-Fiber Distributed Data Interface)*

a teleport. Data is carried between the various locations in packet form or in fixed-bandwitdth channels. Video transmissions facillitate meetings, training, or project management. To safeguard all these transfers, the MAN utilizes mechanisms to ensure the highest degree of network availability and reliability (dual-bus architecture). MAN deployment is being driven by the increasing need for high-speed data services among LAN users. MANs provide an infrastructure allowing telephone companies to better position themselves for the provision of multiple services by means of Broad Integrated Services Digital Network (B-ISDN). MANs are designed to serve several organizations, and traverse public rights-of-way (telephone, cable,...). Emerging technologies like SONET (Synchronous Optical Network) and ATM (Asynchronous Transfer Mode) are interconnecting MANs.

## The Opportunity Infostrada of West Michigan

Consortia of telecommunications providers can utilize market forces in much the same way that coalitions of users can leverage market power to gain access to advanced telecommunications services and technologies. By cooperating or entering joint ventures, telecommunications providers can distribute the high costs and diminish some of the risk of investing in advanced telecommunications technology in rural areas.

Iowa Network Services (INS) illustrates the problems that could arise if telecommunications providers ally to create RANs. INS, a consortium of 128 of Iowa's 150 independent telephone companies, joined forces in 1984 to build a fiber-optic network providing centralized equal access to rural communities across the state (Davidson et al., 1990). Individually, none of the independent companies could afford to provide such sophisticated services to their rural customers. Organizing the consortium and finding financing for their venture proved to be only the first hurdle these independent companies encountered in offering advanced telecommunications capabilities to their customers. The BOC (Bell Operating Companies) serving Iowa brought an anti-trust suit against INS in an effort to block the network. The suit failed, but it took 31/2 years of federal and state regulatory hearings and proceedings before INS could offer its services. (see Figure 8-4).

In contrast to the INS experience, a group of independent telephone companies in South Carolina encountered few regulatory or legal obstacles when they joined together to create a fiber optic network, called PalmettoNet. Each participating independent telephone company builds, operates and maintains the section of the network that passes through their territory. The consortia, PalmettoNet, then leases capacity from the individual companies to create the unified network (Figure 8-5) (Sawhaney, 1990).

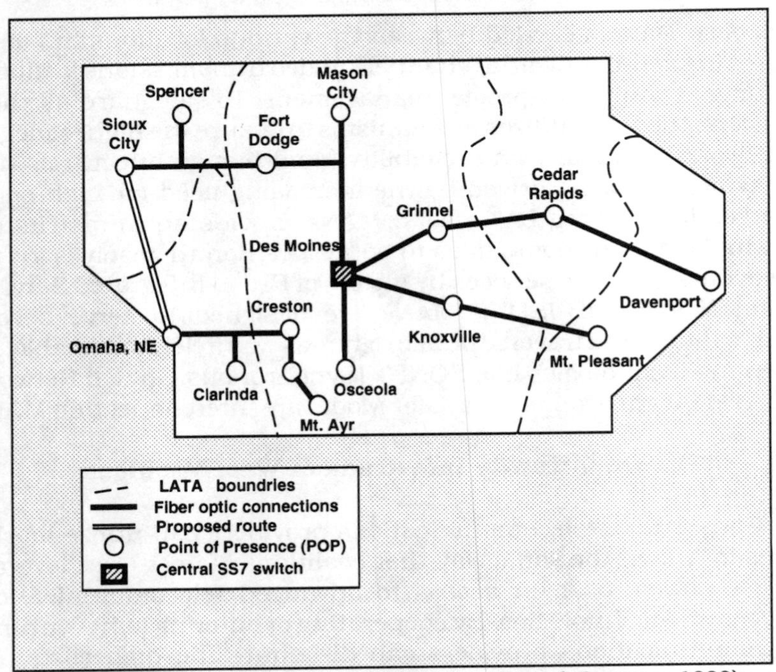


Source: The INS Story (Des Moines, IA: Iowa Network Services, 1990)

*Figure 8-4: Iowa Network Services is a Consortium of Independent Telephone Companies, Who Jointly Invested in a Fiber Optic Network and in SS7 Switch*

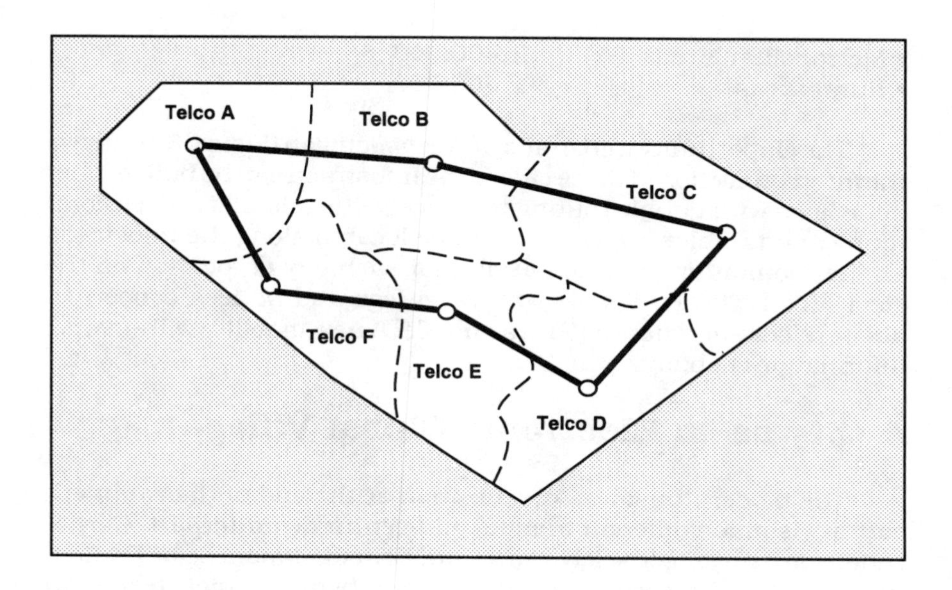


*Figure 8-5: Palmetto Net Configuration in South Carolina*

Arrangements like the Iowa Network Services and PalmettoNet take advantage of synergy to make market forces work to the advantage of rural subscribers. If rural areas are to have economical access to advanced communications technologies, policymakers at the local, state, and federal levels must think about and plan for such arrangements.

The application of telework centers in midsize metropolitan areas makes sense when more cities share the same TWC. West Michigan has seven midsize cities, each about one hour or less from the other. Consider the following West Michigan cities:

| | |
|---|---:|
| • Kalamazoo/Portage | 120,000 |
| • Battle Creek | 36,000 |
| • Grand Rapids | 180,000 |
| • Muskegon | 158,000 |
| • Grand Haven | 12,000 |
| • Holland | 26,000 |
| • Benton Harbor/St. Joseph | 22,000 |
| • Niles | 13,000 |
| Total | 567,000 |
| | |
| • Lansing (State capitol, Central Michigan) | 130,000 |
| Total | 697,000 |

- Metropolitan areas 1,400,000
- Rural areas 2,800,000

The Opportunity Infostrada of West Michigan (Figure 8-6) shows the interconnection of these cities. The Infostrada can be built of fiber optic lines with double routing to increase the reliability of telecommunications. Telework centers can be located along the Infostrada and telecommuting will be possible for each city or location on the electronic highway. People will travel less and be less concerned about relocating where a job is available. A new mobility will emerge: information mobility.

## A City as an Electronic Global Village

The modern business organization at the end of the twentieth century is much different from its mid-century counterpart. Corporations are no longer solely American, British, Japanese or French. Increasingly, the corporation is a facade behind which teems an array of decentralized and empowered groups, subplants, regularly connected with similarly diffuse working units around the world. We

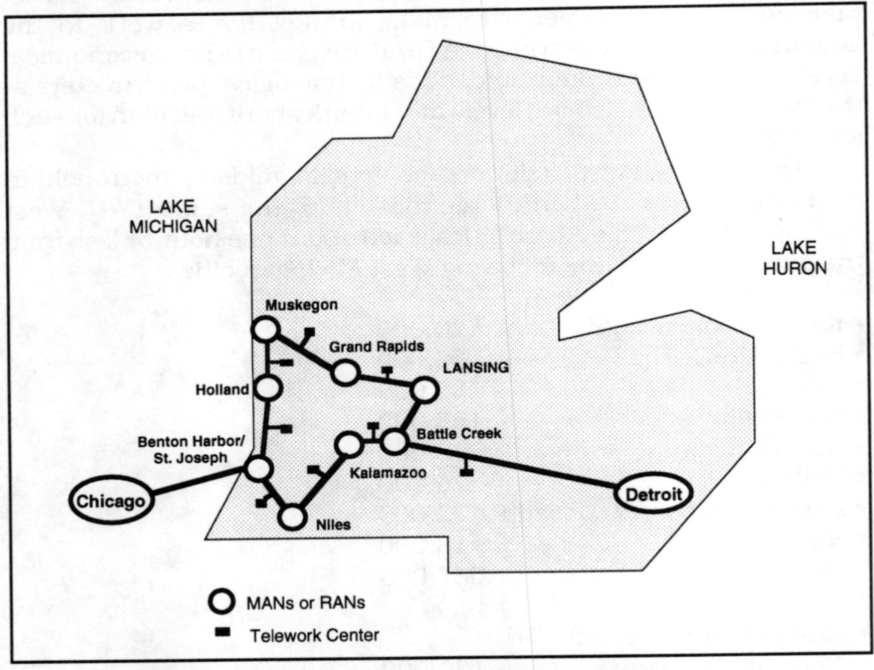


*Figure 8-6: The Opportunity Infostrada of Western Michigan (The Targowski Model)*

live in an increasingly global economy, one in which our lives increasingly intertwine with those of people in other countries.

We have witnessed the emergence not only of global corporations, but also of global citizens. Global citizens get information about almost everything from all corners of the world and make decisions about their personal needs and priorities. During 1988, nearly 90% of all Japanese honeymooners went abroad. For the first time in two thousand years, the Japanese people are revolting against their government and telling it what it must do for them. Omache (1991) argues that when a GNP exceeds $ 10,000 per year per capita, religion and government become declining "industries." Today, when the most developed countries' citizen produces $25,000 of GNP, people want to buy the best and cheapest products, without regard for where in the world they are produced. Japanese leaders used to say American and Australian beef was too lean and too tough to chew. But Japanese travelers have tasted it for themselves and discovered that it is cheap and good. American consumers developed a taste for Japanese cars, despite the Lee Iaccoca's claim about the "Chrysler Advantage."

In 1988, the total value of goods traded between the United States, Japan and Europe amounted to $600 billion annually. In contrast, foreign exchange trading amounted to $600 billion daily (Ohmae, 1991). In other words, trade in "soft" goods exceeded 300 times the trade in "hard" goods. This is a fruit of the Electronic Global Village (EGV). With an electronic highway linking our countries, the world is truly one. Since cargo is less transported than information, a teleport becomes more important than a seaport. A city with a teleport has more chances to thrive than a city with a seaport.

The beneficiaries of these new tools and behaviors will be those communities which share capacity for accessing and processing data, information, knowledge and wisdom.

## Fiber-to-the-Curb/ Video on Demand

A television set can be a futurist's dream. With only a few touches of a remote keypad, a viewer can select any of several dozen first-run movies, exercise videos or instructional tapes - a mini-video store in the living room. Or, with the flick of a button, a TV set can become a video phone. A camera and microphone mounted on the set allow the user to see a similarly equipped caller at the other end of the line and vice versa. The Southern California community of Cerritos is one of handful in the nation with homes equipped with just such experimental telecommunications equipment, installed by a telephone company, GTE Corp. This TV set is more than gee-whiz, state-of-the-art technology. It represents the determination of the telephone industry to play the leading role in the Interactive Age, one in which video programs, phone calls, electronic shopping services, computer data and other information would be brought into and out of the home over high-capacity "optical fiber" transmission lines.

Because a single fiber-optic cable can carry many times the electronic information contained in the typical copper line, customers would have video access to classrooms, shops, and doctors. Fiber-optic cable place the contents of whole libraries at their fingertips and make thousands of movies and TV shows a simple click away. Telecommuting or distance learning at home will become common.

This is the future vision promoted by the nation's phone companies, which would like to spend between $200 and 400 billion to install and control this super conduit. They expect to generate billions of dollars in revenue from information services (every thing from audio horoscopes to computer databases) and entertainment. They have taken dead aim at replacing another industry whose transmission lines already enter America's households, cable TV.

In October of 1991, a federal court removed legal prohibitions that kept the Regional Bell Operating Companies (BOC) from providing information services over their local phone networks. The court effectively decided that the Bell companies can sell everything over fiber-optic lines but television programs and video services. Income from these services would offset the investment required to install fiber-optic lines. If telephone companies are allowed to provide TV and video programming services, the cable industry will go out of business. It is highly unlikely that two parallel lines will supply information from and to a home.

The cable industry is promoting a far different vision of the future. Their "multi-wire" scenario has direct-broadcast TV, satellite transmission, cellular phones, multimedia computer software, and cable TV coexisting. It is this future of competition that the Bell companies fear most. On the other hand, the cable industry is not subject to effective competition either. Even if the regulatory roadblocks were lifted, the phone companies still face the task of replacing their existing copper-wire networks with fiber-optic lines. That job is underway, although slowly. Local phone companies are gradually installing fiber-optic lines where copper lines have worn out. At its current pace, a nation-wide replacement of the copper lines will take about 40 years. If the telephone companies could sell video programming, they would have the proper financial incentive to replace those line by the year 2010.

By extending optical fiber into the home and combining it with the phone companies' sophisticated switching systems, every household could become a part of the Interactive Age. A fiber-optic network will change the way people live their lives. Some critics argue that installing a fiber-optic line into every home is too expensive, like replacing every residential driveway with a superhighway. The

Federal Communications Commission (FCC) supports an idea of "video dial-tone", in which the phone companies are builders and operators of TV-transmission facilities that could be leased by all comers.

While the debate rages, cable companies are touting their own experiments with fiber optics. By adding signal-squeezing digital "compression" technology, and installing fiber-optic lines in the main trunks of a cable system, avoiding the huge expense of wiring every home, in a few years cable systems could be outfitted to offer as many as 400 or 500 channels of service. (Farhi, 1992). Imagine that you are watching a film on a cable TV and the phone rings. You do not reach for a phone, just mute the movie and you take a call through the wireless remote unit which is a subsystem of a cable TV. The CATV companies taking on the phone companies? One of the most significant deals occurred on February 18, 1992 when Tele-Communications Inc., the country's largest cable operator (see Table 8-4) agreed to buy 49% of Teleport Communications Group, Inc. from Merrill Lynch. Teleport has built a $100 million per year business by interconnecting major buildings and corporations in downtown with fiber-optic cable and linking customers to other phone networks. Teleport, in short, brings TCI and Cox (another co-buyer of Teleport) into direct competition with local phone companies. The cable TV industry is changing from being a video entertainment source to being a full-service telecommunications supplier.

The cable TV companies already have a substantial lead over the telephone companies in the race to put smart wires into American households. 60% of American homes are currently served by cable TV, and another 33% can be easily hooked up. By contrast, the telephone companies have installed their wires into 93% of American households. However, these penetration figures are misleading. In terms of the quantity data which can be passed through the wire, the cable TV companies are out in front of the telephone companies.

The arteries of a cable TV system are the coaxial cables that run

| Name | Number of subscribers (millions) | Operating Cash Flow 1991 ($millions) | Cable as percent of operating cash flow 1991 |
|---|---|---|---|
| Telecommunications Inc | 11.3 | 1,540 | 95 |
| Time Warner Inc. | 6.7 | 2,630 | 39 |
| Comcast Corporation | 2.8 | 309 | 92 |
| Jones Intercable | 1.7 | 49 | 100 |
| Cablevision | 1.6 | 269 | 100 |

Source: Moran & Associates

*Table 8-4: The Public Top Five Cable Companies*

from the center of the system, the so-called headend, out to subscribers' homes. Over long distances, about every quarter-mile, coaxial cable requires an amplifier to boost the signal and compensate for line resistance. These amplifiers create electronic noise and virtually prohibit efficient two-way communications on coaxial cable. Therefore, the TV shopping channels require an additional telephone call to place an order. Over short distances, for 300 feet or less into the home, coaxial cable requires no amplifiers. Thus, across the so-called drop, the distance from the curbside cable into the living room. coaxial cable can now handle as much data as fiber-optic cable, and far more than a twisted-pair copper telephone line. Coaxial cable technology on short distances is called "broadband pipe," capable of transmitting at 1 billion hertz (1 gigahertz). One hertz is a wave per second. Compare this with the 4,000 hertz (4 kilohertz) of a twisted pair of a telephone cable. The former can transmit the entire contents of the Library of Congress within eight hours, while the latter would require 500 years. Current telephone system can support transmission of voice and data (narrowband), and coaxial or fiber-optic cable can support far more complex full-motion video, high-resolution medical images, vivid educational simulations and lifelike teleconferences. Comparing these two wires is like comparing a five car ferry with an eight lane bridge.

The cable companies are increasing their fiber-optics base much faster than the telephone companies. The former has doubled fiber mileage every year since 1988 and will have 22,000 miles by the end of 1992. While telephone companies today expend around 7% of their investment budget for fiber optics, this is half the cable companies' expenses (in %) for the same installation. The cable companies install fiber-optic lines from the headend to clusters of homes and then tie the fiber into the neighborhood coaxial system. A cable operator can deliver a 1 gigahertz signal almost without amplifiers. Two-way traffic can be switched at the cable system's headend, just like phone trafiic.

As long as the twisted-pair copper connections remain, phone company fiber does not substantially improve the bandwith of service to the home. There is no point in sending a tidal wave of information down a broadband fiber line and into the neck of a tiny bottle. In short, a cable system can turn itself into a supplier of a wide range of broadband services, including wireless phone and computer video, at relatively low cost. Regardless of the outcome of the cable/ phone war, one certainty remains. The telephone, television and computer are rapidly merging into a single, very intelligent box, a telecomputer. This telecomputer will be linked to the rest of the world by high-capacity smart wires (Gilder, 1992).

Presently, the cable companies have moved these broadband wires closer to homes than have the telephone companies. On the other hand, the telephone companies are better positioned in downtowns, where they install fiber-optic rings. From the consumer point of view, there is a need to integrate the downtown's office-driven fiber-optic ring, as the backbone in a Digital Loop Carrier (DLC), with a home-driven coaxial TV cable into a hybrid architecture for an Integrated Broadband Network (IBN). Technically, this is possible with MAN technology, utilizing a bus broadband pedestal, a set of 4-way passive spliters, radio frequency combiners, and amplifiers. Some third party has to initiate the deal, since discussion between these two utility vendors could be perceived as the cartel conspiracy. The strategy of the third party would be to combine the two way capability of the phone system with the broadband capacity of cable. Such a strategy has already been applied by the Pacific Telesis Group.

During the FCC cross-ownership inquiry, telephone companies have made clear their desire to acquire existing cable TV systems. The First Amendment is likely to be necessary to maintain open access to the medium. Public control (via Community Access Channel committees) is likely to be necessary when the means of communication are concentrated, monopolized, and scarce, as may be the case with telephone and cable TV networks.

A combination of fiber and switching investments are needed to deliver one-way switched video services over either telephone or cable company networks. Switched video services are attractive, enabling television to become a more democratic medium, capable of providing access to literally thousands of video publishers. The most advanced cable systems can carry at most 80 channels per cable. In the future they will carry not more than 150, assuming that channels continue to occupy 6 MHz slots. In contrast, video rental stores stock up to 10,000 titles to satisfy customer demand. Networks providing switched video can provide access to an essentially unlimited number of video programming sources (movie studios, television networks, video brokers, small video publishers) unrestricted by 24 hours or less of broadcasting. Electronic bulletin board systems and audiotext services demonstrate a precedent for such a proliferation of content providers.

## Information Kiosks

While browsing in a suburban San Diego's Parkway Plaza Mall, an unfamiliar contraption catches a shopper's eye. "Touch my screen," implores the video monitor in the seven-foot structure wedged between a T-shirt cart and a lottery ticket booth. Intrigued, the shopper places his/her index finger on the glass. Suddenly, California Governor Pete Wilson appears on the screen and talks about the many services of Info/California. This videocomputer offers: job listings, information on HIV testing, an electronic application for a fishing license, and more. A shopper selects data on area beaches. "Cool" he/she says as he/she tears off a printout.

With familiarity bred from the use of automatic teller machines, many business and government agencies believe that Americans are ready to retrieve information and order products via information kiosks. Housed in wood-and-plastic cabinets and equipped with touch screens and simple menus, info kiosks are popping up everywhere - in supermarkets, auto showrooms, malls, and schools. Market researcher Inteco Corporation indicates there are about 59,400 such kiosks located in the United States today. By 1996, Inteco predicts there will be 2 million, making them more common than gasoline pumps.

That forecast is based on a new breed of multimedia machines. Unlike the electronic kiosks that have been gathering dust in airports for a decade, these machine have colorful graphics and sound and allow consumers to conduct transactions. For example, from the 15 Info/California kiosks around San Diego and Sacramento, it is possible to order a copy of a birth certificate or get health care data. One kiosk in riot-torn South Central Los Angeles is helping citizens find community-assistance programs. The citizens of California will be able to obtain copies of birth certificates, renew their driver's licenses and reserve campsites in state parks all in the same place—a sort of "one-stop shopping" for government services.

According to Martinez (1992), information on any of 90 subjects is available to citizens by following the instructions of the on-screen narrator and the graphic display to select topics. Touching a picture (icon) on the screen activates a videotape that gives the requested information in Spanish or English. The categories currently available are employment, the legal system and business, education, health, environment and natural resources, transportation, family and children, and general assistance. Under transportation, citizens obtain information about California's boating laws and driver's licenses, how to register vehicles, join Rideshare programs, and for applying for license plates and placards for the disabled. A special interactive program on the system allows citizens to access the state's "Job Match" program. By touching a series of screens, individuals define the type of work which interests them. The computer presents the number of jobs currently available in the selected category and location, and the salary range. This job information is updated daily. The individual can use the "keys" on an on-screen keyboard to complete an application form, which the system prints out along with the name and address of the nearest employment office. The applicant can take the completed form to the employment office, thus saving time for both the applicant and the interviewer. This process usually requires a long, face-to-face interview.

Info/California is a product developed by the state in partnership with the IBM Corporation and North Communications. From 1988 to 1992, the state has spent approximately $300,000 on the project, while IBM's out-of-pocket expenses, not including personnel costs, have been $750,000. In addition, for the pilot project IBM and North Communications lent the state software and equipment valued between $3 and $4 million. The network utilizes state of-the-art multimedia technology incorporating video, computer graphics, high-quality stereo sound and computer touch screen operations. IBM first developed this technology for its "24-Hour City/Hall County Courthouse" project in late 1980s. The goal of this project was to make government more accessible to the public by providing access in convenient locations, not just local government offices, and by expanding delivery of government services beyond the 9-to-5 work day.

Working with Public Technology Inc. (PTI), a non-profit technical arm of the national League of Cities, the International City/County Management Association and the national Association of Counties, IBM developed multimedia touch screen access systems for a number of local governments. PTI is developing a Well-Connected Community Catalog that brings together products and services to help cities and counties become well connected. The first installment involves six PTI partners, providing technical assistance for local governments interested in 24-Hours City Hall Applications:

- Floyd Design, Atlanta, GA
- Interactive Design & Development, Blacsburg, VA
- Interactive Information Technologies, Salt Lake City, UT
- Joel Wittkamp Design, Raleigh, NC
- SGX Communications, Baton Rouge, LA
- Technology Applications Group, Troy, MI.

The California system is based on "Hawaii Access" a dramatically successful experimental touch screen network started in 1990. Hawaii's multimedia network provides health, human services and employment information in English, Samoan and Ilocano (spoken in the Philippines). During its six-month pilot project, the Hawaiian system was used by 30,260 people, 216% ahead of projections for use.

Californians seem to like Info/California, too. In its first four months, 54,357 people used 15 kiosks, an average of 52 contacts per day per kiosk, and 10% higher than projections. Half of the users said Info/California saved them a phone call, letter or trip to a government office, and 79% found the system easy to use. Most of the people (57%) used the system outside the regular weekday work hours. The most popular topics were Job Match, student aid, beaches, driver's licenses, the California State University system, in-demand jobs, community colleges, state parks, the University of California system and AIDS/HIV. In the second phase of the project, during the spring of 1992, citizens began to pay for services. For example, they can use credit cards to pay for traffic fines.

Other states are starting to use kiosks for some services:

- In Anne Arundel County, Md. a visitor can touch the "Your Voice in Government" option on the screen, and point out his residence on a map, and a picture of his council representative appears on the screen, delivering a taped message. In Plano, Texas' Municipal Center in the Mall, "Fireworks," explains fireworks regulations through a colorful, dynamic display.
- In February 1992, New Jersey began offering "TAG, the Motor Vehicle Self-Service Helper" to automate automobile registration renewals. Ten machines placed around the state allow users, for a small fee, to connect with the state vehicle database and receive copies of vehicle registration cards on the spot. The TAG system was developed by the NCR Corporation.
- Delaware, Maryland, New York, Pennsylvania, Virginia, West Virginia and the district of Columbia offer job information through Automated labor Exchange (ALEX) kiosks, financed by the United States Department of Labor. Job seekers have access to lists and descriptions of job openings around the country. More than 1,000 people were using the machines in Virginia each week.

Business is also adopting kiosks. Four music stores in California have installed Note Stations, kiosks that sell sheet music, with the scores printed in any key. Music Writer, Inc. in Los Gatos, California, plans to install its kiosks in 1,800 stores. At deli counters in New England, shoppers can order their favorite sandwiches and cold cuts. At Expo '92 in Seville, Spain, visitors used multimedia systems to preview pavilions, find their way around, or make reservations. At auto shows and at some dealers, tire-kickers can get details on new Cadillacs.

While kiosks are a fairly small business today, only about $250 million annually in 1992, giant IBM is aggressively pursuing it. The IBM programmers at the Thomas J. Watson Research Center study the best "human interface" for the masses. The results ar visible on Info/California screens and at the 320 kiosks IBM built for Expo '90 in Seville. If they catch on, kiosks could radically alter the delivery of service. They can replace $6 to $8 per hour workers. Workers can only memorize a limited amount of product and pricing data, but computers can master torrents of it. In Toronto, Allstate Insurance Co. sells policies at Sears and Food City stores. These are convenient locations for people who do not want to run all over the place. Instead, they may just be running down to the corner kiosk (Schwartz et al.,

1992).

Future kiosks will provide access to all levels of government service, from city, county, state, to federal. Integrated systems will allow citizens to resolve problems, obtain quick answers to their questions, and order, pay for and obtain government documents without the need for locating the correct government office and getting there during working hours. With hundreds of kiosks handling routine questions, the municipal bureaucracy can function better with fewer workers. A fee per transaction will recoup some capital investment cost.

Services available through information kiosks will greatly simplify and expedite government service by providing a "single face" to government to all citizens. By relegating to kiosks the repetitive tasks of answering the same questions, such as which office to contact, which document to bring and how to fill out forms, government workers are available to meet people's special needs. This advanced technology can reduce the costs of administering government services and increase the ease with which citizens interact with governments. This technology is a tool to improve government efficiency and productivity while "bringing government directly to people's fingertips" (Martinez 1992).

## High-Tech Business of the Future

The high-tech business of tomorrow can be illustrated by the hotel business. The "new" hotels are responding to the greater sophistication of travelers and increase in business travel through information technology. Jerome E. Klein write that in such hotels, "guests unlock the door of their room with a credit card. Once inside, they can use the television set to order breakfast, keep tabs on their spending in the hotel, receive messages, and even check out. If they have been guests in the past, a computer storage has recorded their 'history,' including personal preferences and special needs, if any."

Guests equipped with a sensing device will enter future hotels without checking in. The sensing device, in the form of a microdisc, could be worn on the lapel, kept in the wallet, or even attached to a watchband. The microdiscs could also be attached to luggage, enabling lost luggage to be located instantly anywhere in the world.

With increased business travel, the guest room may increasingly be equipped to serve as an office as well as sleeping quarters. Many hotels in Asia have already installed executive centers featuring secretarial services and state-of-the-art conference rooms. A Tokyo hotel even offers cocktail evenings at its business center, so local executives can meet foreign business guests of the hotel.

The trend toward more-specialized hotels will continue. Some hotels will be huge "cities" housed within skyscrapers, while the opposite extreme will exist in intimate deluxe hotels, inns, and lodges. Technological innovations will not eliminate hotel jobs, but rather free up hotel staff to spend more quality time with guests. Guests will continue to demand service, the human contact that electronics can never replace Klein (1989).

# The Electronic Town

### Teledemocracy

Teledemocracy is a catchword for the establishment of direct democracy through the use of communications media. The question is not *whether* the new media will influence our politics, but *how* (Arterton1987). Imagine it is 1994, the economy is still stagnating. Japan remains in the doldrums as well, interest rates are rising, and the deficit has reached $600 billion. Something has to be done. The president organizes an electronic town meeting of the whole nation. Even before his presentation is over, the returns begin to pour in - by telephone, fax and two-way interactive cable TV. By morning, the will of the American people is clear; they have decided to cut back on Social Security payments, further slash military spending and raise their own taxes. This is how teledomocracy could work. Fed-up with the paralysis of Congress and the special-interest outrages that characterize politics as usual. the idea that citizens could bypass some of the musty machinery of representative democracy and directly influence the government is enormously attractive.

Cable operators should allocate $20 to $30 billion on digital-compression and fiber optic technology to prepare their systems for interactive audiences. The telephone companies, for their part, would have to invest $300 billion to $500 billion in a fiber-optic network before they could deliver TV-quality pictures into every American's home. However, it is not necessary to have interactive TV or videophones to call a town meeting, just a standard TV and some form of telephone communication, fax, phone or modem. But the long-distance phone networks assume that everybody will not call all at once. If masses of people dial simultaneously, the lines quickly get jammed. After President's Bush's State of the Union address in January 1992, for instance, CBS broadcast a toll-free 800 number and invited viewers to respond to questions posed by Charles Kuralt. Of 25 million calls that were made, only 315,000 got through.

For more freewheeling discussions, the president, or mayor, might plug into an electronic-mail or bulletin-board system, such as Prodigy, CompuServe, or Free-Net. These interconnected matrices of computers allow participants to exchange written opinions at any time and from any place without the need to meet face to face. On such networks, future elected officials could quickly tap the views of ordinary citizens and specialists at universities and think tanks across the United States or a metropolitan area.

There are some drawbacks to electronic communication, some of which do not bode well for teledemocracy. Without the visual cues that are so abundant in personal meetings, people behave much differently. Without access to facial expressions which might indicate that you are hurting someone's feelings, it is easy to drive a point too far. On Prodigy, for instance, there have been outbreaks of anti-Semitism and even mass paranoia, that IBM's central office is spying on users' files. The potential for good and mischief is very high. Indeed, Hitler pioneered the electronic referendum, using radio broadcasts to drum up votes for plebiscites supporting his rise to power. Later, through the same vehicle, President Charles De Gaull lost his presidency.

The Founding Fathers of the United States lacked computers and cable TV. They did have some experience with crowds and mass behavior. From this they concluded that people were to easily swayed by passion to be entrusted with direct democracy. The government they fashioned was not a national town meeting, in which everybody votes on issues, but a representative democracy, in which lawmaking power is entrusted to elected officials and constrained by a system of checks and balances to ensure that decisions are not too hastily made.

However, the United States may eventually adopt some forms of electronic government. The challenge to the nation or metropolitan citizens will be to use this new technology to support representative democracy, and not subvert it.

**Bringing the Local Government to Citizens [2]**

The Kalamazoo - Telecity USA can serve as the foundation of an electronic forum in which people communicate with each other and with city leaders informally by Electronic Mail to facilitate civic service and community improvement projects. Electronic Bulletin Boards and Electronic Conferences can be developed to inform city residents about current and future events, city improvement plans, and ideas for future activities, and to allow each citizen a voice in discussing the relative merits of particular ideas and approaches.

A neighborhood leader, faced with a problem, would be able to use the network to reserve a public meeting room using the Electronic Facilities Reservation service, and then an "Electronic Telephone Pad" to instantly send notices of the meeting to the 100 households in the neighborhood.

Information on ordinance requirements, licensing and zoning regulations, and many others could be routinely communicated to citizens by electronic media. Electronic availability would save the City the printing and postage funds currently required to diseminate this information, while personalizing communication with its' citizens.

## Bringing "the People" to Politics

The TeleCity might include activities registration performed through the network, a volunteer registry, and distribution of town minutes and other civic information without the delays and expense associated with printing and mailing.

A few experiments have been conducted in which a modest number of citizens participated in electronically mediated policy discussions. Usually, public officials at one site converse with citizens at remote locations. For example, a congressman in Washington holds electronic office hours with constituents in California, a legislative hearing in Juneau, Alaska, enables citizens on the North Slope to testify without leaving their villages. The first Electronic Town Meeting (ETM), MINERVA , was conducted in 1973 in New York City, during which the cable system's 3,600 apartment inhabitants joined the ETM without leaving their apartments. Another early ETM took place in Hawaii in 1978, utilizing theater to dramatize issues and capture the citizens' attention. The first Televotes system was applied in California in 1974 and in Hawaii in the 1980s. Both processes increased citizens' participation by about 35%. The agenda, however, was controlled by the organizers.

## Experimenting with Referenda

In 1978, the Qube system was applied to a plebiscite on the future of Upper Arlington in Columbus, Ohio. This Electronic Town Meeting attracted 2,500 out of 32,000 (just 8%), and only 4% actually voted. As a result, the politicians and commissioners rejected the construction of an aerial highway. The work involved in ETM was so exhausting that it was five years before another meeting of this type was suggested. The Health Vote '82 in Des Moines, Iowa, to discuss the options available to limit the rising cost of medical services was publicized in newspapers, television and radio. Almost 24% of the mailed ballots were returned. The public relation campaign was successful, since 76% responders had heard about the issue's options. The options selected were support of HMO-like organizations, and an increase of the deductible amount on health insurance. The public rejected cost-cutting measures, such as cutting support of poor and for medical research. The vote indicated an increased degree of sophistication among the respondents. A similar ETM project was organized from 1974-76, called Alternatives for Washington (State), or AFW.

Governmental plebiscites have a broader reach than those teledemocracy projects aimed at encouraging dialogues. Participation in these plebiscitory contexts appears lower than turnout for candidates (Magleby 1984). When both appear on the same ballot, many voters ignore the referenda questions.

### Empowering the Governed Rather Than Governors

The rapid changes of the Interactive Age upset established institutions and created elites. As instant information and instant markets erode the power of governments, so they erode the power of corporations and elites. The technology of the Information Age empowers the governed rather than the governors. In its constant churning, today's technology has a dark side to be conquered, but its bright side offers the hope of an era of liberation. Certainly technology is not everything; these opportunities must be exploited by the human spirit. We have the ability to build new versions of the institutions of the *Electronic Bell Époque* of the 1990s, and rekindle its spirit.

## General Architecture of Kalamazoo-TeleCity USA

The Kalamazoo TeleCity USA embraces the following goals [1]:

A. Business-intensive:

1. To increase social and business responsiveness through enhanced communication and interconnectivity, both locally and globally: e.g.: Electronic mail, teleconferencing, EDI, business TV provide instant communication between many participants at costs much lower than travel. The Upjohn Company saves about $400,000 annually on trips made unnecessary by teleconferencing. It facilitates quick response to business crises, like the worldwide concern raised recently regarding possible side-effects of Upjohn's Halcion. Company executives used global teleconferencing as a crisis management tool in conveying the urgent business policy and image management.
2. To increase social, business, and individual productivity through user-friendly information processing, handling, and access: e.g.: Electronic mail and word processing significantly increase the speed and scope of communications and documents creation as well as boosting its quality. Office automation accomodates telecommuting.
3. To increase business and individual competitive advantage using techniques for doing business better and in innovative ways: e.g.: A terminal in a customer office provides an edge in existing markets. Linking field units to the head office allows service staff

to keep in touch with the office (and vice versa). New services can be created by network interconnections.

B. Quality of life-intensive ( ILSG-California, 1991):

1. To improve democratic governance via effective participation by residents in municipal decision making; e.g.: Meeting notifications, staff reports, policy outcomes (distributed by E-mail services, such as Santa Monica's PEN, and accessed by computers and telephone (the State of Hawaii), video coverage of city council and commission meetings can be distributed over Community Access Channels.
2. To enhance inter-community cooperation which could diminish the need for regional agencies that frequently infringe on local self-determination: e.g.: Growth management, transportation, crime prevention, and economic development can be facilitated by E-mail, teleconferencing, business TV, and BISDN services. The Portland Metropolitan Communications Commission coordinates a 14 community INET; a five city joint powers agreement for cable TV franchising in the Palo Alto area; Reno and Washoe County project to build a six mile fiber optic network.
3. To ensure delivery of municipal services, especially to low and moderate income constituents, and those with limited mobility: e.g.: Action for the 90's predicted that cities will need to accept increased responsibility for social conditions that involve safe neighborhoods, citizen participation and youth and senior programs. Some school districts keep home bound students "in school" through live, interactive distance learning. In other cities, health clinics are linked to major hospitals through fax networks to improve speed of diagnoses.
4. To reduce air pollution and transportation, traffic congestion and **energy consumption** without stopping the physical development necessary to accommodate the growth projected from natural increase and immigration: e.g.: Innovation is necessary to protect the quality of life. Local governments will have to provide affordable housing, natural resources, infrastructure, waste disposal/recycling, and open spaces while minimizing the environmental costs.
5. To encourage economic development within environmental constraints: e.g.: Start-up and other small businesses, educational institutions including universities, colleges, community colleges, primary and secondary public schools and specialized schools and associations that offer continuing education for:
   - professionals
   - libraries
   - business resource organizations such as chambers of commerce, downtown development associations, or tourist bu-

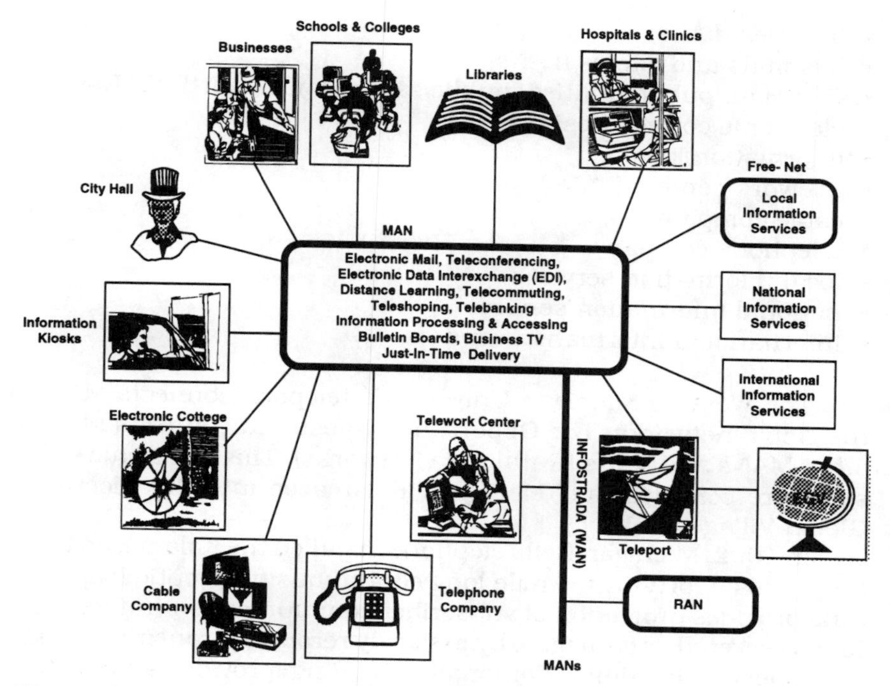


*Figure 8-7: The Architecture of Kalamazoo TeleCity*

reaus

- convention centers, airports or sea ports (found in large cities)
- social services and other nonprofit organizations providing vital community services
- city and county governments, whose success at solving traffic congestion problems, maintaining low taxes with high service levels and reducing per capita energy consumption are prerequisites for economic growth

6. To empower local governments to respond to the information and communications needs entrepreneurially: e.g., A network of info kiosks throughout the metropolitan area could be established. By turning over to the kiosks the repetitive tasks of answering the same questions, such as which office to contact, which documents to bring and how to fill out forms, government workers can attend to people's special needs.

Figure 8-7 shows the general architecture of the Kalamazoo - TeleCity USA.

The main component of the TeleCity is the Municipal Area Network (MAN) which interconnects all metropolitan users where they live, work, shop, play, meet, commute to work:

- businesses (LANs)
- schools and colleges (LANs)

- libraries (LANs)
- hospitals and clinics (LANs)
- City Hall, public utilities, public safety, and so forth (LANs)
- electronic cottage (residences)
- information kiosks
- telework centers (LANs)
- cable company
- telephone company
- local information services
- national information services
- international information services

The Western Michigan University's teleport connects MAN to the global networks. The Opportunity Infostrada connects MAN to other MANs and RANs (Rural Area Networks). This electronic infrastructure converts the TeleCity of Kalamazoo into the Electronic Global Village (EGV).

In 1992, Michigan Bell Telephone installed the Kalamazoo Fiber Ring. It is a forty- three mile long continuous fiber-optic loop. The Ring provides protection of subscriber communication traffic, if the Ring is severed or damaged, by instantly rerouting transmissions in the opposite direction. Additionally, the Ring provides subscribers with a host of enhanced services and features, such as:

- increased capacity
- security
- twenty-four hour network monitoring
- video conferencing
- high-speed, high-volume transmissions
- clarity of transmission
- Synchronous Optical Network (SONET)
- Fiber Distributed Data Interexchange (FDDI)
- Integrated Services Digital Network (ISDN).

The alliance of the telephone, television, and the computer is accelerating the introduction of a vast array of electronic information services operating on the telecommunications network. For example, France Telecom's Teletel provides an on-line gateway to 12,000 different databases in nearly one in five French households. Similar videotext systems exist in the United Kingdom, Germany, Japan, and the United States.

Electronic infrastructure enables the operations of the following telecommunication and computer services, including:

- electronic mail
- teleconferencing

- Electronic Data Interexchange (EDI)
- distance learning
- telecommuting
- teleshopping
- telebanking
- bulletin boards
- just-in-time delivery
- information processing
- information accessing
- payment of taxes, purchasing of permits with credit cards
- book, video, and tapes reservations from library collections
- credit card verification
- automatic health care claims processing
- integration of multidatabases to single patient record
- private network management
- personal communications service
- many others.

With the development of Kalamazoo - TeleCity USA, new applications will emerge and become "irreplaceable."

## Strategy of Kalamazoo - TeleCity USA Implementation

TeleCity Aims are defined as follows [1]:

- Mission: To improve economic and community development, create opportunities for citizens and businesses, and enhance the City's performance into 21st Century.
- Goal: To establish the City's leadership role in defining policy, direction, projects and funds for emerging Telecity of Kalamazoo.
- Spirit: Cooperation to compete globally
- Results: Gradually build components of the telecity in five to ten years.

The strategy of Telecity is based on the following premises:

1. Common electronic infrastructure, shared by citizens and organizations such as:
    a. A common manu to a network of networks
    b. A Metropolitan Area Network (MAN).
2. Set of comprehensive inter-organizational information and communications services.
3. Different user devices for the delivery of same information and communication services.
4. Easy acces to the national and global telematic services (comput-

ers + telecommunications + television) services.
5. Affordable access for all citizens and organizations.
6. Users team work which is prioritized and supported by the local governments.

Most everything proposed by the Telecity task forces can be technologically accomplished today; and the technology improves almost daily. Most proposed services are already available. The technological challenge is to integrate isolated activities into a larger whole for broad community benefits. The business challange is to deliver quality services with high reliability and performance level — at a cost people are willing to pay [4].

The **benefits** will depend on the features available and the amount of usage. For France's Mintel experiment, only around 6% of the citizens make full use of the system. If substantial TeleCity benefits are to be achieved, a massive County-wide training program will be needed to acquaint people with all of its services. The potential benefits are:

1. Improved educational delivery system
2. Wide-spread access to many databases
3. Improved intelligence and skills of people entering and in the work force
4. Increased attraction for companies to locate in the area due to the benefits from EGV and TeleCity and a skilled workforce
5. Additional delivery jobs, if people order groceries and other products through TeleCity
6. Increased administrative efficiency via instant communications
7. Improved democratic governance through teledemocracy
8. Reduced air pollution, traffic congestion and energy consumption through telecommuting
9. Encouragement of economic development within environmental constraints
10. Ensured delivery of municipal services, especially to low and moderate income constituency (Info/Kiosks, Community Access Channel, health care, programs for mothers and children at risk)
11. Closer to balanced budgets from public information utility income
12. Increased inter-city cooperation, through the Opportunity Infostrada and video conferencing, electronic mail, and data and graphics communications capabilities, available to city representatives, business organizations and the general public.

**The premises of the telecity**: Imagine, you had a device that combined a telephone, a TV set, a camcorder, and a personal computer. No matter where you went or what time it was, your child could see you and talk to you, you could watch a reply of your team's last game, you could browse the latest additions to the library, or you could find the best prices in town on groceries, furniture, clothes — where you needed. Imagine further the dramatic changes in your life [5]:

- The best schools, teachers, and courses were available to all students, without thought to geography, distance, resources, or disability;
- The vast resources of art, literature, and science were available everywhere, not just in large institutions or big city libraries and museums;
- Services that improve America's health care system and respond to other important social needs were available on-line, without waiting in line, when and where you needed them;
- You could live in many places without foregoing opportunities for usuful and fulfilling employment, by "telecommuting" to your office through an information highway, instead of by automobile, bus or train;
- Small manufacturers could get orders from all over the world electronically — with detailed specification — in a form that the machines could use to produce the neccessary items;
- You could see the latest movies, play your favorit video games, or bank and shop from the compfort of your home whenever you choose;
- You could obtain government information directly or through local organizations like libraries, apply for and receive government benefits electronically, and get in touch with government officials easy; and
- Individual government agencies, businesses and other entities all could exchange information electronically — reducing paperwork and improving services.

The development of telecity is not an end in itself; it is a means by which Greater Kalamazoo can achieve a broad range of economic and social goals. Although the telecity is not a "silver bullet" for all problems we face, it can make an important contribution to our most pressing economic and social changes.

• • • •

As a result of Telecity Steering Committee and Task Forces' work in 1993-94, the following functional architecture on Figure 8-8 is shown. There are five categories of functional components [1]:

1). **Users** with different interactive devices, such as interactive television, telephone, personal computers, and information kiosks. These input/output media will be in use according to users' affordability;

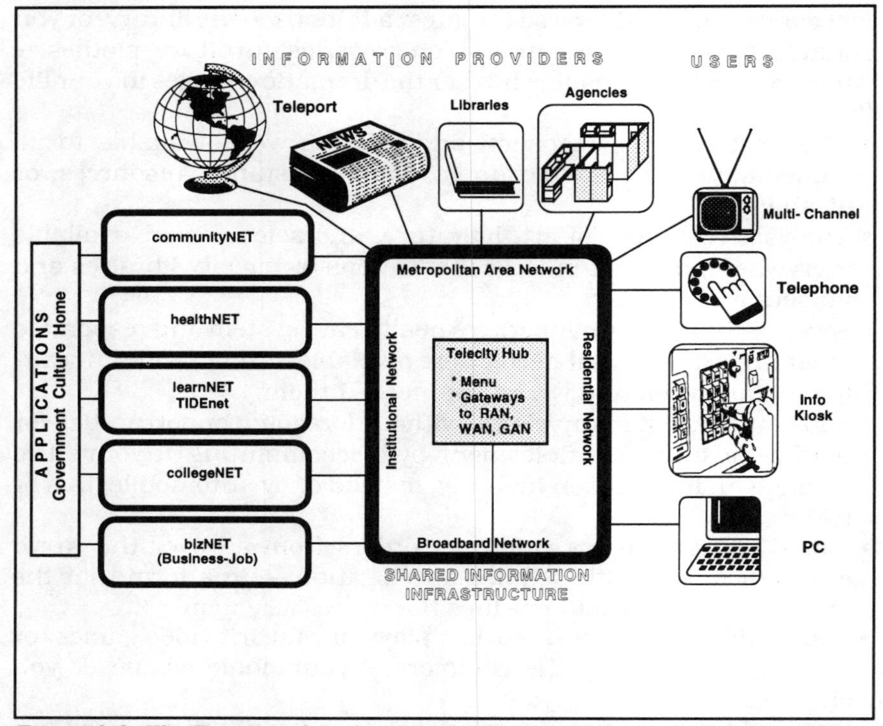


*Figure 8-8: The Functional Architecture of Greater Kalamazoo TeleCity USA (May 1994) (The Targowski Model)*

2). **Information providers** such as global, national, and local information providers (authors, polllsters, educational institutions, and various levels of government), libraries, and agencies, including governmental, schools, and universities;

3). **Information Networks** as: clusters of specialized users, information providers, and information/communication distributors.

4). **Applications** as: pre-network entities which can be implemented within other networks for governmental, cultural, and home services;

5). **Metropolitan Area Network (MAN)**: a broadband residential and institutional network which provides gateways to Regional Area Network, Wide Area Network, and Global Area Network.

The phrase 'telecity" or "local information infrastructure (LII)" has been defined as:

> *A wide and ever-expanding range of equipments including cameras, scanners, keyboards, telephones, fax machines, telcom switches, compact discs, video and audio tape, cable, wire, satellites, optical fiber transmission lines, microvave nets, televisions, monitors, printers, and much more in order to integrate and share communicated information.*

The Telecity will integrate and interconnect these physical components in a technologically neutral manner so that no one industry will be favored over any other. Most importantly, the Telecity requires building foundations for living in the Information Civilization and for making these technological advances useful to the public, business, institutions (e.g. libraries), and other non-governmental entities. That is why, beyond physical components of the information infrastructure, Telecity's value to users and the community will depend in large part on the quality of its other elements:

- The information itself, under the form of video programming, scientific and bysiness databases, sound recordings, images, library archives, and other media;
- The Telecity applications and software at citizens fingerprints to access, manipulate, organize and digest a prolifirating mass of information;
- The network standards and transmission codes that facilitate interconnection and interoperation between networks, and ensure the privacy of persons and the security of the information carried, as well as the security and reliability of the networks.
- The people—largely in the private sector —who create the information, develop applications and services, construct the facilities, and train other to tap its potential. Many of these people will be vendors, operators, and service providers working for private industry.

Every component of the Telecity must be developed and integrated if Kalamazoo is to capture the premise of the Information Civilization. In the 1990s, each Telecity component is in a period of either strong conceptualization or development. These are caused by the bifurcation which begun in the 1990s (globalism vs. tribalism) and by currents of cross-cultural communications and global cultures. A system definition of Telecity is as follows:

> *A Telecity is a set of information and telematic services and systems which are provided through the information infrastructure to virtual, electronic, and interactive organizations as well as electronic citizens in order to integrate and share communication-mediated information any time, anywhere in the synchronicity with events and accordingly with the rules and practice of cross-cultural communication and global culture.*

The first, most general Telecity's community information network is infoNET. As is shown in Figure 8-9.

This network of interactive communication is commonly called a freenet becaused the basic services in some cities are offered to the user without charge. The following list adapted from the National Public Telecomputing Network provides a scope of users and information providers:

- Individual citizens
- Community organizations and institutions
- Public and private schools
- Government
- Small, medium and some large size businesses
- The agricultural community
- Enhancement of existing services

Grass-roots networks are springing up all over the country, providing citizens with a wide range of information services.

The next important set of information and communication services should be develop within the health care sector. The healthNET shown in Figure 8-10 should support the following issues:

- stopping the rising cost of health care services

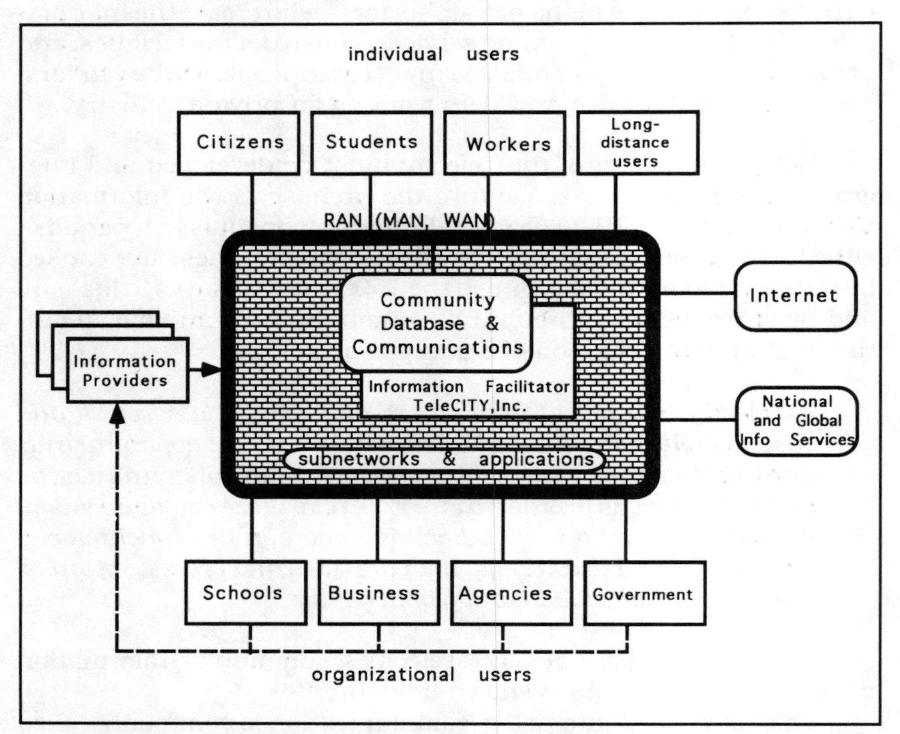


*Figure 8-9: The Architecture ofCommunity Information Network—infoNET (The Targowski Model)*

- improving the quality of care
- minimizing administrative costs
- improving the access to health care
- integration of health care services
- minimize malpractice claims and defensive medicine

The set of healthNET's applications is as follows [6]:

- Computerized Patient/Client Records
- Public Information and Education Resources for self-care, prevention, and early diagnosis. Example components might include:
  - Telephone hotline for refferals and access
  - Directories of health care organizations and resources
  - On-line information services
  - Mechanism for feedback and surveys such as bulletin boards and electronic mail
- Professional Information and Education Resources. Example components might include:
  - On-line databases
  - Clinical guidelines
  - Distance learning
  - Teleconferencing for timely consultations

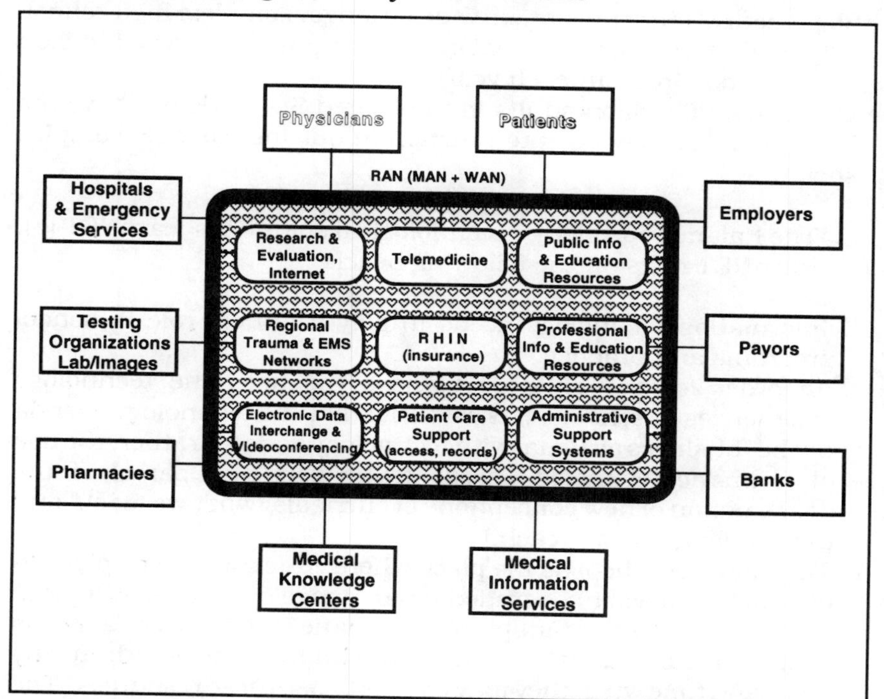


*Figure 8-10: The Architecture of the Southwestern healthNET (RHIN-Regional Health Information Systems: insurance) (The Targowski Model)*

- Alerts and reminers which process data from electronic records
- Telemedicine, it includes the following services:
  - Teleconsultation (remote patient care by video conferencing)
  - Teleradiology (transmission of X-ray images)
  - Telepathology (transmission of microscope slides)
- Medical Data Exchange, instead of using 1500 different claims form, standardized electronic submission and processing claims should limit 20 million health claims submitted annually in the Southwestern Michigan region
- Video conferencing
- Regional Trauma and Emergency Network
- Administrative Support which includes:
- Electronic mail
- Bulletin boards
- Transfer of records
- Electronic data interexchange
- and so forth

Increasingly, what we earn depends on what we learn. Americans must be well-educated and well trained if we are to compete internationally and enjoy a healthy democracy. The magnitude of the challenge we face is well known [5]:

- 25 percent of students nation-wide no longer complete high school, a figure that rises to 57 percent in some large cities (nearly 1 million high- school drop-out each year)
- Currently, 90 million adults in the United States do not have the literacy skills, they need to function in our increasingly complex society.

The Kalamazoo Telcity's technological solution is learnNET [6]. The learnNET rests on the following beliefs:

1). Information technology needs to play a critical role in school programs and practice.

2). In order for our children to have the need the technology experiences in schools, it is essential that the technology agenda not be "business as usual with technology added;" rather, the use of telematic technology needs to be part and parcel of the construction of new conceptions of curricula (what we teach) and instruction (how we teach.)

3. The image of school as *the* place where education takes place in our society needs to be traded for an image of schools as a place —albeit a very important place. Telematic technology enables us to deliver educational resources to children and youth at any place and time when they may choose to use these resources. The learnNET will enable us to avil ourselves of this opportunity.

4. The real benefit of telematic technology in the lives of children is

highly contigent on the quality of the applications. The most powrful network infrastructure can be used to send insipid—or worse—information. Substantial energy and talent will be required to construct powerful and productive applications.

Figure 8-11 illustrates the Targowski-Basco functional architecture of Southwest Michigan learnNET.

As will be evident, some of the applications below are immediately feasible, other will develop. The following applications of learnNET are recognized:

- TIDE —Technology in Distance Education from one site to multiple sites simultaneously
- Education Forum —a set of bulletin boards for parents, students, tewachers, and administrators
- Supercomputing—an access to supercomputers for simulations and visualizations in science and mathematics
- Business Education Link —for the interaction between employers and potential employees
- Internet —for the Electronic Global Village connection
- School-Home Communication; homework hotline, news of schools activities, grades, and so forth

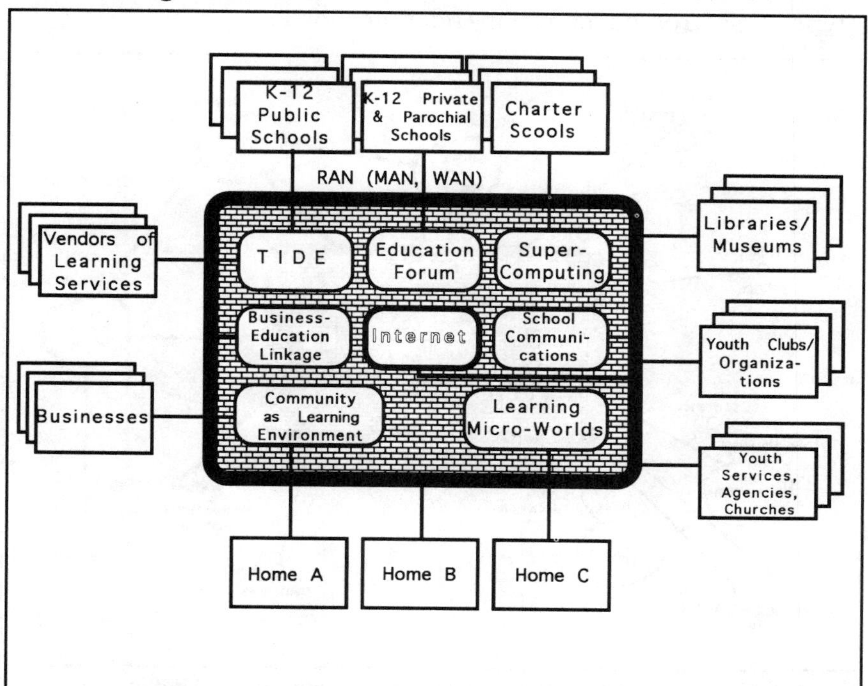


*Figure 8-11: The Architecture of learnNET (TIDE-Technology in Distance Education) (The Targowski Model)*

- Community as Learning Environment —provides information from throughout the community on learning opportunities
- Learning Microworlds—multimedia knowledge environments which will provide students the opportunity to explore information worlds in ways which are similar to Papert's "knowledge machine" concept.

A network which should establish higher education's role in exploring, developing, operating, and improving regional, national, and global knowledge and skills is the collegeNET. Its functional architecture is depicted on Figure 8-12.

The collegeNET will support the folliwng functions [7]:

- Telecommunications Connectivity (RAN: Regional Area Network), it is important that various voice, video, image and data will be transmittable over all connections among regional 27 colleges
- Research and Development (R&D) to solve grand challenges, enabling remote access to scientific instruments, supporting scientific collaboration
- Library Services (LS) to continue the expansion of existing library networks and on-line catalogs, a full-text, an access to "pay-as-you-see" reference databases, on-line ineteractive archives, and a type of electronic discussion/reference forum.

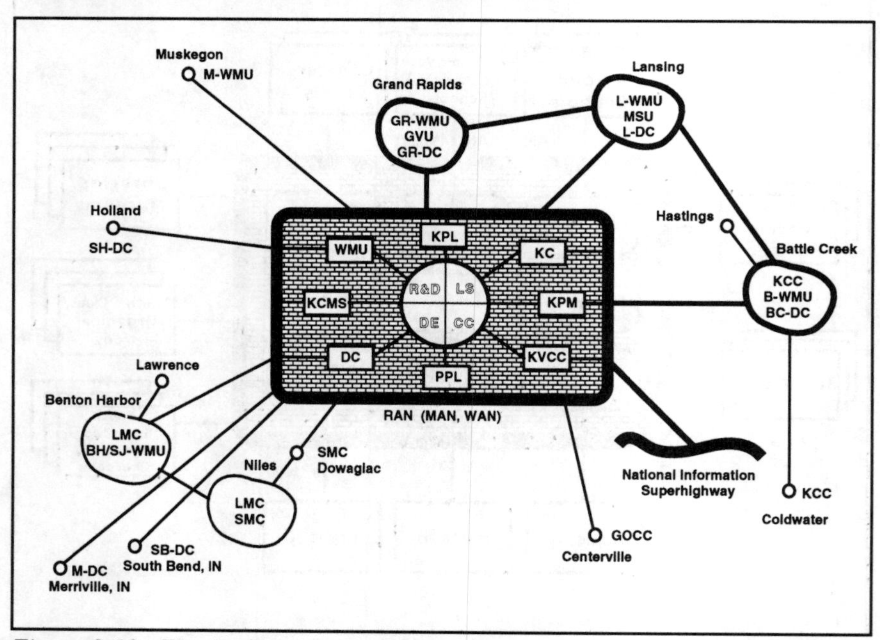


*Figure 8-12: The Architecture of collegeNET - Southwestern Michigan Area, All Colleges (R&D-Research and Development, LS-Library Service, DE-Distance Education, CC-Cable Channel) (The Targowski-Poole-Palchick Model)*

- Higher Education Cable Channel (CC) as an identiable cable television resource for higher education classes and teleconferences.
- Distance Education (DE) for academic degrees in interactive communications systems.

A network integrating business organizations with schools schools and colleges is bizNET. Its functional architecture is shown in Figure 8-13.

The applications of the bizNET are as follows [8]:

- Skills (SK) with the following components:
  - business practice and needs for skilled workers
  - distance video-based trips by pupils and students to business locations to learn about business practices
  - college education programs
  - college registration
- Distance Learning (DL) for companies training:
  - Police and Fire Department training
  - Course taken in the home
  - Remote college credit courses
  - Business initiated vocational/technical training courses

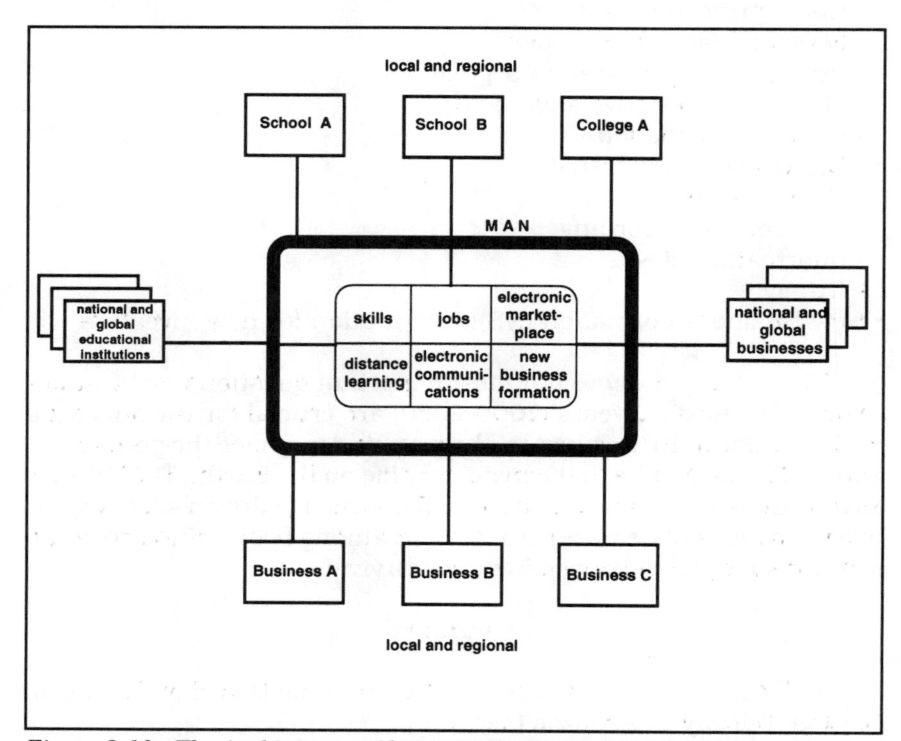


*Figure 8-13: The Architecture of bizNET (The Targowski Model)*

- Adult education
- College recruitment
- Remote medical consultation
- Business teleconferencing
- Law enforcement applications, such as prisoner arraignment
- Remote interviewing

• Jobs (JB) with the following components:
- Employment services
- Job listings
- Jobs seeking persons listings
- Job matching
- Employment status and trends (regional, national, and global)
- Other

• Electronic marketplace (EM). The range of potential applications is virtually infinite. A very small subset is suggested as follows:
- Yellow pages
- On-line catalog shopping
- Consumer advisory services
- Water, electric, and gas meter reading
- Hotel, travel, and theater ticketing
- Plays and movies on dial-tone demand
- Tailored news reports (dedicated newspaper)
- Home protection service
- Personal data management
- Personal tax preparation services
- Mass mail; selective advertising
- Interim transactions
- Electronic funds transfer
- Cashless transactions
- Point of sale recording
- Information kiosks
- Other

• New Business Formation (NB): information for new investors

The Telecity touches our most important questions "to be or not to be." The next 50 years (2000-2050) are crucial for the surviving of the mankind. In these years, the mankind may face the population and ecological bombs that can destroy life on the Earth. The Telecity as information, communication, and knowledge driven set of cognitive tools may create positive awarness among people and providing solutions for global consensus and survival.

**Endnotes**

1.The author was the first Chairman of the Board of Directors, Greater Telecity Kalamazoo (1993-95).

2. Subtitles after Arterton (1987)

3. . Adapted from Andrew S. Targowski and Adrian B. Horton, Greater Kalamazoo GRAND STRATEGY, Kalamazoo: July 1994

4. Adapted from The Health Care Task Force led by Charles Barr, MD (Michigan State University) and contributors: R. Lewis (MSU), Andrew Targowski (Western Michigan University)

5. Adapted from The national Information Infrastructure: Agenda for Action, Federal Information Task Force, September 15, 1993.

6. Adapted from The K-12 Task Force led by James Basco from Western Michigan University

7. Adapted from The Higher Education Task Force led by Howard Poole (Western Michigan University) and Lisa Palchick (Kalamazoo College)

8. Adapted from The Business Task Force led by Robert Bodzianowski (First of America Bank) and Torry Cobb (The Kalamazoo CEO Council)

## References

Adler, R. P., (1988). "Telecommunications, Information technology, and Rural Development," *Proceedings of the Aspen Institute Conference on the Importance of Communications and Information Systems to Rural Development in the United States,"* July 24-27, Menlo Park, CA: Institute for the Future.

Arterton, C.F. (1987). *Teledemocracy, Can Technology Protect Democracy,* Newbury Park, CA.: SAGE Publications, p.14.

Bookchin, M. (1987). *The Rise of Urbanization and the Decline of Citizenship,* San Francisco: Sierra Club Books.

Bloomquist, L. (1988). *Rural Economic Development in the 1980's : Prospects for the Future,* Washington, DC: United States Department of Agriculture, p.52.

Culver, L. (1979). "America's Troubled Cities: Better Times Ahead?" *Habitats Tomorrow,* Bethesda, MD: World Future Society.

Davidge, C. (1987). "America's Talk-Back Television Experiment: QUEBE," in ed. W.H. Dutton, J. G. Blumler, K. L. Kraemer, *Wired Cities, Shaping the Future of Communications,* Boston: G.K.Hall & Co.

Dern, D. (1992). "Business Users Find Internet is more Than E-mail," *Infoworld,* September 14, 1992, p. 62.

Dewey, J. (1915). *Democracy and Education,* New York: Macmillan Co.

Doxiadis, C.A. (1974). *Anthropopolis: A City for Human Development,* Athens: Athens Publishing Center.

Duhl, L. and T. Hanckok (1986). *Promoting Health in the Urban Context,* Copenhagen: The WHO Healthy Cities Project, p. 33.

Dutton, W.H., J. G. Blumler, K. L. Kraemer (1987). "Continuity and Change in Conceptions of the Wired City," in ed. W.H. Dutton, J. G . Blumler, K. L. Kraemer, Wired Cities, *Shaping the Future of Communications,* Boston: G.K.Hall & Co.

Egan, B. (1990). "Bringing Telecommunications to Rural America: The Cost of Technology Adoption," contractor report prepared for the Office of Technology Assessment, October.

Farrel, Ch., M. Galen, M. Mandel, G. McWiliams, M. Schroeder, J. Weber (1992). "The Economic Crisis of Urban America," *Business Week*, May 18, pp. 38-43.

Farhi, P. (1991). "Fighting for a Leading Edge on the Future, Cable TV, Phone Firms Compete for Control of Tomorrow's Technology," *The Washington Post*, January 24.

Gilder, G. (1992). "Cable's Secret Weapon," *Forbes*, April 13.

Harman, W. (1977). "The Coming Transformation," *The Great Transformation, Alternative Futures for Global Society*, Bethesda, MD: World Future Society.

ILSG, (1991). *A Telecommunications Framework for Cities*, Sacramento, CA.: Institute for Local Self Government.

Kemp, R.L. (1990). "Cities in the Year 2000," *The FUTURIST*, September-October, pp.13-15.

Margolis, M. (1992). A Third-World City That Works," March, *World Monitor*, pp.43-50.

Martinez, J (1992). "Getting in Touch with Government," *State Legislature*, p.31

Kielbowicz, R. (1987). The Role of Communication in Building Communities and Markets," contractor report prepared for the Office of Technology Assessment, November.

Klein, J.E. (1989). Welcome to the Future...Resort 2000," *Tours $ Resort*, December/January.

Magleby, D (1984). *Directing Legislation: Voting On Ballot Propositions in the United States*. Baltimore: John Hopkins University Press.

Mokhtarian, P.L. (1992). "Telecommuting in the United States: Letting our Fingers Do the Commuting," *TR News 158*, January-February, pp.2-7

Murata, T. (1987). "Competition for Shaping the New Utopia," in ed. W.H. Dutton, J. G. Blumler, K. L. Kraemer, Wired Cities, *Shapingthe Future of Communications*, Boston: G.K.Hall & Co.

Nazem, S.M., "Telecommunications Policy and Rural Economic Development," University of Nebraska at Omaha, International Center for Telecommunications Management, nd.

Pelton, J. N. (1990). *Future Talk*, Boulder, CO.: Cross Communication Company, p.92.

Reich, R. B. (1992). *The Work of Nations*, New York: Vintage Books, p. 113.

Royston, R. (1985). *Cities 2000*, New York: Facts On File Publications.

Saffo, P. (1991). "*Widespread Telecommuting Will Change the Urban Landscape*," Menlo Park, CA: Institute for the Future.

Segal, T. (1992). "The Riots: 'Just As Much About Class As About Race," *Business Week*, May 18, p.47.

Sims, C. (1989). "Global Communications Net Planned by GE for Its Staff," *The New York Times*, May 31, 1989.

Schmid, G. (1991). What Will Make the 1990s Global Economy Global?, 1991 Ten-Year Forcast, Menlo Park, CA.: Institute for the Future.

SMS Research (1991). "Final Evaluation Report On Year One of The Hawaii Telwork Center Demonstration Project," Honolulu: SMS Research.

Schwartz, E. I., P.M. Eng, S.L. Walker, A. Cueno (1992). "The Kiosks are Coming, The Kiosks are Coming," *Business Week*, June 22, p. 122.

Quaid, M., L. Heifez, M. Farley, D. Christensen (1992). "Puget Sounding

Telecommuting Demonstration," Olympia: Washington State Energy Office.

Qvortrup, L. (1989). "The Nordic Telecottages: Community Teleservice Centers for Rural Regions," *Telecommunications Policy,* March, pp. 59-68.

Weingarten, F.W. (1987). "The New R&D Push in Communications Technology," in ed. W.H. Dutton, J. G. Blumler, K. L. Kraemer, *Wired Cities, Shaping the Future of Communications,* Boston: G.K.Hall & Co.

United States Department of Commerce, (1988). "Rural and Rural Farm Population: 1988", Washington, DC: United States Government Printing.

Van Till, J. (1979). "A New Type of a City for an Energy-Short World, " *Habitats Tomorrow*, Bethesda, MD: World Future Society.

# Index